Eric M...

Useful Properties of Common Areas

Rectangle $\dfrac{VQ}{Ib} = \dfrac{3V}{2A}$ (center only)	**Half of thin tube**
$A = bh$ $I_x = \dfrac{bh^3}{12}$ $I_y = \dfrac{hb^3}{12}$	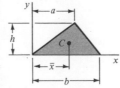 $A = \pi rt$ $I_x \approx 0.095\,\pi r^3 t$ $I_y = 0.5\,\pi r^3 t$
Triangle	**Triangle**
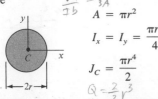 $A = \dfrac{bh}{2}$ $I_x = \dfrac{bh^3}{36}$ $I_y = \dfrac{hb^3}{48}$	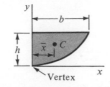 $A = \dfrac{bh}{2}$ $\bar{x} = \dfrac{(a+b)}{3}$
Circle $\dfrac{VQ}{Ib} = \dfrac{4V}{3A}$	**Parabola** ($y = kx^2$)
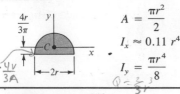 $A = \pi r^2$ $I_x = I_y = \dfrac{\pi r^4}{4}$ $J_C = \dfrac{\pi r^4}{2}$ $Q = \dfrac{2}{3}r^3$	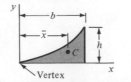 $A = \dfrac{2bh}{3}$ $\bar{x} = \dfrac{3b}{8}$
Semicircle	**Parabolic spandrel** ($y = kx^2$)
$x = \dfrac{-VQ}{Ib} = \dfrac{4V}{3A}$ 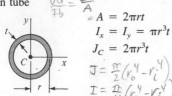 $A = \dfrac{\pi r^2}{2}$ $I_x \approx 0.11\,r^4$ $I_y = \dfrac{\pi r^4}{8}$ $Q = \dfrac{2}{3}r^3$	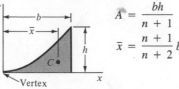 $A = \dfrac{bh}{3}$ $\bar{x} = \dfrac{3b}{4}$
Thin tube $\dfrac{VQ}{Ib} = \dfrac{2V}{A}$	**General spandrel** ($y = kx^n$)
$A = 2\pi rt$ $I_x = I_y = \pi r^3 t$ $J_C = 2\pi r^3 t$ $J = \dfrac{\pi}{2}(r_o^4 - r_i^4)$ $I = \dfrac{\pi}{4}(r_o^4 - r_i^4)$	$A = \dfrac{bh}{n+1}$ $\bar{x} = \dfrac{n+1}{n+2}b$

MECHANICS OF MATERIALS

MECHANICS OF MATERIALS

A. C. UGURAL, Ph.D.

Adjunct Professor of Mechanics and Materials Science
College of Engineering
Rutgers University

McGRAW-HILL, INC.

New York St. Louis San Francisco Auckland Bogotá Caracas
Hamburg Lisbon London Madrid Mexico Milan Montreal New Delhi
Paris San Juan São Paulo Singapore Sydney Tokyo Toronto

MECHANICS OF MATERIALS
Copyright © 1991 by McGraw-Hill, Inc. All rights reserved. Printed in the United States of
America. Except as permitted under the United States Copyright Act of 1976, no part of this
publication may be reproduced or distributed in any form or by any means, or stored in a data
base or retrieval system, without the prior written permission of the publisher.

1 2 3 4 5 6 7 8 9 0 DOH DOH 9 5 4 3 2 1 0

ISBN 0-07-065737-8

This book was set in Serif by Waldman Graphics, Inc.
The editors were John J. Corrigan and James W. Bradley;
the designer was Amy E. Becker;
the production supervisor was Denise L. Puryear.
The cover illustrator was Joe Gillians.
R. R. Donnelley & Sons Company was printer and binder.

Library of Congress Cataloging-in-Publication Data

Ugural, A. C.
    Mechanics of Materials / A.C. Ugural.
        p.   cm.
    Includes bibliographical references.
    ISBN 0-07-065737-8
    1. Strength of materials.   I. Title.
    TA405.U43   1991   89-13257
    620.1'12—dc20

ABOUT THE AUTHOR

Ansel C. Ugural is Adjunct Professor of Mechanics and Materials Science at the College of Engineering at Rutgers University. He received his Ph.D. in engineering mechanics from the University of Wisconsin—Madison. Dr. Ugural has taught at the University of Wisconsin as well as at Fairleigh Dickinson University, where he also served as Professor and Chairman of Mechanical Engineering from 1966 to 1990. He has considerable and varied industrial experience in both full-time and consulting capacities. A member of several professional societies, including the American Society of Mechanical Engineers and the American Society of Engineering Education, he is currently listed in *Who's Who in Engineering*. Dr. Ugural is the author of *Stresses in Plates and Shells* (McGraw-Hill, 1981), a coauthor (with S. K. Fenster) of *Advanced Strength and Applied Elasticity* (Elsevier, 1987), and has published numerous articles in the trade and professional journals.

CONTENTS

13   Buckling of Columns 362

Appendixes 391

A   Moments of Areas 391

B   Tables 403

PREFACE

This volume is designed for an undergraduate-level engineering course in strength, or mechanics, of materials, although the author has endeavored to make it useful as a reference for engineering professionals as well. Fundamentals of the subject, the three aspects of solid mechanics problems, and the applications necessary to prepare students for more advanced study and for engineering practice are emphasized throughout.

The text offers a simple, comprehensive, and methodical presentation of the basic concepts in the analysis of members subjected to axial loads, torsion, bending, and pressure. The coverage presumes a knowledge of statics; however, topics that are particularly significant to an understanding of the subject are reviewed as they are taken up.

To enhance the student's overall understanding of the multiple aspects of structural strength, the behavior of various types of members is discussed, using design criteria such as yielding, plastic collapse, fatigue, and fracture. Failure criteria are employed to predict the behavior of structures under combined loadings. Analyses of members made of isotropic as well as composite and inelastic materials under ordinary and/or high-temperature loads are presented. Applications of the equilibrium, numerical, and energy methods are described in detail, and a treatment of the column-stability problem is included.

The rational design procedure—and its applications to axially loaded and twisted bars and beams—is described in order to make clear the relation of mechanics of materials to design. Equations of elasticity are presented to acquaint the reader with the essentially simple nature of the foundations of an important field in the mechanics of solids. These formulations are used to make critical evaluations of the force-deformation relations of the strength of materials. The text attempts to fill what the author believes to be a void in this regard.

The material presented here strikes a balance between the theory necessary to gain insight into mechanics and numerical solutions, both so useful in performing stress analysis in a realistic setting. Above all, an effort has been made to provide a visual interpretation of the basic equations and of the means by which the loads are resisted in typical members. Finally, attention is given to analytical procedures, formulations, and numerical techniques suitable for computer programming.

Emphasis is placed upon the illustration of important principles and applications through examples, and a broad range of practical problems is provided for solution by the student. This volume offers more than 135 illustrative examples, fully worked out; more than 935 problem sets, many of which are drawn from engineering practice; and a multitude of formulas and tabulations (Appendix B) from which preliminary design calculations can be made. The properties of moments and centroids of areas are also described (Appendix A).

Both the International System of Units (SI) and the U.S. Customary System of units are used; but since in practice the former is replacing the latter, this book places a greater emphasis on SI units. (The conversion factors for common SI units and U.S. customary units are listed in a table inside the front cover of this text.) A sign convention that is consistent with vector mechanics is employed throughout for loads, internal forces, and stresses. This convention has been carefully chosen to conform to that used in most classical strength of materials texts as well as to that most often employed in numerical analysis of complex structures.

Most chapters are independent of one another and are self-contained. Hence the order of presentation can be smoothly altered to meet the instructor's preference. Optional sections that can be deleted without destroying the continuity of the text are identified by an asterisk.

The text is accompanied by a supporting package of instructional materials: a solutions manual to the problems and three diskettes containing the TK Solver designed for the IBM-PC, XT, AT, PS/2, or any compatible computer. The TK Solver is a powerful tool that builds and solves mathematical equations and models. It plots, uses graphics and charts, and has spreadsheet functions. Its purpose is to provide students with a tool for analyzing and solving a wide variety of problems in mechanics of materials. Thirty-one documented models address specific problems from all but the first chapter of the text. They were prepared by the author, programmed by Universal Technical Systems, Inc., and should be easy to use and understand.

Instructors should contact their local McGraw-Hill sales representatives to obtain the complete supplementary package. Students can purchase their TK Solver through the college bookstore.

ACKNOWLEDGMENTS

Thanks are due to the many students who offered constructive suggestions and checked the solutions to the problems when drafts of this work were used as a text. Dr. Saul K. Fenster of New Jersey Institute of Technology read an early draft of the manuscript and made numerous significant corrections, for which I am very grateful. I am also pleased to acknowledge the helpful recommendations and valuable perspectives offered by the reviewers of this text, particularly George G. Adams, Northeastern University; Nicholas Altiero, Michigan State University; Leon Bahar, Drexel University; J. R. Barber, University of Michigan; Harry Conway, Cornell University; Herb Corten, University of Illinois–Urbana–Champaign; Akhtar S. Kahn, University of Oklahoma; John Kennedy, Clemson University; V. J. Meyers, Purdue University; Donald Pierce, University of Nebraska–Lincoln; Michael H. Santare, University of Delaware; and John G. Thacker, University of Virginia. The author is indebted to William H. Frank of UTS, who programmed the TK Solver models for this book.

A. C. Ugural

LIST OF SYMBOLS

A	area	R	reaction, force
b	breath, width	r	radius, radius of gyration
C	centroid	S	elastic section modulus
c	distance from neutral axis to extreme fiber, radius	s	distance, length along a line
D	diameter	T	torque, temperature
d	diameter, distance, depth, dimension	t	thickness, width, tangential deviation
E	modulus of elasticity	U	strain energy
e	eccentricity, dilatation, distance	u, v, w	displacement components
F	force	V	shearing force, volume
f	frequency, flexibility, shape factor	v	velocity
f_s	factor of safety	W	work, weight
G	modulus of rigidity, shear modulus of elasticity	w	load per unit length
		Z	plastic section modulus
g	acceleration of gravity (≈ 9.81 m/s^2)	x, y, z	rectangular coordinates, distances
h	height, depth of beam	α (alpha)	angle, coefficient of thermal expansion, form factor for shear
I	moment of inertia of area		
J	polar moment of inertia of area	γ (gamma)	shearing strain, specific weight
K	stress concentration factor, impact factor	δ, Δ (delta)	deformation, displacement, finite difference
k	spring constant, stiffness, shear coefficient	ε (epsilon)	normal strain
L	length, span	θ (theta)	angle, slope
M	bending moment, couple	κ (kappa)	curvature
m	mass	μ (mu)	micro
N	number of cycles	ν (nu)	Poisson's ratio
n	modular ratio, number of coils	ρ (rho)	radius, radius of curvature, density
P	force, concentrated load	σ (sigma)	normal stress
p	pressure	τ (tau)	shearing stress
Q	first moment of area, force	ϕ (phi)	angle, total angle of twist
q	shear force per unit length, shear flow	ω (omega)	angular velocity

MECHANICS OF MATERIALS

CHAPTER

1

FUNDAMENTAL PRINCIPLES

1.1 INTRODUCTION

Mechanics of materials is the branch of applied mechanics that deals with the internal behavior of variously loaded solid bodies. The ''solid bodies'' referred to here include shafts, bars, beams, plates, shells, and columns, as well as structures and machines that are assemblies of these components. Also called *strength of materials* or *mechanics of deformable bodies,* mechanics of materials focuses primarily on stress analysis and on the mechanical properties of materials.

The study of mechanics of materials is based upon an understanding of the equilibrium of bodies under the action of forces. While *statics* treats the external behavior of bodies that are assumed to be ideally rigid and at rest, mechanics of materials is concerned with the relationships between external loads (forces and moments) and internal forces and deformations induced in the body. *Stress* and *strain* are fundamental quantities connected with the former and the latter, respectively.

Complete analysis of a structure under load requires the determination of stress, strain, and deformation through the use of three fundamental principles, which will be outlined in Sec. 1.3: the laws of forces, the laws of material deformation, and the conditions of geometric compatibility. We consider here the first principle, static equilibrium, and its application to a loaded body. The remaining principles are studied in Chap. 3.

Investigation of the behavior of solids under loads began with Galileo Galilei (1564–1642), though Robert Hooke (1635–1703) was the first to point out that a body is deformed if a force acts upon it. Since then many engineers, scientists, and mathematicians in the field of stress analysis have developed the basic knowledge on which modern methods are based, and the literature related to the strength of materials is voluminous. A number of selected references are identified at the end of the text for those seeking more extensive treatment.

1.2   SCOPE OF TREATMENT

The usual objective of mechanics of materials is the examination of the load-carrying capacity of a body from three standpoints: *strength, stiffness,* and *stability*. These qualities relate, respectively, to the ability of a member to resist permanent deformation or fracture, to resist deflection, and to retain its equilibrium configuration. The *stress level,* sometimes expressed through failure theories relating the complex stresses in a structure with the experimentally obtained axial stress, is used as a measure of strength. *Failure* can be defined, in very general terms, as any action that results in an inability on the part of the structure to function in the manner intended. For instance, when loading produces an abrupt shape change of a member, instability occurs; in like manner, an inelastic deformation or an excessive magnitude of deflection in a member will cause malfunction in normal service. Each of these examples indicates a type of failure.

The main concerns in the study of mechanics of materials may be summarized as follows:

1. Analysis of stress and deformation within a loaded body, which is accomplished by application of one of the methods to be described in Sec. 1.3.
2. Determination by analysis (or by experiment) of the largest load a structure can sustain without suffering damage, failure, or compromise of function.
3. Determination of the body shape and selection of those materials which are most efficient for resisting a prescribed system of forces under specified environmental conditions of operation. This is called the *design function* and will be discussed further in Sec. 1.9.

Clearly, the last item cited relies upon the performance of the first two, and it is to these primary concerns that this book is largely directed.

The ever-increasing industrial demand for more sophisticated structural and machine components calls for a good grasp of the concepts of stress and strain and of the behavior of materials—and for a considerable degree of ingenuity. This text will, in the very least, provide the reader with the ideas and information necessary for a basic understanding of the mechanics of deformable bodies and will encourage the creative process based on that understanding. Very few basic formulas are actually derived in this volume; rather the student will master these formulas through their repeated application. It is important, however, that the reader visualize the nature of the quantities being computed. Complete, carefully drawn, free-body diagrams facilitate visualization, and these we have provided, all the while knowing that the subject matter can be learned best by solving practical problems.

1.3   METHODS OF ANALYSIS

The approaches in widespread employment for determining the influence of loads upon deformable bodies are the *mechanics of materials theory* (also known as *technical theory*), which is presented in this text, and the *theory of elasticity*. The difference between these theories lies primarily in the extent to which strains are described and in the nature of simplifications used.

The mechanics of materials approach uses assumptions, based upon experimental evidence and the lessons of engineering practice, to make a reasonable solution of the basic problem possible. On the other hand, the theory of elasticity establishes every step rigorously from the mathematical point of view and hence seeks to verify the validity of the assumptions introduced. This technique can provide ''exact'' results where configurations of loading and shape are simple. In general, however, the theory of elasticity yields solutions with considerable difficulty.

The complete analysis of structural members by the so-called *method of equilibrium* requires consideration of a number of conditions relating to certain laws of forces, laws of material deformation, and geometry. These essential relationships, referred to as *basic principles of analysis,* should be outlined in summary form before we proceed:

1. *Statics.*   The equations of equilibrium of forces must be satisfied throughout the member.
2. *Deformations.*   The stress-strain or force-deformation relations (for example, Hooke's law) must apply to the behavior of the material of which the member is constructed.
3. *Geometry.*   The conditions of geometric fit or compatibility of deformations must be satisfied; that is, each deformed portion of the member must fit together with adjacent portions.

The stress and strain distribution obtained by applying these principles must be such as to conform to the conditions of loading imposed at the boundaries of a member. This is known as *satisfying the boundary conditions*. Applications of the foregoing procedure will be shown in the problems presented as the subject unfolds. We note here, however, that it is not always necessary to execute an analysis in the exact order of steps listed above.

As an alternative to the equilibrium methods, the analysis of stress and deformation can be accomplished through the use of *energy methods,* which are based upon the concept of strain energy. The role of both the equilibrium and the energy methods is twofold. These methods can provide solutions of acceptable accuracy where configurations of loading and member shape are regular, and they can be employed as the basis of *numerical methods* in the solution of more complex problems.

In conclusion, it should be noted that a degree of caution is necessary when employing formulas for which there is uncertainty in applicability and restriction of use. The relatively simple form of many formulas often results from idealizations made in their derivations—idealizations such as simplified boundary conditions and loading on a member and approximation of shape or material. Designers and stress analysts must be aware of such limitations.

1.4 FORCE AND LOAD CLASSIFICATIONS

All forces acting on a body, including the reactive forces caused by supports, are considered *external forces*. These forces are classified as surface forces and body forces. A *surface force* is of concentrated type when it acts at a point, but it may also be distributed over a finite area. A *body force* acts on a volumetric element rather than on a surface and is attributable to fields such as gravity and magnetism. The force of the earth on an object at or near the surface is termed the *weight* of the object. *Internal forces* in a body can be considered as forces of interaction between the constituent material particles of the body.

The loads on bodies may be concentrated and distributed forces, and couples. Any force applied to an area that is relatively small compared with the size of the loaded structural member is assumed to be a *concentrated load*. A load slowly and steadily applied is regarded as a *static load*, while a rapidly applied load is called an *impact load*. Multiple applications and removals of load, usually measured in thousands of episodes or more, is referred to as *repeated loading*. Having said all that, we add that, unless otherwise stated, we assume in this text that the weight of the body can be neglected and that the load is static.

In the International System of Units (SI), force is measured in newtons (N), but because the newton is a small quantity, the kilonewton (kN) is often used in practice. In the U.S. Customary System, force is expressed in pounds (lb) or kilopounds (kip). Both systems of units are used here. However, greater emphasis is placed on SI units in line with international convention. (The table inside the front cover of this book compares the two systems.)

1.5 CONDITIONS FOR STATIC EQUILIBRIUM

The analysis and design of structural and machine components requires a knowledge of the distribution of the internal forces throughout such members. Fundamental concepts and conditions of static equilibrium provide the necessary background for the determination of internal as well as external forces. In Sec. 1.7 we shall see that the components of internal-force resultants have special meaning in terms of the type of deformations they cause, as applied, for example, to slender members.

When a system of forces acting upon a body has zero resultant, the body is said to be in *force equilibrium*. From another viewpoint, equilibrium of forces is the state in which the forces applied on a body are in balance. Newton's first law states that if the resultant force acting on a particle (the simplest body) is zero, the particle will remain at rest or will move with constant velocity. Statics, as its name implies, deals essentially with the case in which the particle or body remains at rest.

Consider the equilibrium of a body in space. The *equations of statics* require that the following conditions be satisfied:

$$\Sigma F_x = 0 \qquad \Sigma F_y = 0 \qquad \Sigma F_z = 0$$
$$\Sigma M_x = 0 \qquad \Sigma M_y = 0 \qquad \Sigma M_z = 0 \tag{1.1}$$

In words we are saying that the sum of all forces acting upon a body in any direction must be zero and that the sum of all moments about any axis must be zero.

If the forces act on a body in equilibrium in a single (xy) plane, three expressions—$\Sigma F_z = 0$, $\Sigma M_x = 0$, and $\Sigma M_y = 0$—in Eqs. (1.1), while still valid, are trivial. This leaves only *three* independent conditions of equilibrium for planar problems:

$$\Sigma F_x = 0 \qquad \Sigma F_y = 0 \qquad \Sigma M_A = 0 \tag{1.2}$$

In words then, the sum of all forces in any two (x, y) directions must be zero, and the resultant moment with respect to any axis z or any point A in the plane must be zero.

Alternative sets of conditions can imply Eqs. (1.2). That is,

$$\Sigma F_x = 0 \qquad \Sigma M_A = 0 \qquad \Sigma M_B = 0 \tag{1.3a}$$

provided that the line connecting the points A and B is *not perpendicular* to the x axis, and

$$\Sigma M_A = 0 \qquad \Sigma M_B = 0 \qquad \Sigma M_C = 0 \tag{1.3b}$$

where points A, B, and C are *not collinear*. Clearly the alternative sets are obtained by replacing a force summation by an equivalent moment summation. We often find it convenient to employ Eqs. (1.3a) or even Eqs. (1.3b). The judicious selection of points for taking moments can generally simplify algebraic computations.

It is noted that, for a body in *accelerated* motion, additional forces must be included for the equations of statics to be applicable. These additional forces are the *inertia forces*; for purposes of structural analysis, incorporating them allows us to consider the body as being acted upon by a set of forces in equilibrium. This is the so-called *d'Alembert principle*.

Common engineering problems involve machines and structures in equilibrium. Certain forces—usually loads—are specified, and the problem is to obtain the unknown forces—usually reactions—balancing the loads. If it is possible to determine all forces by using the equilibrium conditions alone, the system is called *statically determinate*. However, there are problems where equations of statics are *not* sufficient to ascertain the unknown forces on the member. Such problems are called *statically indeterminate*. The *degree* of static indeterminacy is equal to the difference between the number of unknown forces and the number of pertinent equilibrium equations. Any reaction that is in excess of those that can be obtained by statics alone is said to be *redundant*. Thus the number of redundants is the same as the degree of indeterminacy.

1.6 ANALYSIS OF INTERNAL FORCES: METHOD OF SECTIONS

A body responds to the application of external forces by deforming and by developing internal forces which hold together the particles forming the body. We shall now deal with one of the principal problems of mechanics of materials: the investigation of internal forces by the familiar approach of statics, or the *method of sections*. The steps involved in applying the method of sections may be summarized as follows:

1. Isolate the bodies. The sketch of the isolated body and all external forces acting on it is called a *free-body diagram*. Since in practice the allowable deformations are negligible as compared with the size of the member, free-body diagrams contain the *initial* dimensions and ignore the deformations.
2. Apply the equations of statics to the diagram to determine the unknown external forces.
3. Cut the body at a section of interest by an *imaginary* plane (Fig. 1.1a), isolate one of the segments, and repeat step 2 for that segment. If the entire body is in equilibrium, any part of it must be in equilibrium. That is, there must be internal forces transmitted across the cut sections (Fig. 1.1b).

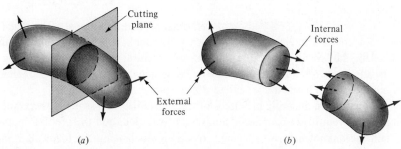

(a)   (b)

Figure 1.1   Application of the method of sections to a loaded body.

It can be concluded from the foregoing that the external forces are balanced by internal forces. The former can thus be regarded as a continuation of the distribution of the latter, though the precise distribution of the forces within a member depends upon the imaginary plane selected to separate the two parts.

The section on which internal forces produce the largest stress is called the *critical section* of the loaded member. When a single load is acting on a member, the critical section and its orientation are usually evident by inspection. With combined loads, however, the angular position of the cutting plane is determined by approaches that will be given in Secs. 4.2 and 4.4.

1.7   COMPONENTS OF INTERNAL-FORCE RESULTANTS

Many structural elements can be classified as *slender members*. According to the criterion often used to define a slender member, the length of such a member should be at least 5 times greater than its largest cross-sectional dimension. In general, forces within a slender member can be represented by a statically equivalent set consisting of a force vector and a moment or couple vector acting *at the centroid C* of the cross section. These *internal-force resultants*, also called *stress resultants,* are usually resolved into components normal and tangent to the section. This is seen in Fig. 1.2*a,* and the *right-handed coordinate system* shown in the figure will be used throughout this text. It is observed that the x axis coincides with the longitudinal axis of the member; the y axis is taken as the vertical upward axis, and the z axis points toward the reader. Note that the sense of the moment components follows the right-hand screw rule—that is, a right-hand screw advances with the sense of the vector when it is twisted in the sense indicated by the moment couple. For convenience, the moment components are also represented by double-headed vectors, as depicted in Fig. 1.2*a.*

Each internal force and moment component reflects a different effect of the applied loading on the member. These effects can be described as follows (Fig. 1.2*a*):

The *axial force* F_x tends to elongate (or contract) the member and is often identified by the letter P. If the force acts away from the cut, it is termed axial *tension*; if toward the cut, it is called axial *compression*.

The *shear forces* F_y and F_z tend to shear one part of the member relative to the adjacent part and are often designated by the letters V_y, V_z, or V.

The *twisting moment* or *torque* M_x is responsible for twisting the member about its axis and is identified by the letter T.

The *bending moments* M_y and M_z cause the member to bend and are often designated by the letter M.

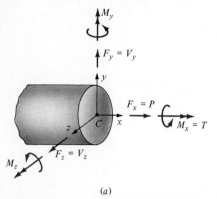

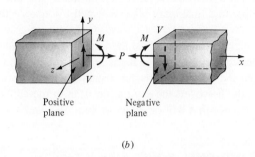

Figure 1.2 Definitions of positive internal forces and moments: (a) the general case and (b) components in two dimensions.

A structural element may be subject to any combination of or all of these four modes of force transmission simultaneously, though the modes are usually treated separately, and, if appropriate, the results are combined to obtain the final solution. Therefore the method of sections is counted on as a first step in all problems where the internal forces, and thus the corresponding stresses and strains, are being investigated.

In practice the problem is often considerably simpler in that all forces act in a single plane, here taken as *xy*. In *planar problems*, we find only three components acting across a section: the axial force $F_x = P$, the shear force $F_y = V$, and the bending moment $M_z = M$. The diagrammatic representations of these components, as used in this text, are shown in Fig. 1.2*b*. It is noted that the cross-sectional *face*, or *plane*, is defined as *positive* when its outward normal points in a positive coordinate direction and as negative when its outward normal points in the negative coordinate direction. According to Newton's third law, the axial forces, shear forces, and bending moments acting on these faces at a cut section are equal and opposite.

Sign Convention.   To assure consistency among the various analytical approaches, the following sign convention is established for the axial force, shear force, twisting moment, and bending moment. When *both* the outer normal and the internal force or moment vector component point in a *positive* (or negative) *coordinate direction*, the force or moment is defined as *positive*. When a *negatively directed* component acts on a *positive* face (or vice versa), the force or moment is negative. Thus Fig. 1.2 depicts positive internal force and moment components. Accordingly, the tensile force at a section is positive. Note that the sense of the positive twisting moment vector is identical with that of the positive axial-force vector.

This sign convention adopted for the internal forces and moments is also associated with the behavior under load of a member. For example, a straight bar elongates when subjected to positive axial forces at its ends, a point that shall be made clear in Sec. 7.3 when we consider beam deformation.

1.8 CONVENTIONAL DIAGRAMS FOR SUPPORTS

Figure 1.3 illustrates the conventional diagrams for three types of support in common usage for structural members loaded with forces acting in the same plane. These representations are employed when constructing the free-body diagrams of bars and beams. For convenience, the members are depicted in horizontal positions.

A *hinge,* or *pin, support* (Fig. 1.3*a*) restrains the member from translating (moving) in *any* direction of the plane, but it does not prevent rotation. At a *roller support* (Fig. 1.3*b*), translation is prevented in the vertical direction but not in the horizontal direction. Just as at a pin, there is no restraint offered to rotation at a roller. Thus a pinned support resists tangential and normal forces to the surface, while a roller support withstands forces normal to the surface only. Both are called *simple supports.*

At the *fixed,* or *clamped, support* (Fig. 1.3*c*), the bar can neither translate nor rotate. Therefore a fixed support resists both a moment and a force in any direction. This force may have components normal and tangential to the surface.

The idealized diagrams also indicate the nature of the reactive forces R and moment M at each support. The analysis of stress resultants in members generally begins with the determination of the reactions.

*1.9 RATIONAL DESIGN PROCEDURE

The main objective of a design process is to determine proper materials, dimensions, and shapes of the components of a structure or machine so that they will support given loads and function without failure. This problem is more particularly that of *optimum design.* Efficiency may be gaged by such criteria as minimum weight or volume, minimum cost, and/or any other standard deemed appropriate.

The role of analysis in design is observed in examining the following *rational procedure* in the design of a load-carrying member:

1. Evaluate the mode of possible failure of the member.
2. Determine a relationship between the applied load and the resulting effect such as stress or deformation.

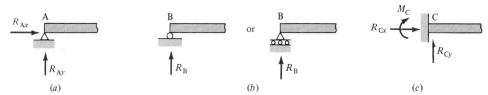

Figure 1.3 Common supports: (*a*) hinge, or pin; (*b*) roller; and (*c*) fixed, or clamped, end.

3. Determine the maximum usable value of a significant quantity such as stress or deformation that could conceivably cause failure. Employ this value in connection with the equation found in step 2 or, if required, in any of the formulas associated with the various theories of failure, which will be discussed in Sec. 8.9.

4. Select the factor of safety, as outlined in Sec. 2.9.

The foregoing procedure must be tailored somewhat to each individual case, since some steps may be regarded as unnecessary or as obvious for a certain member. Suffice it to say here that complete design solutions are not unique, involve a consideration of many factors, and often require a trial-and-error process.

This text provides an elementary treatment of the concept of "design to meet strength requirements" as those requirements relate to individual component parts. That is, the geometrical configuration and material of a member are preselected and the applied loads are specified. Then the basic formulas, to be developed in Chaps. 5, 6, and 7, are used to select members of adequate size in each case.

Other aspects of the design of members are the prediction of the deformation of a given component under prescribed loading and the consideration of buckling. The methods of determining deformation are discussed in several chapters throughout the text. In Chap. 13 we shall be concerned with the buckling of slender members loaded axially in compression.

To conclude, we note that there is a very close relationship between analysis and design, and the examples and problems that appear throughout this book illustrate that connection.

CONCEPT OF STRESS

2.1 INTRODUCTION

Stress and strain are concepts of paramount importance to a comprehension of the mechanics of materials. They permit the mechanical behavior of load-carrying members to be described in terms essential to the engineer. The study of stress is begun in this chapter, while the concept of strain will be introduced in Chap. 3.

We shall apply the method of sections to isolate an infinitesimal element in defining stress and its components (Secs. 2.2 and 2.3). In Secs. 2.4 through 2.6 we shall outline procedures for determining the internal axial forces, corresponding normal stresses, and average shearing stresses. Then applications to various structures, the allowable stress, and the preliminary design of prismatic bars for axial loads will be discussed. We emphasize that the definitions and procedures presented here can be applied in all types of loading situations.

2.2 STRESS DEFINED

As discussed in Chap. 1, a body subjected to external forces develops an associated system of internal forces. We shall now describe the intensity of those internal forces, which represents a particularly significant quantity.

Reconsider one of the isolated segments of a body in equilibrium under the action of a system of forces, as first shown in Fig. 1.1b and now redrawn in Fig. 2.1. An element of area ΔA, positioned on an interior surface passing through a point O, is acted upon by force $\Delta \mathbf{F}$. Let the origin of the coordinate axes be located at O, with x normal and y, z tangent to ΔA. Generally $\Delta \mathbf{F}$ does not lie along x, y, or z. Components of $\Delta \mathbf{F}$ parallel to x, y, and z are also indicated in the figure. The *normal stress* σ (sigma) and the *shear*, or *shearing stress*, τ (tau) are then defined as

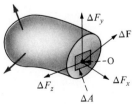

Figure 2.1 Components of an internal force $\Delta \mathbf{F}$ acting on a small area centered on point O.

$$\sigma_{xx} = \sigma_x = \lim_{\Delta A \to 0} \frac{\Delta F_x}{\Delta A} = \frac{dF_x}{dA}$$

$$\tau_{xy} = \lim_{\Delta A \to 0} \frac{\Delta F_y}{\Delta A} = \frac{dF_y}{dA} \tag{2.1}$$

$$\tau_{xz} = \lim_{\Delta A \to 0} \frac{\Delta F_z}{\Delta A} = \frac{dF_z}{dA}$$

These relations represent the stress components at the point O to which area ΔA is reduced in the limit. Note that the intensity of the force *perpendicular, or normal, to the surface* is called the *normal stress* at a point. On the other hand, the intensity of the force *parallel to the surface* is termed the *shear stress* at a point. Hence the primary distinction between normal and shearing stress is one of direction. It is important that the reader have a clear mental picture of the stresses that are normal and those that are shear.

From the foregoing we observe that two indices are needed to denote the components of stress. For the normal stress component the indices are identical, while for the shear stress component they are mixed. The two indices are given in double subscript notation: The first subscript indicates the direction of a normal to the plane, or face, on which the stress component acts; the second subscript relates to the direction of the stress itself. Repetitive subscripts will be avoided in this text, so that the normal stress will be designated σ_x, as seen in Eqs. (2.1). Note that a plane is defined by the axis normal to it; for example, the x face is perpendicular to the x axis.

The limit $\Delta A \to 0$ in Eqs. (2.1) is of course an idealization, since the area itself is not continuous on an atomic scale. Our consideration is with the *average stress* on areas where size, while small as compared with the size of the body, are large as compared with the distance between atoms in the solid body. Therefore stress is an adequate definition for engineering purposes. Note that the values obtained from Eqs. (2.1) differ from point to point on the surface as $\Delta \mathbf{F}$ varies. The components of stress depend not only upon $\Delta \mathbf{F}$, however, but also upon the orientation of the plane on which it acts at point O. Thus, even at a specified point, the stresses will differ as different planes are considered. The complete description of stress at a point therefore requires the specification of stress on all planes passing through the point.

We observe from the foregoing definitions that the units of stress (σ or τ) consist of units of force divided by units of area. In SI units, stress is measured in *newtons per square meter* (N/m^2) or in *pascals* (Pa). Since the pascal is a very small quantity, the megapascal (MPa) is commonly used. The U.S. Customary System uses the pound per square inch (psi)—or the kilopound per square inch (ksi)—as the unit for stress.

2.3   COMPONENTS OF STRESS

In order to be able to determine stresses on an infinite number of planes passing through a point O (Fig. 2.1), thus defining the *state of stress* at that point, we need only specify the stress components on three mutually perpendicular planes passing through the point. These planes, perpendicular to the coordinate axes, contain three sides of an infinitesimal cubic element.

This *three-dimensional* state of stress acting on an isolated element within a body is shown in Fig. 2.2. Stresses are considered to be identical at points O and O′ and are *uniformly* distributed on each face. They are indicated by a single vector acting at the center of each face. We observe a total of nine *components of stress* that compose three groups of stresses acting on the mutually perpendicular planes passing through O. This representation of state of stress is called a *stress tensor*. It is a tensor of second rank, requiring two indices to identify its elements or components. (A vector is a tensor of first rank; a scalar is a tensor of zero rank.)

We now consider the property of shear stress from an examination of the equilibrium of forces acting on the cubic element shown in Fig. 2.2. It is clear that the first three of Eqs. (1.1) are satisfied. Taking moments of the *x*- and *y*-directed forces about point O, we find that $\Sigma M_z = 0$ results in

$$(-\tau_{xy}\, dy\, dz)\, dx + (\tau_{yx}\, dx\, dz)\, dy = 0$$

from which

$$\tau_{xy} = \tau_{yx} \tag{2.2}$$

Similarly, from $\Sigma M_x = 0$ and $\Sigma M_y = 0$, we obtain

$$\tau_{xz} = \tau_{zx} \quad \text{and} \quad \tau_{yz} = \tau_{zy}$$

The subscripts defining the shear stresses are commutative, and the stress tensor is symmetric. This means that each pair of equal *shear stresses acts on mutually perpendicular planes*. Because of this, no distinction will hereafter be made between the stress components τ_{xy} and τ_{yx}, τ_{xz} and τ_{zx}, or τ_{yz}

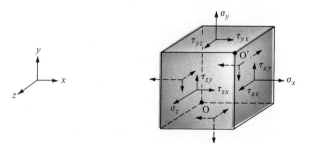

Figure 2.2   Three-dimensional state of stress.

and τ_{zy}. It is verified rigorously in Sec. 4.5 that the foregoing is valid even when stress components vary from one point to another.

We shall employ here a *sign convention* that applies to both normal and shear stresses and that is based upon the relationship between the direction of an *outward* normal drawn to surface and the direction of the stress components on the same surface. When *both* the outer normal and the stress component point in a positive (or negative) direction relative to the coordinate axes, the stress is *positive*. When the normal points in a positive direction while the stress points in a negative direction (or vice versa), the stress is negative. Accordingly, tensile stresses are always positive and compressive stresses always negative.

It is clear that the same sign and the same notation apply no matter which face of a stress element we choose to work with. Figure 2.2 depicts positive normal and shear stresses. This sign convention for stress, which agrees with that adopted for internal forces and moments as discussed in Sec. 1.7, will be used throughout the text.

A study of the three-dimensional state of stress is beyond the scope of this volume, but fortunately for us, ample justification may be found in many engineering applications, for simplifying assumptions with respect to the distribution of stress (and strain). Of special importance, because of the resulting decrease in complexity, are those cases justifying a reduction of a three-dimensional problem to one involving two dimensions or even one dimension. In this regard, we shall describe some special cases for states of stress.

Consider the projection on the xy plane of a thin element and assume that σ_x, σ_y, τ_{xy} do not vary throughout the thickness and that other stress components are zero. When only one normal stress exists, the stress is referred to as a *uniaxial,* or *one-dimensional,* stress (Fig. 2.3a); when only two normal stresses occur, the state of stress is called *biaxial* (Fig. 2.3b). An element subjected to shearing stresses alone (Fig. 2.3c) is said to be in *pure shear*. The combinations of these stress situations, *two-dimensional* stress (Fig. 2.3d), will be analyzed in Chap. 4.

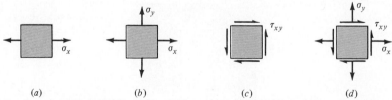

(a)   (b)   (c)   (d)

Figure 2.3 Special cases of state of stress: (a) uniaxial; (b) biaxial; (c) pure shear; and (d) two-dimensional.

2.4 INTERNAL AXIAL FORCES

We are concerned here with the internal forces at a section of a straight bar loaded by forces along its longitudinal axis. The method of sections is employed to calculate these forces. (We shall observe in Chap. 6 that there is a direct analogy between the analysis of axially loaded members and twisted members. In one case we have axial forces and elongations or contractions, in the other, torques and rotational effects.)

The *applied* axial forces and torques are *positive* if their vectors are in the direction of a positive coordinate axis. Hence in Fig. 2.4, loadings 2*P* and 3*P* are positive but *P* and 4*P* are negative. When a bar is subjected to loads at several points along its length, the internal axial forces or torques will vary from section to section. Graphical representation of these variations is often useful for the determination of maximum axial force or torque. A graph showing the variation of axial force along the bar axis is called the *axial-force diagram*. A similar graph for the twisting moment is referred to as the *torque diagram*. These diagrams are usually located below the free-body diagram, or *load diagram,* for the member.

The reactions at the supports of a loaded bar are usually determined first. The axial-force and torque diagrams can be readily constructed after determining the internal forces at various sections in a bar at which the loading conditions change. These sections are called *change-of-load points*. According to the sign convention, internal tensile forces are plotted as positive and compressive forces as negative. The sense of the positive internal-torque vector coincides with that of the tensile force. Hence the axial-force and torque diagrams are constructed by proceeding continuously along a horizontal bar *from the left end,* taking the sum of the forces and torques. We note that the foregoing diagrams are *not* used as commonly as shear and moment diagrams, which will be discussed in Sec. 7.5. This is because, in practice, variations in axial force and twisting moment in a given member are generally of lesser magnitude than variations in shear and moment.

A check on the accuracy of the internal force and torque diagrams can be made by noting whether or not they close. Closure of these diagrams serves to demonstrate that the sum of the forces and moments acting on the member are zero, as they must be for equilibrium. When any of the diagrams fails to close, you will know that there is a construction error or an error in the calculation of the reactions. The following example illustrates the procedure.

EXAMPLE 2.1

An aluminum bar carries the axial loads as shown in Fig. 2.4*a*. Draw the axial-force diagram.

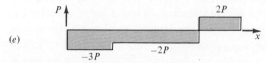

Figure 2.4   Example 2.1: (a) load diagram; (b–d) free-body diagrams of the bar segment; (e) axial-force diagram.

Solution   The change-of-load points are designated A, B, C, and D in the figure. To determine the variations of internal forces, we must pass sections through the segments AB, BC, and CD of the bar. In so doing, we obtain the free-body diagrams shown in Fig. 2.4b, c, and d. Note that a *positive* force is placed on each exposed face. From the equilibrium condition $\Sigma F_x = 0$, we have

$$P_{AB} = -3P \qquad P_{BC} = -2P \qquad P_{CD} = 2P$$

The axial-force diagram can now be plotted by continuously summing the forces from the left end of the bar, as shown in Fig. 2.4e.

Observe that just to the right of A, $P_{AB} = -3P$ must be plotted downward. No other forces are applied until B; hence there is no change in the value of the axial force. To the right of B, tension force decreases the axial force to $-2P$. Similarly, the value of the axial force is increased to $2P$ at C. This remains constant until point D, where the end force $2P$ closes the diagram, proving the accuracy of our work.

2.5   NORMAL STRESS

The condition under which the stress is *constant* or *uniform* at a section within a body is known as *simple stress*. In many load-carrying members, the internal actions on an imaginary cutting plane consist of either only the axial force or only the shear force. Examples of such elements include cables, simple truss members, centrally loaded brace rods and bars, and bolts, pins, and rivets connecting two members. In these bodies, the values of the simple normal stress and shearing stress associated with each action can be approximated directly from the definition of stress and the conditions of equilibrium. However, to learn the "exact" stress distribution, it is necessary to consider the deformations resulting from the particular mode of application of the loads.

Consider, for example, the extension of a prismatic bar subject to an axial force P. A *prismatic bar* or rod is a straight member having constant cross section throughout its length. The front and top views of such a rod are shown in Fig. 2.5a. To obtain the algebraic expression for the normal stress, we make an imaginary cut (section *a-a*) through the member at right angles to its axis. The free-body diagram of the isolated segment of the bar is shown in Fig. 2.5b. Note that the stress is substituted on the cut section as a replacement for the effect of the removed portion. The equilibrium of axial forces requires that $P = A\sigma$, where $A = bh$ is the cross-sectional area of the rod. That is, the system of stress distribution in the rod is statically equivalent to the force P. The *normal stress* is thus

$$\sigma = \frac{P}{A} \tag{2.3}$$

When the rod is being stretched as shown in the figure, the resulting stress is a *uniaxial tensile* stress; if the direction of the forces is reversed, the rod is in compression and *uniaxial compressive* stress occurs. In the latter case, Eq. (2.3) is applicable only to chunky or short members.*

Equation (2.3) represents the value of the *uniform* stress over the cross section rather than the stress at a specified point of the cross section. When a nonuniform stress distribution occurs, then we must deal instead with the *average* stress. A uniform distribution of stress is possible only if three conditions coexist:

1. The axial force P acts through the centroid of the cross section.
2. The rod is straight and made of a homogeneous material.
3. The cross section is remote from the ends of the rod, a situation we shall look at in Sec. 5.8.

---

*Further discussion of uniaxial compressive stress will be found in Sec. 13.5 where we take up the classification of columns.

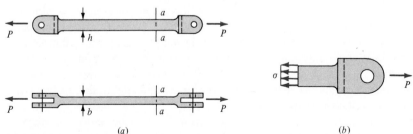

(a)   *(b)*

Figure 2.5   (a) Prismatic bar with clevised, or forked, ends in tension and (b) free-body diagram of the bar segment.

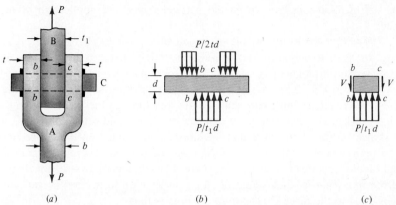

Figure 2.6 (a) A clevis-pin connection; (b) pin in bearing; and (c) pin in double shear.

As can be demonstrated by the equilibrium requirements, bending of the rod will result when the first situation cited above is not realized. In that case, a more accurate analysis is necessary, and we shall take that up in Chap. 7.

In practice, the force P is applied to clevised—forked—ends of the rod through a connection such as shown in Fig. 2.6a. This joint consists of a *clevis* A, a *bracket* B, and a *pin* C. As the force P is applied, the bracket and the clevis press against the rivet in bearing, and a nonuniform pressure develops against the pin (Fig. 2.6b). The average value of this pressure is determined by dividing the force transmitted by the *projected area* of the pin into the bracket (or clevis). This is called the *bearing stress*. The bearing stress in the bracket then equals $\sigma_b = P/(t_1 d)$. Here t_1 is the thickness of the bracket and d is the diameter of the pin. Similarly, the bearing stress in the clevis is given by $\sigma_b = P/(2td)$.

2.6 AVERAGE SHEARING STRESS

A *shearing stress* is produced whenever the applied forces cause one section of a body to tend to slide past its adjacent section. An example is shown in Fig. 2.6, where the pin resists the shear across the two cross-sectional areas at *b-b* and *c-c*. This rivet is said to be in *double shear*. Since the pin as a whole is in equilibrium, any part of it is also in equilibrium. At each cut section, a *shear force V* equivalent to $P/2$, as shown in Fig. 2.6c, must be developed. Thus the shear occurs over an area parallel to the applied load. This condition is termed *direct shear*.

Unlike normal stress, the distribution of shearing stresses τ across a section cannot be taken as uniform. Dividing the total shear force V by the cross-sectional area A over which it acts, we can determine the *average*

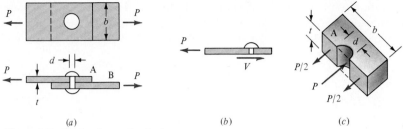

(a)   (b)   (c)

Figure 2.7   Single shear of a rivet.

shearing stress in the section:

$$\tau_{\text{avg}} = \frac{V}{A} = \frac{P}{A} \tag{2.4}$$

Two other examples of direct shear are depicted in Figs. 2.7 and 2.8. Figure 2.7*a* illustrates a connection where plates A and B are joined by a rivet. The rivet resists the shear across its cross-sectional area, a case of *single shear*. The shear force V in the section of the rivet is equal to P (Fig. 2.7*b*). The average shearing stress is therefore $\tau_{\text{avg}} = P/(\pi d^2/4)$.

Note that the rivet exerts a force P on plate A equal and opposite to the total force exerted by the plate on the rivet (Fig. 2.7*c*). The *bearing stress* in the plate is obtained by dividing the force P by the area of the rectangle representing the projection of the rivet on the plate section. As this area is equal to td, we have $\sigma_b = P/(td)$. On the other hand, the average normal stress in the plate on the section through the hole is $\sigma = P/[t(b - d)]$, while on any other section $\sigma = P/(bt)$.

Shown in Fig. 2.8*a* is a typical way in which direct shear stresses are applied to a plate specimen. A hole is to be punched in the plate. The force applied to the punch is designated P. Equilibrium of vertically directed forces requires that $V = P$ (Fig. 2.8*b*). The area resisting the shear force V is analogous to the edge of a coin and equals πtd. Equation (2.4) then yields $\tau_{\text{avg}} = P/(\pi td)$.

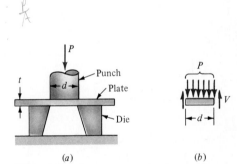

(a)   (b)

Figure 2.8   Direct-shear testing in a cutting fixture.

2.7   APPLICATION TO SIMPLE STRUCTURES

There is an extensive variety of structures that are used in the many fields of engineering. We define a *structure* to be a unit composed of interconnected members supported in such a manner that it is capable of resisting applied forces in static equilibrium. *Machines* are a class of structure designed to transmit and modify forces.

We now illustrate the application of the equilibrium requirement to determine the forces acting within the members and connections of a simple structure (Example 2.2) and a machine (Example 2.3). Once these internal axial and shear forces are obtained, Eqs. (2.3) and (2.4) can readily be applied to compute the average stresses. (Calculation of the displacements in load-carrying members is considered in the chapters to follow, after we develop the stress-strain relationships in Chap. 3.)

EXAMPLE 2.2

A pin-connected truss composed of members AB and BC is subjected to a vertical force $P = 40$ kN at joint B (Fig. 2.9a). Each member is of constant cross-sectional area: $A_{AB} = 0.004$ m² and $A_{BC} = 0.002$ m². The diameter d of all pins is 20 mm, clevis thickness t is 10 mm, and the thickness t_1 of the bracket is 15 mm. Determine the normal stress acting in each member and the shearing and bearing stresses at joint C.

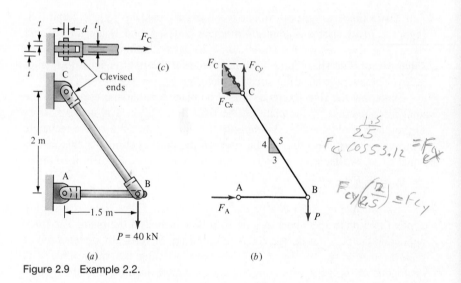

Figure 2.9 Example 2.2.

Solution A free-body diagram of the truss is shown in Fig. 2.9b. The magnitudes of the axially directed end forces of members AB and BC, which are equal to the support reactions at A and C, are labeled F_A and F_C, respectively. For computational convenience the x and y components of the inclined forces are used rather than the forces themselves. Hence force F_C is resolved into F_{Cx} and F_{Cy}, as shown.

Calculation of Support Reactions. Relative dimensions are shown by a small triangle on the member BC in Fig. 2.9b. From the similarity of force and relative-dimension triangles,

$$F_{Cx} = \tfrac{3}{5}F_C \qquad F_{Cy} = \tfrac{4}{5}F_C \qquad (a)$$

$$F_{CA} = \tfrac{3}{4} F_{Cy}$$

It follows then that $F_{Cx} = \tfrac{4}{3} F_{Cy}$. Application of equilibrium conditions to the free-body diagram in Fig. 2.9b leads to

$$\curvearrowleft \Sigma M_C = 0: \quad P(1.5) - F_A(2) = 0 \quad F_A = \tfrac{3}{4}P = 30 \text{ kN} \rightarrow$$

$$\uparrow \Sigma F_y = 0: \quad F_{Cy} - P = 0 \quad F_{Cy} = P = 40 \text{ kN} \uparrow$$

$$\rightarrow \Sigma F_x = 0: \quad -F_{Cx} + F_A = 0 \quad F_{Cx} = F_A = 30 \text{ kN} \leftarrow$$

We thus have

$$F_C = \tfrac{5}{4}P = 50 \text{ kN}$$

Attention is called to the algebraic signs of F_A and F_C. The positive sign means that the sense of each of the forces was assumed correctly in the free-body diagram.

Calculation of Internal Forces.   If imaginary cutting planes are passed perpendicular to the axes of the members AB and BC, separating each into two parts, it is observed that each portion is a two-force member. Therefore the internal forces in each member are the axial forces $F_A = 30$ kN and $F_C = 50$ kN.

Calculation of Stresses.   The normal stresses in each member are

$$\sigma_{AB} = -\frac{F_A}{A_{AB}} = -\frac{30 \times 10^3}{0.004} = -7.5 \text{ MPa}$$

$$\sigma_{BC} = \frac{F_C}{A_{BC}} = \frac{50 \times 10^3}{0.002} = 25 \text{ MPa}$$

where the minus sign indicates compression. Referring to Fig. 2.9c, we see that the double shear in the pin C is

$$\tau_C = \frac{\tfrac{1}{2}F_C}{\pi d^2/4} = \frac{25 \times 10^3}{\pi (0.02)^2/4} = 79.6 \text{ MPa}$$

$$A = \tfrac{\pi}{4} d^2 \qquad \text{pin.}$$

For the bearing stress in the bracket at joint C, we have

$$\sigma_b = \frac{F_C}{t_1 d} = \frac{50 \times 10^3}{(0.015)(0.02)} = 166.7 \text{ MPa}$$

Area $= td.$
where diam pin is d

while the bearing stress in the clevis at joint C is given by

$$\sigma_b = \frac{F_C}{2td} = \frac{50 \times 10^3}{2(0.01)(0.02)} = 125 \text{ MPa}$$

Area (dt)

Area (dt)

The shear and bearing stresses in the other joints are determined in a like manner.

EXAMPLE 2.3

A force P of magnitude 200 N is applied to the handles of the bolt cutter shown in Fig. 2.10a. Compute (a) the force exerted on the bolt and rivets at joints A, B, and C and (b) the normal stress in member AD, which has a uniform cross-sectional area of 2×10^{-4} m^2. Dimensions are given in millimeters.

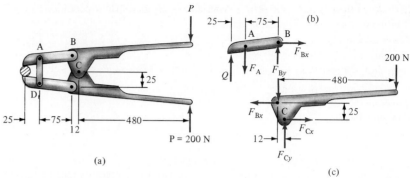

Figure 2.10 Example 2.3.

Solution The conditions of equilibrium must be satisfied by the entire cutter. To determine the unknown forces, we consider component parts. Let the force between the bolt and the jaw be Q. The free-body diagrams for the jaw and the handle are shown in Fig. 2.10b and c. Since AD is a two-force member, the orientation of force F_A is known. It is noted that the force components on the two members at joint B must be equal and opposite, as indicated in the diagrams.
(a)   Referring to the free-body diagram in Fig. 2.10b, we have

$$\rightarrow \Sigma F_x = 0: \qquad\qquad\qquad\qquad\qquad F_{Bx} = 0$$

$$\uparrow \Sigma F_y = 0: \quad Q - F_A + F_{By} = 0 \qquad F_A = Q + F_{By}$$

$$\circlearrowleft \Sigma M_B = 0: \quad Q(0.1) - F_A(0.075) = 0 \qquad F_A = \frac{Q}{0.75}$$

from which $Q = 3F_{By}$. Using the free-body diagram in Fig. 2.10c, we obtain

$$\rightarrow \Sigma F_x = 0: \quad -F_{Bx} + F_{Cx} = 0 \qquad\qquad F_{Cx} = 0$$

$$\uparrow \Sigma F_y = 0: \quad -F_{By} + F_{Cy} - 0.2 = 0 \qquad F_{Cy} = \frac{Q}{3} + 0.2$$

$$\circlearrowleft \Sigma M_C = 0: \quad F_{Bx}(0.025) - F_{By}(0.012) + 0.2(0.48) \qquad F_{By} = 8 \text{ kN}$$

It follows that $Q = 3(8) = 24$ kN. Therefore the shear forces on the rivet at the joints A, B, and C are $F_A = 32$ kN, $F_B = F_{By} = 8$ kN, and $F_C = F_{Cy} = 8.2$ kN, respectively.
(b)   The normal stress in the member AD is given by

$$\sigma = \frac{F_A}{A} = \frac{32 \times 10^3}{2 \times 10^{-4}} = 160 \text{ MPa}$$

The shear stress in the pins of the cutter is investigated as described in Example 2.2. Let us note here that the handles and jaws are subject to combined flexural and shearing stresses, which will be discussed in Chap. 8.

*2.8 APPLICATION TO THIN-WALLED PRESSURE VESSELS

In various applications curved sheets, termed *shells,* are used as structural components. Examples include *pressure vessels* such as pipes, automobile tires, and a variety of containers. The pressure vessels of special importance in engineering are generally classified as thin-walled and thick-walled vessels. A *thin-walled vessel* is one in which the distribution of stress is essentially constant throughout the thickness, while in *thick-walled vessels,* the normal stress varies over the wall thickness. If the ratio of the wall thickness to the inner radius is less than about $1/10$, the vessel is classified as thin-walled. In fact, in thin-walled vessels, there is often no distinction made between the inside and outside radii because they are so nearly equal.

We shall limit our treatment here to the commonly encountered thin-walled cylindrical pressure vessels and spherical pressure vessels. The walls of these shells act as a *membrane*—that is, no bending of the walls takes place. Because of the axisymmetry of the vessel and its contents, the vessel is free to deform under pressure, and hence only uniform normal stresses, so-called *membrane stresses,* exist. This supposition is found to be quite reasonable for thin shells in regions away from external constraints. However, the formulas derived are *not* valid near the ends or supports of the vessel. Any incompatibility at the junction of the cylinder and its end produces bending moments and shearing forces. The stresses associated with this bending and shear are termed *discontinuity stresses,* whose values can be reduced by proper curvature of the ends of the vessel.* It is noted that the equations given in this section apply to cases of *internal* pressure p exerted by a liquid or gas. They pertain equally to vessels under external pressure (for example, vacuum tanks and submarines) with the algebraic sign of p changed. The stresses thus found must, of course, be lower than the critical stresses at which buckling of the walls might occur.

Cylindrical Vessels.   Consider a cylindrical vessel such as an air tank or boiler of inner radius r and wall thickness t containing a fluid under gage pressure p, as shown in Fig. 2.11a. The *gage pressure* refers to the internal pressure in excess of the external pressure. We seek to determine the stresses exerted on a small element of the shell whose sides are parallel and perpendicular to the cylindrical axis. The stress component σ_c is often referred to as the *circumferential,* tangential, or hoop stress, and the stress σ_a is known as the *axial,* or longitudinal, stress.

To evaluate the circumferential stress, we isolate a semicylindrical element of length L of the vessel and the fluid contained therein, as shown in Fig. 2.11b. The weight of the vessel and its contents are assumed to be negligible. Note that for simplicity the pressure and stresses in the axial direction are omitted in this free-body diagram. Referring to the figure, we

*See A.C. Ugural, *Stresses in Plates and Shells,* McGraw-Hill, New York, 1981, secs. 12.7 to 12.11.

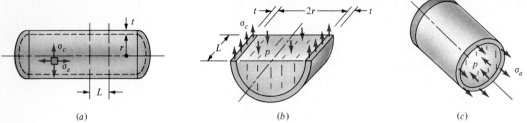

(a) (b) (c)

Figure 2.11 Thin-walled cylindrical pressure vessel.

see that the total force resisted by the walls of the element is $\sigma_c(2tL)$. The resultant fluid force acting on the fluid remaining within the portion of the vessel acting on a section passing through the axis is $p(2rL)$. Thus the condition that the sum of the vertical forces be equal to zero yields $\sigma_c(2tL) = p(2rL)$. After simplifying, we obtain the following expression for circumferential stress:

$$\sigma_c = \frac{pr}{t} \tag{2.5}$$

The axial stress can be determined using the free-body diagram shown in Fig. 2.11c, isolated from the cylinder by a transverse section. Acting on this free body are the force $\sigma_a(2\pi rt)$ on the wall section and the pressure force $p(\pi r^2)$ exerted on the portion of fluid contained therein. Equilibrium of the axial forces leads to $\sigma_a(2\pi rt) = p(\pi r^2)$. From the foregoing, we find the axial stress to be

$$\sigma_a = \frac{pr}{2t} \tag{2.6}$$

From Eqs. (2.5) and (2.6) it is observed that the circumferential stress is twice the axial stress, that is, $\sigma_c = 2\sigma_a$.

Spherical Vessels. The stresses in a spherical vessel can be obtained by using a procedure similar to that described above for cylindrical vessels. Because of symmetry, the stresses exerted on the face of a small element of the wall must be equal; this is shown in Fig. 2.12a. We employ a section through the center of the shell to isolate a hemisphere, as shown in Fig. 2.12b. The equation of equilibrium for this free-body diagram is identical with that of Fig. 2.11c. The circumferential stress is thus

$$\sigma = \frac{pr}{2t} \tag{2.7}$$

This equation yields the same result in *all* directions at any diametral section of the sphere.

(a)   (b)

Figure 2.12   Thin-walled spherical pressure vessel.

Note that the stress acting in the *radial* direction on the wall of a cylinder or sphere varies from $-p$ at the inner surface of the shell to zero at the outer surface. For thin-walled shells, this stress is much smaller than σ_c and σ_a and is generally omitted. The state of stress for an element of a thin-walled vessel as given by Eqs. (2.5), (2.6), and (2.7) is therefore considered *biaxial*.

EXAMPLE 2.4

A steel cylindrical tank with hemispherical ends, as shown in Fig. 2.11a, contains air at a pressure of 200 psi. Calculate the stresses in the vessel if each portion has the same radius of 10 in. and thickness t of $\frac{5}{16}$ in.

Solution   The axial stress in the cylinder and the circumferential stresses in the ends are the same. Applying Eq. (2.6) or (2.7), we have

$$\sigma_a = \sigma = \frac{pr}{2t} = \frac{200(10)}{2(5/16)} = 3200 \text{ psi}$$

Applying Eq. (2.5), we find that the circumferential stress in the cylinder is

$$\sigma_c = \frac{pr}{t} = 2\sigma_a = 6400 \text{ psi}$$

The largest radial stress, occurring on the inner surface of the tank, is $\sigma_r = -200$ psi.

2.9 ALLOWABLE STRESS: FACTOR OF SAFETY

To account for *uncertainties* in various aspects of analysis and design of structures—including those related to service loads, material properties, maintenance, and environmental factors—it is of practical importance to select an adequate *factor of safety*. A significant area of uncertainty is connected with the assumptions made in the analysis of stress and deformation. In addition, one is not likely to have sure knowledge of the stresses that may be introduced during the manufacturing and shipment of a part. For the above-mentioned reasons, the factor of safety is sometimes referred to as the ''factor of ignorance.''

Under certain circumstances, the deformation of a material may continue with time while the *load remains constant*. This deformation, beyond that experienced when the material is initially loaded, is called *creep*. On the other hand, a loss of stress is also observed with time even though the strain level remains constant in a member. Such loss is called *relaxation*; it is basically a relief of stress through the mechanism of internal creep. In materials such as lead, rubber, and certain plastics, creep may occur at ordinary temperatures. Most metals, on the other hand, manifest appreciable creep only when the absolute temperature is roughly 35 to 50 percent of the melting temperature. The rate at which creep proceeds in a given material is dependent not only on temperature but on stress and history of loading as well. In any event, stresses must be kept low in order to prevent intolerable deformations caused by creep.

The factor of safety is thus used to provide assurance that the load applied to a member does not exceed the largest load it can carry. This factor is the *ratio* of the maximum load the member can sustain under testing without failure to the load allowed under service conditions. When a linear relationship exists between the load and the stress caused by the load, the factor of safety f_s may be expressed as*

$$\text{Factor of safety} = \frac{\text{maximum usable stress}}{\text{allowable stress}}$$

or

$$f_s = \frac{\sigma_{\max}}{\sigma_{\text{all}}} \tag{2.8}$$

The *maximum usable stress* $\sigma_{\max}$ represents either the yield stress or the ultimate stress, terms we explore in Sec. 3.6. The *allowable stress* σ_{all} is the working stress. If the factor of safety used is too low and the allowable stress is too high, the structure may prove weak in service. On the other hand, when the working stress is relatively low and the factor of safety relatively high, the structure becomes unnecessarily heavy and uneconomical.

Values of the factor of safety are usually 1.5 or greater. The value is selected by the designer on the basis of experience and judgment. For the majority of applications, pertinent factors of safety are found in various construction and manufacturing codes.

---

*In the field of aeronautical engineering, the *margin of safety* is used instead of the factor of safety. The margin of safety is defined as the factor of safety minus 1, or $f_s - 1$.

EXAMPLE 2.5

Determine the pressure that can be carried by a 3-mm-thick steel cylinder 1.5 m in diameter if the maximum usable strength is $\sigma_{max} = 240$ MPa. Let $f_s = 2$.

Solution We have $\sigma_{all} = 240/2 = 120$ MPa. From Eq. (2.5) we find that the limiting value of pressure for circumferential stress is

$$p = \frac{\sigma_{all}t}{r} = \frac{120 \times 10^6 (0.003)}{0.75} = 480 \text{ kPa}$$

On the other hand, the axial-stress formula, Eq. (2.6), requires

$$p = 2\frac{\sigma_{all}t}{r} = 960 \text{ kPa}$$

Thus the gage pressure may not exceed 480 kPa.

*2.10 DESIGN OF TENSILE AND SHORT COMPRESSION MEMBERS

Of special concern is the design of axially loaded prismatic bars for strength. The application to this case of the rational design outline given in Sec. 1.9 is rather simple:

1. *Evaluate the mode of possible failure:* It is usually assumed that the normal stress is the quantity most closely associated with failure. This assumption applies regardless of the type of failure that may actually take place on a plane of the member, as we shall see in Sec. 5.6.

2. *Determine the relationship between load and stress:* The significant value of the normal stress is given by $\sigma = P/A$.

3. *Determine the maximum usable value of stress:* The maximum usable value of σ without failure is denoted by σ_{max}.

4. *Select the factor of safety:* A factor of safety f_s is applied to σ_{max} to obtain the allowable stress $\sigma_{all} = \sigma_{max}/f_s$. The *required* cross-sectional area of the member is thus

$$A = \frac{P}{\sigma_{all}} \tag{2.9}$$

The foregoing is applicable to prismatic bars in tension and to *short* compression members. Note that the design of *slender* compression members using Eq. (2.9) requires trial-and-error calculations, which will be considered in Sec. 13.7. If the bar contains an abrupt change of cross-sectional area, the procedure outlined above is repeated, using a stress-concentration factor to determine the normal stress (step 2); we shall look at this in Sec. 5.7.

The following examples demonstrate the application of Eq. (2.9) and at the same time provide additional review of statics.

EXAMPLE 2.6

Calculate the cross-sectional areas of the square aluminum post AB and the round steel eyebar AC of the hoist shown in Fig. 2.13a. The required load P is 14 kips. The maximum usable stress in aluminum and steel is approximately 40 and 70 ksi, respectively. Use a factor of safety of 2.5.

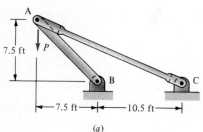

(a)

Figure 2.13 Example 2.6.

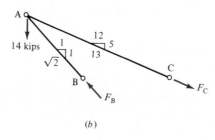

(b)

Solution Members AB and AC are subjected to axial loading. Application of the equations of equilibrium to the free-body diagram of Fig. 2.13b yields

$$\circlearrowleft \Sigma M_B = 0: \quad -14(7.5) + \tfrac{5}{13}F_C(10.5) = 0 \qquad F_C = 26 \text{ kips}$$

$$\circlearrowleft \Sigma M_C = 0: \quad -14(18) + \frac{1}{\sqrt{2}}F_B(10.5) = 0 \qquad F_B = 33.94 \text{ kips}$$

The results can be verified as follows:

$$\rightarrow \Sigma F_x = 0: \quad -\frac{1}{\sqrt{2}}(33.94) + \tfrac{12}{13}(26) = 0$$

The allowable stresses, from Eq. (2.8), are

$$(\sigma_{\text{all}})_{AB} = \frac{40}{2.5} = 16 \text{ ksi} \qquad (\sigma_{\text{all}})_{AC} = \frac{70}{2.5} = 28 \text{ ksi}$$

Applying Eq. (2.9), we find that the required areas of the members are

$$A_{AB} = \frac{33.94 \times 10^3}{16} = 2.12 \text{ in}^2 \qquad A_{AC} = \frac{26}{28} = 0.93 \text{ in}^2$$

Thus a $1\tfrac{1}{2}$-in. by $1\tfrac{1}{2}$-in. commercial-size aluminum post and a $1\tfrac{1}{8}$-in. round steel eyebar are used.

EXAMPLE 2.7

The pin-connected truss shown in Fig. 2.14a is subjected to 50- and 20-kN vertical loads at joints C and E, respectively. Determine the areas of the cross sections of members AC, AD, CD, and CE if the allowable stress is set at 140 MPa in tension and 100 MPa in compression.

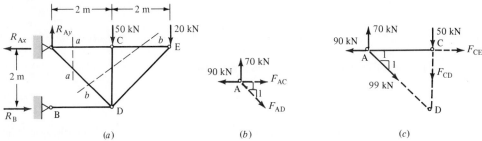

Figure 2.14   Example 2.7.

Solution   The free-body diagram of the entire truss is shown in Fig. 2.14a, where the reactions are denoted R_{Ax}, R_{Ay}, and R_B. Equilibrium of the forces indicated requires that

$$\uparrow \Sigma F_y = 0: \qquad R_{Ay} - 50 - 20 = 0 \qquad\qquad R_{Ay} = 70 \text{ kN} \uparrow$$

$$\circlearrowleft \Sigma M_A = 0: \qquad 50(2) + 20(4) - R_B(2) = 0 \qquad R_B = 90 \text{ kN} \rightarrow$$

$$\rightarrow \Sigma F_x = 0: \qquad -R_{Ax} + R_B = 0 \qquad\qquad R_{Ax} = 90 \text{ kN} \leftarrow$$

Now we pass imaginary cutting planes at a-a and b-b to expose the forces in the members to be designed. Referring to the free-body diagram in Fig. 2.14b, we find:

$$\uparrow \Sigma F_y = 0: \qquad 70 - \frac{1}{\sqrt{2}} F_{AD} = 0 \qquad\qquad F_{AD} = 99 \text{ kN}$$

$$\rightarrow \Sigma F_x = 0: \quad -90 + \frac{1}{\sqrt{2}} (99) + F_{AC} = 0 \qquad F_{AC} = 20 \text{ kN}$$

Using the free-body diagram of Fig. 2.14c, we have

$$\circlearrowleft \Sigma M_D = 0: \quad -90(2) + 70(2) + F_{CE}(2) = 0 \qquad F_{CE} = 20 \text{ kN}$$

$$\uparrow \Sigma F_y = 0: \qquad 70 - 50 - \frac{1}{\sqrt{2}} (99) + F_{CD} = 0 \qquad F_{CD} = -50 \text{ kN}$$

Note as a check that

$$\rightarrow \Sigma F_x = 0: \quad -90 + \frac{1}{\sqrt{2}} (99) + 20 = 0$$

The minus sign means that the direction of F_{CD} is opposite to that assumed in Fig. 2.14c.

Applying Eq. (2.9), we can see that the cross-sectional areas of the members are therefore

$$A_{AC} = A_{CE} = \frac{20 \times 10^3}{140} = 142.9 \text{ mm}^2$$

$$A_{AD} = \frac{99 \times 10^3}{140} = 707.1 \text{ mm}^2$$

and

$$A_{CD} = \frac{50 \times 10^3}{100} = 500 \text{ mm}^2$$

Bars BD and DE are treated in a like manner.

The approach of analyzing a single joint of the truss, as illustrated in Fig. 2.14b, is called the *method of joints*. The analysis of a section of the truss composed of two or more joints, as shown in Fig. 2.14c, is referred to as the *method of sections,* which was set forth in Sec. 1.6.

PROBLEMS

Secs. 2.1 to 2.7

2.1 Determine the normal stress in each segment of the stepped bar shown in Fig. P2.1. Load $P = 20$ kN.

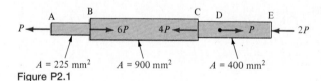

Figure P2.1

2.2 A rod is subjected to the five forces shown in Fig. P2.1. What is the maximum value of P for the stresses not to exceed 100 MPa in tension and 140 MPa in compression?

2.3 Redo Prob. 2.2, assuming that the force $4P$ at C is directed to the right and that the left end A of the rod is fixed.

2.4 The bell-crank mechanism shown in Fig. P2.4 is in equilibrium. Determine (*a*) the diameter of the connecting rod CD, given that the normal stress is limited to 140

MPa; (b) the shearing stress in the 8-mm-diameter pin at B; (c) the bearing stress in
the bracket supports at B; and (d) the bearing stress in the crank at B.

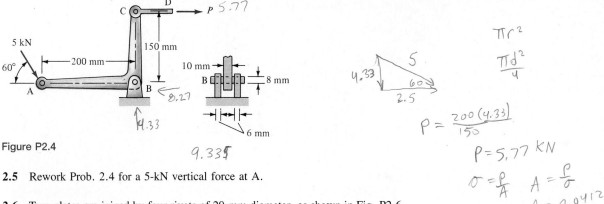

Figure P2.4

2.5 Rework Prob. 2.4 for a 5-kN vertical force at A.

2.6 Two plates are joined by four rivets of 20-mm diameter, as shown in Fig. P2.6.
Determine the maximum load P if the shearing, tensile, and bearing stresses are
limited to 80, 100, and 140 MPa, respectively. Assume that the load is equally divided
among the rivets.

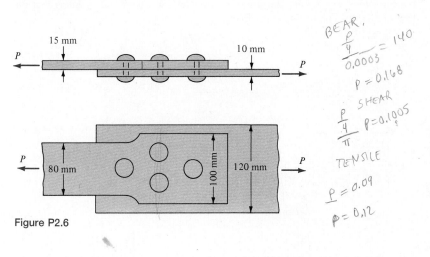

Figure P2.6

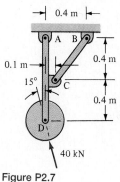

Figure P2.7

2.7 Calculate the shearing stresses produced in the pins at A and B for the landing
gear shown in Fig. P2.7. Assume that each pin has a diameter of 25 mm and is in
double shear.

2.8 The piston, connecting rod, and crank of an engine system are depicted in Fig.
P2.8. Assuming that a force $P = 2.4$ kips acts as indicated, determine (a) the torque
T required to hold the system in equilibrium and (b) the normal stress in the rod AB
if its cross-sectional area is 0.8 in.²

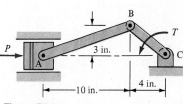

Figure P2.8

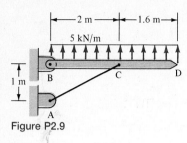

Figure P2.9

2.9 The wing of a monoplane is shown in Fig. P2.9. Determine the normal stress in rod AC of the wing if it has a uniform cross section of 18×10^{-5} m^2.

2.10 A 150-mm pulley subjected to the loads shown in Fig. P2.10 is keyed to a shaft of 25-mm diameter. Calculate the shear stress in the key.

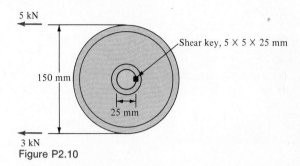

Figure P2.10

2.11 The lap joint seen in Fig. P2.11 is fastened by five 1-in.-diameter rivets. For $P = 10$ kips, determine (a) the maximum shear stress in the rivets; (b) the maximum bearing stress; and (c) the maximum tensile stress at section a-a. Assume that the load is divided equally among the rivets.

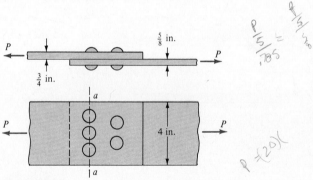

Figure P2.11

2.12 A punch having a diameter d of 1 in. is used to punch a hole in a steel plate with a thickness t of $\frac{1}{16}$ in., so illustrated in Fig. 2.8a. Calculate (a) the force P required if the shear stress in steel is 20 ksi and (b) the corresponding normal stress in the punch.

2.13 Two plates are fastened by a bolt as shown in Fig. P2.13. The nut is tightened to cause a tensile load in the shank of the bolt of 60 kN. Determine (a) the shearing stress in the threads; (b) the shearing stress in the head of the bolt; (c) the bearing stress between the head of the bolt and the plate; and (d) the normal stress in the bolt shank.

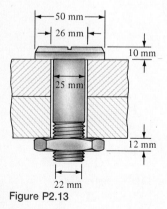

Figure P2.13

2.14 Determine the stresses in members BE and CE of the pin-connected truss of Fig. P2.14. Each bar has a uniform cross-sectional area of 5×10^{-3} m^2.

2.15 The frame of Fig. P2.15 consists of three pin-connected, 1-in.-diameter bars. Calculate the normal stresses in bars AC and AD for $P = 1.2$ kips.

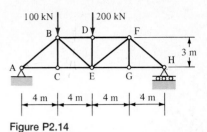

Figure P2.14

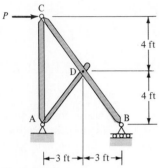

Figure P2.15

2.16 Solve Prob. 2.15 for $P = 3$ kips directed leftward at joint C.

2.17 The two tubes shown in Fig. P2.17 are joined with an adhesive of shear strength $\tau = 2$ MPa. Determine (*a*) the maximum axial load P the joint can transmit and (*b*) the maximum torque T the joint can transmit.

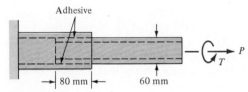

Figure P2.17

2.18 Redo Prob. 2.17, given that two 10-mm-diameter rivets, each having a shear strength of $\tau = 70$ MPa, are used instead of the adhesive. (See Fig. P2.18.)

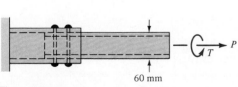

Figure P2.18

$$140\, MPa = \frac{P}{4}$$

$$2\,\frac{\pi d^2}{4} \qquad \frac{\pi d^2}{2}$$

2.19 The connection shown in Fig. P2.19 is subjected to a load $P = 4$ kips. Calculate (*a*) the shear stress in the pin at C; (*b*) the maximum tensile stress in the clevis; and (*c*) the bearing stress in the clevis at C.

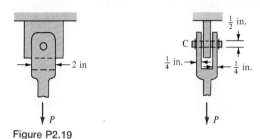

Figure P2.19

2.20 Two rods AC and BC are connected by pins to form a mechanism for supporting a vertical load P at C, as shown in Fig. P2.20. The normal stresses σ in both rods are to be equal. Determine the angle α if the frame is to be of minimum weight.

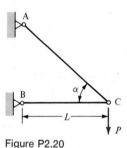

Figure P2.20

2.21 The butt joint of Fig. P2.21 is fastened by four 15-mm-diameter rivets. Determine the maximum load P if the stresses are not to exceed 100 MPa in shear, 140 MPa in tension, and 210 MPa in bearing. Assume that the load is equally divided among the rivets.

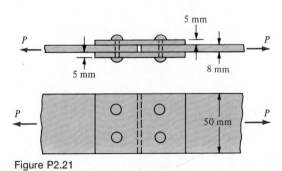

Figure P2.21

2.22 The truss shown in Fig. P2.22 consists of seven pin-connected 50-mm-diameter rods. Calculate (*a*) the normal stress in members AB, AE, and DE and (*b*) the shearing stress in the pin at D if it has a diameter of 25 mm and is in double shear.

2.23 The pin-connected frame shown in Fig. P2.23 supports the loads $Q = 5$ kN and $P = 10$ kN. Determine, for $\alpha = 0°$, (*a*) the normal stress in the bar CE of uniform cross-sectional area 15×10^{-5} m^2 and (*b*) the shearing stresses in the 10-mm-diameter pins at D and E if both are in double shear.

2.24 Rework Prob. 2.23 for the case in which $\alpha = 30°$ and $Q = 0$.

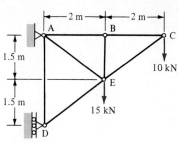
Figure P2.22

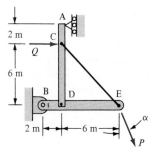
Figure P2.23

2.25 For the frame of Fig. P2.25, calculate (*a*) the normal stress in member BD for a cross-sectional area $A_{BD} = 6 \times 10^{-4}$ m^2 and (*b*) the shearing stress in the pin at A for a diameter of 24 mm in double shear.

2.26 Calculate the shearing stress in the pin at support B of the frame seen in Fig. P2.26. It has a diameter of 20 mm and is in single shear.

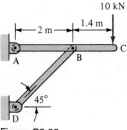
Figure P2.25

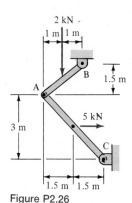
Figure P2.26

2.27 Determine the shearing stress in the pin at hinge B of the frame shown in Fig. P2.27. It has a diameter of 25 mm and is in double shear.

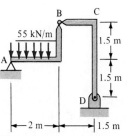
Figure P2.27

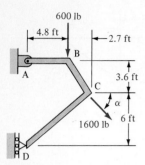

Figure P2.28

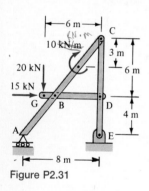

Figure P2.31

2.28 For the frame shown in Fig. P2.28, calculate the shearing stress in the pin at support A if it has a diameter of $\frac{3}{4}$ in. and is in single shear. Use $\alpha = 0°$.

2.29 Rework Prob. 2.28 for the case in which $\alpha = 90°$.

2.30 For the frame shown in Fig. P2.30, determine the shearing stress in the pin at hinge D. Assume that the pin is of 25-mm diameter and acts in double shear.

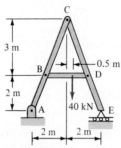

Figure P2.30

2.31 Calculate the shearing stress in the pin at hinge D of the frame shown in Fig. P2.31. It has a diameter of 30 mm and is in single shear.

Secs. 2.8 to 2.10

2.32 What is the required cross-sectional area of rod AC of the monoplane wing of Fig. P2.9 for an allowable stress of 50 MPa?

2.33 A spherical vessel of 2.5 ft radius and $1\frac{1}{2}$-in. wall thickness is submerged in water (density $\gamma = 62.4$ lb/ft³). Based upon a factor of safety of $f_s = 1.5$, calculate the water depth at which the circumferential stress in the sphere would be 4200 psi.

2.34 A compressed-air tank of uniform 5-mm thickness is subjected to an internal pressure of 1.4 MPa, as shown in Fig. P2.34. Determine the maximum axial and circumferential stresses.

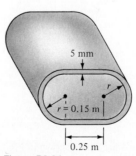

Figure P2.34

2.35 Calculate the required diameter of strut CB of the landing gear of Fig. P2.7. The maximum usable stress of the strut is 140 MPa in tension, and the factor of safety is to be 4.

2.36 Rod AC in the mechanism shown in Fig. P2.20 is to be designed with a factor of safety of 3 with respect to its maximum usable stress of 30 ksi. Calculate the necessary cross-sectional area of the rod. Use $\alpha = 30°$ and $P = 6$ kips.

2.37 The closed-ended cylindrical steel tank seen in Fig. P2.37 has a radius r of 5 m and a height h of 20 m. It is completely filled with a liquid of density $\gamma = 15$ kN/m³ and is subjected to an additional internal gas pressure of $p_0 = 400$ kPa. Based upon an allowable stress of 150 MPa, calculate the wall thickness needed (a) at the top of the tank; (b) at quarter-height; and (c) at mid-height.

2.38 Solve Prob. 2.37, assuming that the gas pressure is 100 kPa.

2.39 Determine the required thickness of a cylindrical vessel 3.6 ft in diameter under a gage pressure of 150 psi. Use a maximum usable stress of 25 ksi and an f_s of 1.5.

2.40 What is the required cross-sectional area of member BC of the frame shown in Fig. P2.40 for a maximum usable stress of 80 MPa? Use $f_s = 2$.

2.41 Calculate the allowable load P for the connection of Fig. P2.19 if the maximum usable stress in the clevis is 30 ksi and the maximum usable shear stress in the pin at C is 22 ksi. Use $f_s = 1.6$.

2.42 A pin-connected truss is shown in Fig. P2.42. If it is subjected to a vertical force P of 10 kips, and the maximum usable stresses are 20 ksi in tension and 14 ksi in compression, what is the required cross-sectional area of each member? Let $f_s = 2$.

2.43 Determine the pin diameter in the metal gate-valve operating shaft of Fig. P2.43 that is needed to resist the torque created by force P of 400 N. The maximum usable stress for the pin is 90 MPa in shear, and the factor of safety is 1.5.

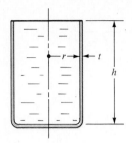

Figure P2.37

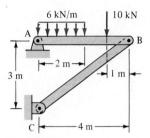

Figure P2.40

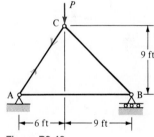

Figure P2.42

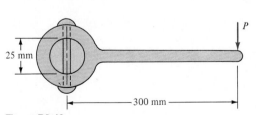

Figure P2.43

2.44 Rework Prob. 2.43 for the case in which $P = 600$ N.

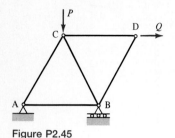

Figure P2.45

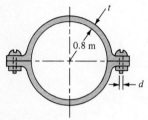

Figure P2.47

2.45 A pin-connected truss in which all members have the same length L is subjected to the forces $P = 10$ kN and $Q = 15$ kN, as shown in Fig. P2.45. Calculate the cross-sectional area of each member for allowable stresses of 120 MPa in tension and 80 MPa in compression.

2.46 Redo Prob. 2.45 for the case in which $P = 0$ and $Q = 25$ kN.

2.47 A spherical vessel of 1.6-m inner diameter is constructed by joining two hemispheres with 40 equally spaced bolts, as shown in Fig. P2.47. The vessel will operate at an internal pressure of 600 kPa. Calculate the bolt diameter d and the vessel thickness t. Use allowable stresses for the bolts and sphere wall of 100 and 50 MPa, respectively.

2.48 A penstock, a pipe for conveying water (specific weight $\gamma = 9.81$ kN/m^3) to a turbine, operates at a head of 120 m. The penstock has a 0.9-m diameter and a wall thickness of t. Determine the minimum required value of t for a maximum usable stress of 100 MPa. Let $f_s = 1.6$.

2.49 Redo Prob. 2.48 for the case in which the allowable stress is 80 MPa.

2.50 Calculate the required pin diameter at support D of the frame shown in Fig. P2.27. Assume that a steel having an allowable strength of 80 MPa is used in the pin, which acts in double shear.

2.51 What is the required pin diameter at support D of the frame shown in Fig. P2.28 for a maximum usable shearing stress of 24 ksi? Use $f_s = 2$ and $\alpha = 0°$. Assume that the pin acts in single shear.

2.52 Rework Prob. 2.51 for the case in which $f_s = 1.5$ and $\alpha = 90°$.

2.53 For the frame shown in Fig. P2.30, determine the required pin diameter at hinge B for an allowable shearing stress of 40 MPa. Assume that the pin acts in double shear.

2.54 Determine the required pin diameter at hinge B of the frame shown in Fig. P2.31 for an allowable shearing stress of 50 MPa. Assume that the pin acts in single shear.

3

STRAIN AND MATERIAL RELATIONS

3.1 INTRODUCTION

In Chap. 2 our concern was with stress within a structure or machine element. We now turn to deformation, the analysis of which is as important as that of stress. The analysis of deformation requires a description of strain, since the latter provides a measure of the intensity of the former (Secs. 3.2 through 3.4).

Section 3.5 defines the significant characteristics of materials. The mechanical properties of engineering materials, as determined from the tension test, are considered in Sec. 3.6. Following this, there is a discussion of the relation between strain and stress under uniaxial, multiaxial, and shear loading conditions. The concept of strain energy is taken up in Sec. 3.9, and the chapter concludes with an introduction to the phenomenon of fracture due to repeated loadings.

3.2 DEFORMATIONS

Consider a body subjected to external forces, as shown in Fig. 1.1a. Assume that, owing to the loading, all points in the body are displaced to new positions. The *displacement* of any point may be a consequence of *deformation,* rigid-body motion (translation and rotation), or some combination of the two. If the *relative positions* of points in the body are altered, the body has experienced deformation. If the distance between any two points in the body remains fixed, yet displacement is evident, the displacement is attributable to rigid-body motion. We shall not treat rigid-body displacements in this book; only *small displacements* by deformation, commonly found in engineering structures, will be considered here.

Extension, contraction, or change of shape of a body may occur as a result of deformation. In order to determine the actual stress distribution within a member, it is necessary to understand the type of the deformation taking place in that member. Examination of the deformations caused by loading, or by a change in temperature in various members within a structure, makes it possible to compute statically indeterminate forces.

We shall designate the total axial deformations by δ (delta). The components of displacement at a point within a body in the x, y, z directions are denoted by u, v, and w, respectively. The strains resulting from small deformations are small compared with unity, and their products (higher-order terms) are omitted. This assumption leads to one of the fundamentals of solid mechanics, the *principle of superposition*. It is valid whenever the quantity (deformation or stress) to be determined is directly proportional to the applied loads, for example, as expressed in Eq. (2.3). In such cases, the total quantity owing to all the loads acting simultaneously on a member may be found by determining *separately* the quantity due to each load and then *combining* the results obtained. The superposition principle permits a complex loading to be replaced by two or more simpler loads and therefore renders a problem more amenable to solution; it will be used repeatedly in this volume.

3.3 STRAIN DEFINED

The concept of normal strain is illustrated by considering the deformation of a prismatic bar (Fig. 3.1a). The initial length of the member is L. After application of a load, the length increases an amount δ (Fig. 3.1b). Defining the *normal strain* ε (epsilon) as the unit change in length, we obtain

$$\varepsilon = \frac{\delta}{L} \tag{3.1}$$

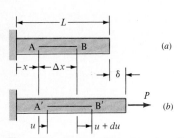

Figure 3.1 Deformation of a prismatic bar.

A *positive* sign applies to *elongation*, a *negative* sign to *contraction*.

The *shearing strain* is the tangent of the total change in angle occurring between two perpendicular lines in a body during deformation. To illustrate, consider the deformation involving a change in shape (distortion) of a rectangular plate (Fig. 3.2). Note that the deformed state is shown by the dashed lines in the figure, where θ' represents the angle between the two rotated edges. Since the displacements considered are small, we can set the tangent of the angle of distortion equal to the angle. Thus the *shearing strain* γ (gamma), measured in radians, is defined as

Figure 3.2 Distortion of a rectangular plate.

$$\gamma = \frac{\pi}{2} - \theta' \tag{3.2}$$

The shearing strain is *positive* if the right angle between the reference lines *decreases,* as shown in the figure; otherwise, the shearing strain is negative.

When uniform changes in angle and length occur, Eqs. (3.1) and (3.2) yield results of acceptable accuracy. In cases of nonuniform deformation, the strains are defined at a point. This state of strain at a point will be discussed in the next section.

In this book, both normal and shear strains are indicated as *dimensionless quantities.* In practice, the normal strains are also frequently expressed in terms of meter (or micrometer) per meter or inch (or microinch) per inch, while shear strains are expressed in radians (or microradians). For most engineering materials, strains seldom exceed values of 0.002 or 2000 μ in the elastic range.

EXAMPLE 3.1

A thin, triangular plate ABC is uniformly deformed into a shape ABC′, as shown by the dashed lines in Fig. 3.3. Calculate (*a*) the normal strain along the centerline OC; (*b*) the normal strain along the edge AC; and (*c*) the shearing strain between the edges AC and BC.

Solution Referring to Fig. 3.3, we have $L_{OC} = b$ and $L_{AC} = L_{BC} = b\sqrt{2} = 1.41421b$.

(*a*) As the change in length OC is $\Delta b = 0.001b$, Eq. (3.1) yields

$$\varepsilon_{OC} = \frac{0.001b}{b} = 0.001 = 1000\ \mu$$

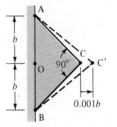

Figure 3.3   Example 3.1.

Note that the answer is read as "1000 micros."

(*b*) The lengths of the deformed edges are $L_{AC'} = L_{BC'} = [b^2 + (1.001b)^2]^{1/2} = 1.41492b$. Thus

$$\varepsilon_{AC} = \varepsilon_{BC} = \frac{1.41492 - 1.41421}{1.41421} = 502\ \mu$$

(*c*) Subsequent to deformation, angle ACB becomes

$$AC'B = 2\ \tan^{-1}\left(\frac{b}{1.001b}\right) = 89.943°$$

The change in the right angle is then $90 - 89.943 = 0.057°$. The corresponding shearing strain (in radians) is

$$\gamma = 0.057\left(\frac{\pi}{180}\right) = 995\ \mu$$

Since the angle ACB is decreased, the shear strain is positive.

*3.4 COMPONENTS OF STRAIN

When uniform deformation does *not* occur, the strains vary from point to point in a body. Then the expressions for uniform strain must relate to a line AB of length Δx (Fig. 3.1*a*). Under the axial load, the end point of the line experiences displacements u and $u + \Delta u$ to become A′ and B′, respectively (Fig. 3.1*b*). That is, an elongation Δu takes place. The definition of normal strain is thus

$$\varepsilon_x = \lim_{\Delta x \to 0} \frac{\Delta u}{\Delta x} = \frac{du}{dx} \tag{3.3}$$

In view of the limit, the foregoing represents the *strain at a point* to which Δx shrinks.

In the case of two-dimensional, or *plane, strain,* all points in the body, before and after application of load, remain in the same plane. Thus the deformation of an element of dimensions dx, dy and of unit thickness can contain linear strains (Fig. 3.4a) and a shear strain (Fig. 3.4b). For instance, the rate of change of u in the y direction is $\partial u / \partial y$, and the increment of u becomes $(\partial u / \partial y) \, dy$. Here $\partial u / \partial y$ represents the slope of the initially vertical side of the infinitesimal element. Similarly, the horizontal side tilts through an angle $\partial v / \partial x$. The partial derivative notation must be used since u or v is a function of x and y. Recalling the basis of Eqs. (3.3) and (3.2), we can use Fig. 3.4 to come to

$$\varepsilon_x = \frac{\partial u}{\partial x} \qquad \varepsilon_y = \frac{\partial v}{\partial y} \qquad \gamma_{xy} = \frac{\partial v}{\partial x} + \frac{\partial u}{\partial y} \tag{3.4a}$$

Clearly, γ_{xy} represents the shearing strain between the x and y (or y and x) axes. Hence we have $\gamma_{xy} = \gamma_{yx}$.

Strains at a point in a rectangular prismatic element of sides dx, dy, and dz are obtained in a like manner. The *three-dimensional strain* components are ε_x, ε_y, γ_{xy}, and

$$\varepsilon_z = \frac{\partial w}{\partial z} \qquad \gamma_{yz} = \frac{\partial v}{\partial z} + \frac{\partial w}{\partial y} \qquad \gamma_{xz} = \frac{\partial u}{\partial z} + \frac{\partial w}{\partial x} \tag{3.4b}$$

where $\gamma_{xz} = \gamma_{zx}$ and $\gamma_{yz} = \gamma_{zy}$. (The sign convention for the strains is given in Sec. 3.3.) Equations (3.4) express the strain tensor in a manner like that of the stress tensor of Sec. 2.3. If the values of the above strains are known at a point, the increase in size and the change of shape of an element at that point are completely determined.

We observe from the foregoing definitions that the six strain components depend *linearly* on the derivatives of the three displacement components. Therefore the strains cannot be independent of one another. Six expressions,

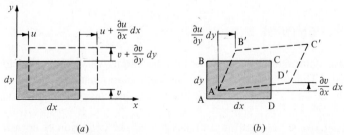

Figure 3.4 Deformations of an element: (a) linear strain and (b) shear strain.

known as the *equations of compatibility,* can be derived to show the inter-relationships among ε_x, ε_y, ε_z, γ_{xy}, γ_{yz}, and γ_{xz}. The number of such equations becomes *one* for a two-dimensional problem. The expressions of compatibility assert that the deformation of a body is continuous. Physically, this means that *no voids are created in the body.* The approach of the theory of elasticity is based upon the requirement of strain compatibility as well as on stress equilibrium and on the general relationships between stresses and strains.*

In the method of mechanics of materials, basic assumptions are made concerning the *distribution of strains* in the body as a whole so that the difficult task of solving Eqs. (3.4) and of satisfying the equations of compatibility is avoided. The assumptions regarding the strains are usually based upon the measured strains. Thus our purpose in studying the nature of the deformations is only twofold: to satisfy the requirements of geometric compatibility with restraints in statically indeterminate structures and to calculate the average values of the deflections and the strains.

EXAMPLE 3.2

A 0.4-m by 0.4-m square ABCD is drawn on a thin plate prior to loading. Subsequent to loading, the square has the dimensions shown by the dashed lines in Fig. 3.5. Determine the average values of the plane-strain components at corner A.

Solution   Let the original lengths of a rectangular element of unit thickness be Δx and Δy. An *approximate version* of Eqs. (3.4a), representing Eqs. (3.1) and (3.2), is then

$$\varepsilon_x = \frac{\Delta u}{\Delta x} \qquad \varepsilon_y = \frac{\Delta v}{\Delta y} \qquad \gamma_{xy} = \frac{\Delta u}{\Delta y} + \frac{\Delta v}{\Delta x} \qquad (3.5)$$

where u and v are, respectively, the x- and y-directed displacements of a point.

For the square under consideration, we have $\Delta x = \Delta y = 400$ mm. Application of Eqs. (3.5) to Fig. 3.5 yields

$$\varepsilon_x = \frac{u_D - u_A}{\Delta x} = \frac{0.7 - 0.3}{400} = 1000\ \mu$$

$$\varepsilon_y = \frac{v_B - v_A}{\Delta y} = \frac{-0.25 - 0}{400} = -625\ \mu$$

Similarly,

$$\gamma_{xy} = \frac{u_B - u_A}{\Delta y} + \frac{v_D - v_A}{\Delta x} = \frac{0 - 0.3}{400} + \frac{0.1 - 0}{400} = -500\ \mu$$

The negative sign indicates that angle BAD has increased.

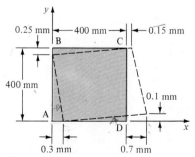

Figure 3.5   Example 3.2.

*A. C. Ugural and S. K. Fenster, *Advanced Strength and Applied Elasticity,* 2d ed., Elsevier, New York, 1987, chap. 3.

3.5   ENGINEERING MATERIALS

In the case of the one-dimensional problem of an axially loaded member, stress-load and strain-displacement relations represent two equations involving three unknowns—stress σ_x, strain ε_x, and displacement u. The insufficient number of available expressions is compensated for by a material-dependent relationship connecting stress and strain. Hence the loads acting on a member, the resulting displacements, and the *mechanical properties of the materials* can be associated.

We now define some important characteristics of commonly used engineering materials (for example, various metals, plastics, wood, ceramics, glass, and concrete). In the next section we shall discuss the tensile test that provides information basic to material behavior, and typical properties of some materials are listed in Table B.1 in App. B.*

An *elastic material* is one which returns to its original (unloaded) size and shape after the removal of applied forces. The elastic property, or so-called *elasticity,* thus precludes permanent deformation. Usually the elastic range includes a region throughout which stress and strain bear a linear relationship. This portion of the stress-strain variation ends at a point called the *proportional limit*. Materials having this elastic range are said to be *linearly elastic*.

In the case of *plasticity,* total recovery of the size and shape of a material does not occur. With the exception of our discussion in Chap. 12, consideration in this text will be limited to elastic materials.

A material is said to behave in a *ductile* manner if it can undergo large strains prior to fracture. Ductile materials, which include structural steel (a low-carbon steel or mild steel), many alloys of other metals, and nylon, are characterized by their ability to yield at normal temperatures. The converse applies to *brittle* materials. That is, a brittle material (for example, cast iron and concrete) exhibits little deformation before rupture and, as a result, fails suddenly without visible warning. A member that ruptures is said to *fracture*. It should be noted that the distinction between ductile and brittle materials is not as simple as might be inferred from the above discussion. The nature of stress, the temperature, and the rate of loading all play a role in defining the boundary between ductility and brittleness.

A *composite* material is made up of two or more distinct constituents. Composites usually consist of a high-strength material (for example, fibers made of steel, glass, graphite, or polymers) embedded in a surrounding material (for example, resin, concrete, or nylon), which is termed a *matrix*. Thus a composite material exhibits a relatively large strength-to-weight ratio compared with a homogeneous material; composite materials generally have

---

*For details, see T. Baumeister, *Mark's Mechanical Engineering Handbook,* 8th ed., McGraw-Hill, New York, 1978, and *Annual Book of ASTM,* American Society for Testing Materials, Philadelphia.

other desirable characteristics and are widely used in various structures, pressure vessels, and machine components.

We assume in this volume that materials are homogeneous and isotropic. A *homogeneous* solid displays identical properties throughout. If the properties are identical in all directions at a point, the material is *isotropic*. A nonisotropic, or *anisotropic,* material displays direction-dependent properties. Simplest among these are those in which the material properties differ in *two* mutually perpendicular directions. A material so described (for example, wood) is *orthotropic*. Mechanical processing operations such as cold-rolling may contribute to minor anisotropy, which in practice is often ignored. Mechanical processes and/or heat treatment may also cause high (as large as 70 or 105 MPa) internal stress within the material. This is termed *residual stress* or *initial stress*. In the cases treated in this text, materials are assumed to be entirely free of such stress. Another kind of residual stress caused by nonuniform plastic deformation will be discussed in Chap. 12.

3.6 STRESS-STRAIN DIAGRAM

The *stress-strain diagram* is used to explain a number of material properties useful in the study of mechanics of materials. Data for this diagram is usually obtained from a *tensile test*. In such a test, a *specimen* of the material, usually in the form of a round bar (Fig. 3.6*a*), is mounted in the grips of a testing machine (Fig. 3.6*b*) and subjected to centric tensile loading, applied *slowly*

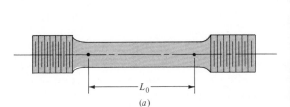

L_0

(*a*)

(*b*)

Figure 3.6 (a) Typical tensile test specimen and (b) testing machine with 60,000-lb capacity load frame, shown with manual control and the DS-50 Data System. (*Photo courtesy of Tinius Olsen Testing Machine Co., Inc., Willow Grove, Pa.*)

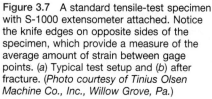

(a)

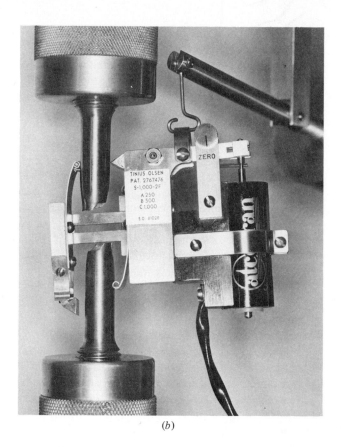

(b)

Figure 3.7 A standard tensile-test specimen with S-1000 extensometer attached. Notice the knife edges on opposite sides of the specimen, which provide a measure of the average amount of strain between gage points. (a) Typical test setup and (b) after fracture. (*Photo courtesy of Tinius Olsen Machine Co., Inc., Willow Grove, Pa.*)

and steadily at room temperature. The ends of the bar may be of any geometric form to fit the test machine in such a way that the load is axial. A gage, called an *extensometer,* is attached to the specimen (Fig. 3.7a). The testing procedure consists of applying successive increments of load while taking corresponding extensometer readings of the elongation between two gage marks on the bar. In this way, a complete stress-strain diagram, a plot with strain as abscissa and stress as ordinate, can be determined for the material.

The typical shapes of two stress-strain plots for a structural steel are shown in Fig. 3.8a. The stress-strain curve for the initial portion of the diagram is replotted using an *enlarged* strain scale M in the figure. Curve OABCDE represents the engineering stress-engineering strain relationship. It is called the *conventional stress-strain diagram.* The other curve, OABCF, represents the true stress-true strain relationships. The *true stress* refers to the load divided by the *actual* instantaneous cross-sectional area of the bar. On the other hand, *engineering stress* equals the load divided by the *initial* cross-sectional area. Note that engineering strain is found by dividing the elongation by the distance L_0, which is known as the *gage length* of the specimen. Clearly, the sum of the elongation increments divided by the corresponding *actual* momentary length L of the bar is the true strain. This strain, also termed the *natural*, or *logarithmic, strain,* is equal to $\int_{L_0}^{L}(dL/L)$ $= \ln(1+\varepsilon)$, where ε is the engineering strain.

For the small elastic strains usually encountered in engineering applications, conventional-strain and true-strain formulas give nearly the same results. Thus the curve of true stress versus true strain is more informative in studying plastic behavior of an axially loaded member; and true stress and true strain are, in general, useful in conducting research on materials. The following discussion is limited to engineering stress and engineering strain.

The portion OA of the diagram in Fig. 3.8a is the *elastic range*. The *linear* variation of stress-strain ends at the *proportional limit* (point A). When the load on the bar is increased beyond point A, a stress level is reached at which the material continues to elongate without an increase in applied stress. For most cases, the proportional limit and the *yield point* (A and B) are *taken as one*. That is, $\sigma_{pl} \approx \sigma_{yp}$, where the subscripts "pl" and "yp" denote the proportional limit and yield point, respectively.

That portion of the stress-strain curve extending from the proportional limit to the *point of fracture* (E) is the *plastic range*. The characteristic of the material in the range CD is such that an increase in stress is required for a continued increase in strain; this is referred to as *strain hardening*. The conventional stress curve for the material when strained beyond C shows a typical maximum called the *ultimate stress* (D), σ_u, and a lower value, the fracture stress (E), σ_f, at which failure takes place by separation of the bar into two parts. Note that in Fig. 3.7b the ductile fracture occurs along a cone-shaped surface forming an angle of approximately 45° with the axis of the specimen.

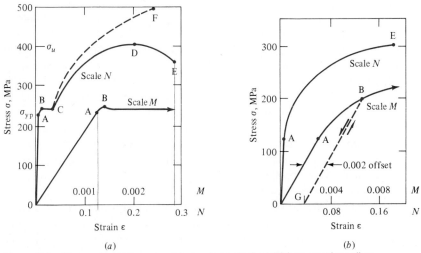

Figure 3.8 Stress-strain diagrams: (a) structural steel and (b) magnesium alloy.

The large disparity between the conventional stress and the true stress curves in the plastic range is attributable to the reduction in the cross-sectional area of the bar in tension. The pronounced lateral contraction that occurs near the breaking point is called *necking* (Fig. 3.9). It is difficult to determine the onset of the necking and to differentiate it from the uniform decrease in diameter of the bar. For practical purposes, however, the conventional stress-strain curve, based upon the initial dimensions of the specimen, provides satisfactory information for structural design.

The standard measures of ductility of a material are defined as follows:

$$\text{Percent elongation} = \frac{L_f - L_0}{L_0} (100) \qquad (3.6a)$$

and

$$\text{Percent reduction in area} = \frac{A_0 - A_f}{A_0} (100) \qquad (3.6b)$$

Here A_0 and L_0 denote, respectively, the initial cross-sectional area and gage length of the specimen. The ruptured bar must be pieced together in order to measure the final gage length L_f. Similarly, the final area A_f is measured at the fracture site where the cross section is minimum. For structural steel, about 25 percent elongation (for a 50-mm gage length) and 50 percent reduction in area are common.

Certain materials, such as aluminum, magnesium, and copper do not exhibit a distinctive yield point, and it is usual to employ a *nominal* yield stress. Figure 3.8b shows the stress-strain diagram for a magnesium alloy. As before, the elastic limit (or proportional limit) and the fracture point are designated by A and E, respectively, in the figure. According to the so-called *0.2 percent offset method,* a line is drawn through a strain of 0.002, parallel to the initial slope at point 0 of the curve. The intersection of this line with the stress-strain curve (point B) defines the yield stress, which is commonly referred to as the *yield strength.*

If a material stressed into the plastic range is unloaded, its stress-strain diagram (BG) is parallel to the loading portion OA, as shown in Fig. 3.8b. A permanent *residual strain* (so-called permanent set) remains in the material. On reloading (GB), the unloading path is retraced and a further loading leads to a continuation of the original stress-strain curve.

Diagrams similar to those in tension may also be determined for a variety of materials in *compression*. For some ductile materials (for example, steel), it is found that the yield-point stress is about the same in tension as in compression. Many brittle materials have ultimate stresses in compression that are much greater than in tension.

Finally, it should be noted that the properties of a material can be determined from direct-shear tests or torsion tests, which will be discussed in Sec. 6.5. *Shear stress-strain diagrams* can be obtained from the results of

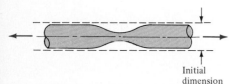

Initial
dimension

Figure 3.9 Necking of a specimen of ductile material in tension.

these tests. These diagrams of τ versus γ are analogous to those seen in Fig. 3.8 for the same materials. However, properties such as yield stress and ultimate stress are usually half as large in shear as they are in tension. For instance, for ductile materials yield stress in shear is about 0.5 to 0.6 times the yield stress in tension.

3.7 HOOKE'S LAW AND POISSON'S RATIO

As has been pointed out in the preceding section, most structural materials have an initial region of the stress-strain curve in which the material behaves both elastically and linearly. This *linear elasticity* is a highly significant property of many engineering materials. For the straight-line portion of the diagram (Fig. 3.8), the stress is directly proportional to the strain:

$$\sigma = E\varepsilon \tag{3.7}$$

The above relationship is known as *Hooke's law* after R. Hooke. The constant E is called the *modulus of elasticity,* or *Young's modulus,* in honor of T. Young (1773–1829). As ε is a dimensionless quantity, E has the units of σ. Hence E is expressed in pascals (or one of its multiples) in SI units and in pounds (or kilopounds) per square inch in the U.S. Customary System.

The modulus of elasticity is observed to be the *slope* of the stress-strain diagram in the linearly elastic region and is different for various materials. For most steels, E lies between 200 and 210 GPa or 29×10^6 and 30×10^6 psi, equal in tension and compression. The slope of the stress-strain diagram beyond the proportional limit is defined as the *tangent modulus E_t,* that is, $E_t = d\sigma/d\varepsilon$. The ratio of stress to strain at any point on the curve above the proportional limit is called the *secant modulus E_s,* that is, $E_s = \sigma/\varepsilon$. Below the proportional limit, both E_t and E_s are equal to the modulus of elasticity E (Fig. 3.10). These quantities are used as measures of the *stiffness* of material in tension or compression.

Elasticity can similarly be measured in a member subjected to shear loading. Referring to Eq. (3.7) for the linearly elastic part of the shear stress-strain diagram, we write

$$\tau = G\gamma \tag{3.8}$$

This is called the *Hooke's law for shear stress and shear strain.* The constant G is termed the *modulus of rigidity,* or *shear modulus of elasticity,* of the material and is expressed in the same units as E—that is, in pascals (Pa) or in pounds per square inch (psi).

It was stated in Sec. 3.6 that the axial (longitudinal) tensile load induces reduction in the cross-sectional area of the specimen (lateral contraction). In a like manner, a contraction owing to a compressive axial load is accompanied by a lateral extension. In the *elastic* range, the ratio of the lateral

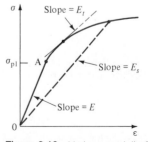

Figure 3.10 Various moduli of elasticity.

strain to the axial strain is constant and is known as *Poisson's ratio,* denoted by ν (nu), after S. D. Poisson (1781–1840):

$$\nu = -\frac{\text{lateral strain}}{\text{axial strain}} \tag{3.9}$$

where the minus sign indicates that the lateral strain is of sense opposite to that of the axial strain. We note that this definition is valid only for a uniaxial state of stress. In the cases of multiaxial stress, the separate strain effects caused by each stress are summed in the required direction, as will be shown in the next section. It is found experimentally that the values of ν are usually in the range 0.25 to 0.35. For steels, Poisson's ratio is taken as 0.3. Extreme cases include a ν of 0.1 for some concretes and a ν of 0.5 for rubber.

Volume Change. The lateral contraction of a cubic element in tension is illustrated in Fig. 3.11, where it is assumed that the faces of the cube at the origin are fixed in position. The deformations shown are greatly exaggerated. For the loading condition represented in the figure, we have $\sigma_y = \sigma_z = 0$, and σ_x is the axial stress. Thus the transverse strains are connected to the axial strain by Eqs. (3.7) and (3.9) as follows:

$$\varepsilon_y = \varepsilon_z = -\nu\varepsilon_x = -\nu\frac{\sigma_x}{E} \tag{a}$$

We observe from the figure that subsequent to deformation, the final volume of the element is

$$V_f = (1 + \varepsilon_x)\,dx \cdot (1 + \varepsilon_y)\,dy \cdot (1 + \varepsilon_z)\,dz$$

Expanding the right side and neglecting higher-order terms involving ε_x^2 and ε_x^3, we obtain

$$V_f = [1 + (\varepsilon_x + \varepsilon_y + \varepsilon_z)]\,dx\,dy\,dz = V_0 + \Delta V$$

in which V_0 is the initial volume $dx\,dy\,dz$ and ΔV is the change in volume. The *unit volume change e* is therefore defined as

$$e = \frac{\Delta V}{V_0} = \varepsilon_x + \varepsilon_y + \varepsilon_z \tag{3.10}$$

Substitution of Eq. (*a*) into this expression yields

$$e = (1 - 2\nu)\,\varepsilon_x = \frac{1 - 2\nu}{E}\sigma_x \tag{b}$$

The quantity e is also referred to as the *dilatation.* It is observed from the foregoing result that a tensile force increases and a compressive force decreases the volume of the element.

Interestingly, in the case of an incompressible material, we have $e = 0$ and Eq. (*b*) shows that $1 - 2\nu = 0$, or $\nu = 0.5$. For most materials in the

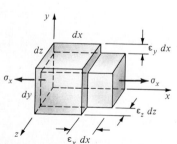

Figure 3.11 Lateral contraction of an element in tension.

linear elastic range $\nu < 0.5$, since some change in volume occurs. In the plastic region, however, the volume remains nearly constant and hence ν is taken as 0.5.

EXAMPLE 3.3

A steel rectangular block, $a = 1$ in. wide and $b = \frac{1}{2}$ in. deep, is subjected to an axial tensile load as shown in Fig. 3.12. Measurements show the block to increase in length by $\delta_x = 2.8 \times 10^{-3}$ in. (initial length $L = 4$ in.) and to decrease in width by $\delta_z = 0.21 \times 10^{-3}$ in., when P is 10.5 kips. Calculate the modulus of elasticity and Poisson's ratio for the material.

Solution   The cross-sectional area of the block is $A = 1 \times \frac{1}{2} = 0.5$ in.^{2}. The axial stress and strain are

$$\sigma_x = \frac{P}{A} = \frac{10.5 \times 10^3}{0.5} = 21 \text{ ksi}$$

$$\varepsilon_x = \frac{\delta_x}{L} = \frac{2.8 \times 10^{-3}}{4} = 700 \ \mu$$

The transverse strain in the z direction is

$$\varepsilon_z = \frac{\delta_z}{a} = -\frac{0.21 \times 10^{-3}}{1} = -210 \ \mu$$

The depth of the block contracts by $\delta_y = 210(10^{-6})(0.5) = 0.105 \times 10^{-3}$ in., since $\varepsilon_y = \varepsilon_z$. Formulas (3.7) and (3.9) result in the values

$$E = \frac{\sigma_x}{\varepsilon_x} = \frac{21 \times 10^3}{700 \times 10^{-6}} = 30 \times 10^6 \text{ psi}$$

and

$$\nu = -\frac{\varepsilon_y}{\varepsilon_x} = \frac{210 \ \mu}{700 \ \mu} = 0.3$$

for the modulus of elasticity and Poisson's ratio, respectively.

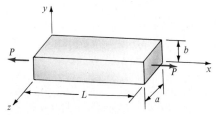

Figure 3.12   Example 3.3.

3.8 GENERALIZED HOOKE'S LAW

The uniaxial stress-strain relation [Eq. (3.7)] may be extended to include biaxial and triaxial states of stress often encountered in engineering applications. In the derivations which follow, we rely upon certain experimental evidence: A normal stress causes no shearing strain whatever, and a shearing stress (τ_{xy}) produces only a shearing strain (γ_{xy}). Also, the principle of superposition applies under multiaxial loading, since the strain components are small quantities. These assumptions are limited to isotropic materials stressed in the linearly elastic range.

Consider a structural element of unit thickness subjected to a biaxial state of stress (Fig. 3.13a). Under the action of the stress σ_x, not only would the direct strain σ_x/E occur, but a y contraction as well, $-\nu\sigma_x/E$, indicated by the dashed lines in the figure. Similarly, were σ_y to act alone, an x contraction $-\nu\sigma_y/E$ and a y strain σ_y/E would result. Thus simultaneous action of both stresses σ_x and σ_y results in the following strains in the x and y directions:

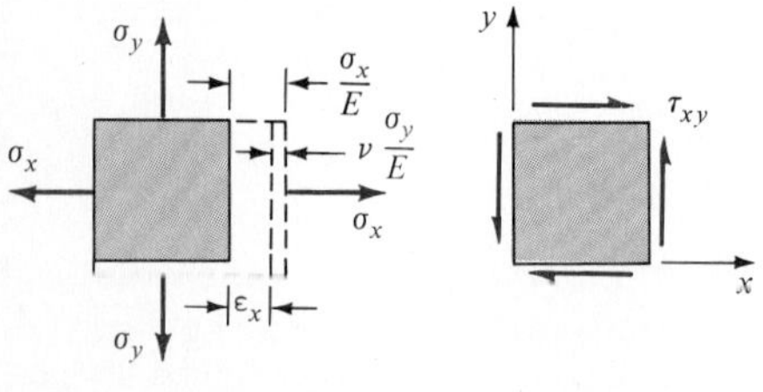

$$\varepsilon_x = \frac{\sigma_x}{E} - \nu\frac{\sigma_y}{E} \tag{3.11a}$$

$$\varepsilon_y = \frac{\sigma_y}{E} - \nu\frac{\sigma_x}{E} \tag{3.11b}$$

The elastic stress-strain relation [Eq. (3.8)], under the state of *pure shear* (Fig. 3.13b), is of the form

$$\gamma_{xy} = \frac{\tau_{xy}}{G} \tag{3.11c}$$

That is, τ_{xy} produces only its corresponding shearing strain γ_{xy}. From Eqs. (3.11) we obtain the following stress-strain relations:

$$\sigma_x = \frac{E}{1 - \nu^2}(\varepsilon_x + \nu\varepsilon_y)$$

$$\sigma_y = \frac{E}{1 - \nu^2}(\varepsilon_y + \nu\varepsilon_x) \tag{3.12}$$

$$\tau_{xy} = G\gamma_{xy}$$

Equations (3.11) or (3.12) represent *Hooke's law for two-dimensional stress*.

An identical analysis enables one to connect the components ε_z, γ_{yz}, γ_{xz} of strain with stress and material properties. The foregoing procedure is readily extended to a *three-dimensional* state of *stress* (see Fig. 2.2). Then the strain-stress relations, known as the *generalized Hooke's law*, consist of the expressions:

(a)   (b)

Figure 3.13 Element in state of (a) biaxial stress and (b) pure shear.

$$\varepsilon_x = \frac{1}{E}[\sigma_x - \nu(\sigma_y + \sigma_z)]$$

$$\varepsilon_y = \frac{1}{E}[\sigma_y - \nu(\sigma_x + \sigma_z)]$$

$$\varepsilon_z = \frac{1}{E}[\sigma_z - \nu(\sigma_x + \sigma_y)]$$

(3.13)

$$\gamma_{xy} = \frac{\tau_{xy}}{G} \qquad \gamma_{yz} = \frac{\tau_{yz}}{G} \qquad \gamma_{xz} = \frac{\tau_{xz}}{G}$$

Recall that a positive value for a stress (strain) component signifies tension (extension), and a negative value compression (contraction). Thus if a particular normal stress is compressive, the sign of the corresponding term in Eqs. (3.13) changes.

We observe from Fig. 3.4b that changes in the lengths of the diagonals of the element are related geometrically to the shearing strain γ_{xy} as well as to the normal strains ε_x and ε_y. We may thus conclude that the shearing modulus G is related to the modulus of elasticity E and Poisson's ratio ν. It will be demonstrated in Sec. 4.8, after we have studied the variation of strain around a point, that the connecting expression is

$$G = \frac{E}{2(1 + \nu)}$$

(3.14)

Therefore, for an isotropic material there are only *two independent* elastic constants. Provided that the constants E and ν are obtained from a tensile test for a given material, G can be calculated from the foregoing basic relationship. Note that G is always less than E, since ν is a positive constant.

EXAMPLE 3.4

Reconsider Fig. 3.12, with the block subjected to a uniform pressure of $p = 150$ MPa acting on all faces and $P = 0$. Calculate the change in volume and dimensions for $a = 40$ mm, $b = 30$ mm, and $L = 100$ mm. Use $E = 200$ GPa and $\nu = 0.3$.

Solution Inserting $\sigma_x = \sigma_y = \sigma_z = -p$ into Eqs. (3.13) and setting $\varepsilon_x = \varepsilon_y = \varepsilon_z = \varepsilon$, we obtain

$$\varepsilon = -\frac{p}{E}(1 - 2\nu)$$

(3.15)

The given numerical values are substituted into the above to yield

$$\varepsilon = -\frac{150}{200(10^3)}(1 - 0.6) = -300\ \mu$$

Thus the change in the volume of the block, using Eq. (3.10), is $\Delta V = -3\varepsilon(abL) = -108$ mm^3. Deformations in the x, y, and z directions are, respectively,

$$\delta_x = (-300 \times 10^{-6})(100) = -0.03 \text{ mm}$$

$$\delta_y = (-300 \times 10^{-6})(30) = -0.009 \text{ mm}$$

$$\delta_z = (-300 \times 10^{-6})(40) = -0.012 \text{ mm}$$

where the minus sign indicates contraction.

EXAMPLE 3.5

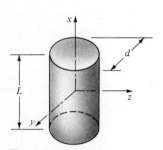

Figure 3.14   Example 3.5.

A short, solid cast-iron cylinder, as seen in Fig. 3.14, is subjected to axial and radial compressive stresses 50 and 10 MPa, respectively. For $E = 100$ GPa, $\nu = 1/4$, $d = 120$ mm, and $L = 200$ mm, determine the change in (a) the length ΔL and diameter Δd and (b) the volume of the cylinder ΔV.

Solution Note that $\sigma_x = -40$ MPa and along any diameter $\sigma_y = \sigma_z = \sigma = -10$ MPa. The corresponding axial and radial strains, using Eqs. (3.13), are

$$\varepsilon_x = -\frac{1}{E}[\sigma_x - \nu(\sigma + \sigma)]$$

$$= -\frac{10^6}{100 \times 10^9}\left[40 - \frac{1}{4}(10 + 10)\right] = -350 \ \mu$$

$$\varepsilon_y = \varepsilon_z = \varepsilon = -\frac{1}{E}[\sigma - \nu(\sigma + \sigma_x)]$$

$$= -\frac{1}{100 \times 10^3}\left[10 - \frac{1}{4}(10 + 40)\right] = 25 \ \mu$$

(a) The decrease in length and increase in diameter are

$$\Delta L = \varepsilon_x L = (-350 \times 10^{-6})(200) = -0.07 \text{ mm}$$

$$\Delta d = \varepsilon d = (25 \times 10^{-6})(120) = 0.003 \text{ mm}$$

(b) From Eq. (3.10),

$$e = \varepsilon_x + 2\varepsilon = (-350 + 2 \times 25)10^{-6} = -300 \times 10^{-6}$$

Therefore

$$\Delta V = eV_0 = (-300 \times 10^{-6})[\pi(60)^2(200)] = -679 \text{ mm}^3$$

where the negative sign connotes a decrease in the volume of the cylinder.

EXAMPLE 3.6

A long, thin plate of thickness t, width b, and length L carries an axial load P, as shown in Fig. 3.15. The edges at $y = \pm b/2$ are placed between the two smooth, rigid walls so that lateral expansion in the y direction is prevented. Determine the components of stress and strain.

Solution For the situation described, we have $\gamma_{xy} = \gamma_{yz} = \gamma_{xz} = 0$, $\varepsilon_y = 0$, $\sigma_z = 0$, and $\sigma_x = -P/bt$. Equations (3.13) then reduce to

$$\varepsilon_x = \frac{1}{E}(\sigma_x - \nu\sigma_y) \qquad (a)$$

$$0 = \frac{1}{E}(\sigma_y - \nu\sigma_x) \qquad (b)$$

$$\varepsilon_z = -\frac{\nu}{E}(\sigma_x + \nu\sigma_y) \qquad (c)$$

from which

$$\sigma_y = \nu\sigma_x \qquad \varepsilon_x = \frac{1 - \nu^2}{E}\sigma_x$$

Substitution of the above into Eq. (c) results in $\varepsilon_z = -\nu\varepsilon_x/(1 - \nu)$. We thus have

$$\sigma_x = -\frac{P}{bt} \qquad\qquad \sigma_y = -\nu\frac{P}{bt}$$

$$\varepsilon_x = -\frac{1 - \nu^2}{E}\frac{P}{bt} \qquad \varepsilon_z = \frac{\nu(1 + \nu)}{E}\frac{P}{bt} \qquad (3.16)$$

It is interesting to note that the following ratios may now be formed:

$$\frac{\sigma_x}{\varepsilon_x} = \frac{E}{1 - \nu^2} \qquad -\frac{\varepsilon_z}{\varepsilon_x} = \frac{\nu}{1 - \nu} \qquad (3.17)$$

The quantities $E/(1 - \nu^2)$ and $\nu/(1 - \nu)$ are called the *effective modulus of elasticity* and the *effective value of Poisson's ratio*, respectively. The former is useful in the theory of wide beams and plates.

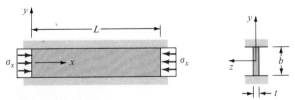

Figure 3.15 Example 3.6.

3.9 STRAIN ENERGY

The work done by external forces in producing deformation is stored within the body as *strain energy*. For a perfectly elastic body, no dissipation of energy occurs, and all the stored energy is recoverable upon unloading. The concept of strain energy, introduced in this section, is useful as applied to the solution of problems involving both static and dynamic loads. It is particularly important for predicting failure in members and for treating the effect of impact loading upon materials, as we shall see in Chaps. 8 and 11.

We shall first express the strain energy owing to uniaxial loading by considering an element subjected to a slowly increasing *normal stress* σ_x (Fig. 3.16a). The element is assumed to be initially free of stress. The force acting on each x face, $\sigma_x\, dy\, dz$, elongates the element an amount $\varepsilon_x\, dx$, where ε_x is the x-directed strain. In the case of a *linearly elastic material*, $\sigma_x = E\varepsilon_x$ (Fig. 3.16b). The average force acting on the element during the straining is $\frac{1}{2}\sigma_x\, dy\, dz$. Thus the strain energy U corresponding to the work done by this force, $\frac{1}{2}\sigma_x\, dy\, dz\,\varepsilon_x\, dx$, is expressed as

$$dU = \tfrac{1}{2}\,\sigma_x\varepsilon_x\,(dx\,dy\,dz) = \tfrac{1}{2}\,\sigma_x\varepsilon_x\,dV \qquad (a)$$

where dV is the volume of the element. The unit of work and energy in SI is the joule (J), equal to a newton-meter (N·m). If U.S. customary units are used, the work and energy are expressed in foot-pounds (ft·lb) or in inch-pounds (in.·lb).

The strain energy per unit volume, dU/dV, is referred to as the *strain-energy density*, designated U_0. From the foregoing, we have

$$U_0 = \tfrac{1}{2}\,\sigma_x\varepsilon_x = \frac{\sigma_x^2}{2E} \qquad (3.18)$$

This quantity represents the shaded area in Fig. 3.16b. The area above the stress-strain curve is known as the *complementary* energy density, denoted U_0^*, as seen in the figure. In SI units the strain-energy density is expressed in joules per cubic meter (J/m³) or in pascals; in U.S. customary units it is measured in inch-pounds per cubic inch (in.·lb/in.³) or pounds per square inch (psi). As the stress (σ_x) is squared, the strain energy is always a positive quantity, and Eq. (3.18) applies for a member in *tension* or *compression*.

Figure 3.16 (a) An element in tension; (b) stress-strain diagram; and (c) an element in shear.

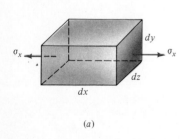

(a)

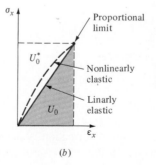

(b)

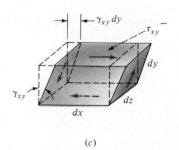

(c)

The strain-energy density when the material is stressed to the yield strength is called the *modulus of resilience*. It is equal to the area under the straight-line portion of the stress-strain diagram (Fig. 3.17a) and measures the ability of the material to store or absorb energy without permanent deformation. For example, the value of the modulus of resilience for a mild steel is $\sigma_{yp}^2/2E = (250 \times 10^6)^2/2(200 \times 10^9) = 156$ kJ/m³ or 22.6 in.·lb/in.³. (See Table B.1.) Another important quantity is known as the *modulus of toughness*; it is equal to the area under a complete stress-strain curve (Fig. 3.17b). Toughness is a measure of the material's ability to absorb energy without fracturing. Clearly, it is related to the ductility as well as to the ultimate strength of a material.

The strain energy for uniaxial normal stress is obtained by integrating the strain-energy density, Eq. (3.18), over the volume of the member:

$$U = \int \frac{\sigma_x^2}{2E} \, dV \tag{3.19}$$

The foregoing can be used for axial loading and bending of beams.

Now consider the element under the action of shearing stress τ_{xy} (Fig. 3.16c). From the figure we observe that force $\tau_{xy} \, dx \, dz$ causes a displacement $\gamma_{xy} \, dy$. The stress varies linearly from zero to its final value as before, and therefore the average force equals $\frac{1}{2}\tau_{xy} \, dx \, dz$. The strain-energy density in pure shear is then

$$U_0 = \tfrac{1}{2} \, \tau_{xy}\gamma_{xy} = \frac{\tau_{xy}^2}{2G} \tag{3.20}$$

This quantity may be visualized as the area under the shear stress-strain diagram. The *strain energy for shear stress* is expressed as

$$U = \int \frac{\tau_{xy}^2}{2G} \, dV \tag{3.21}$$

The integration is over the volume of the member. Equation (3.21) can be used for bars in torsion and transverse shear in beams.

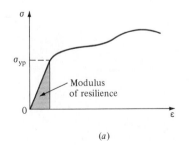

(a)

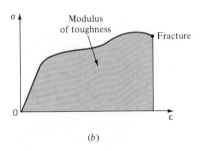

(b)

Figure 3.17 Typical stress-strain diagram: (a) modulus of resilience and (b) modulus of toughness.

*3.10 REPEATED LOADING: FATIGUE

A structural element may fail at stress levels substantially below the ultimate stress of a static test if the element is subjected to thousands of *repeated loadings* rather than to a continuously applied loading. The phenomenon of fracture owing to repeated loading is termed *fatigue*. In general, fatigue is importantly influenced by minor structural discontinuities, the quality of surface finish, and the chemical nature of the environment. A fatigue crack is most often initiated at a point of high stress concentration—for example, at the edge of a notch or hole. Fatigue failure is of a brittle nature even for materials that are normally ductile; the usual failure occurs under tensile stress.

The *fatigue life* or *endurance* of a material is measured by the number of stress repetitions or cycles prior to fracture. Experimental determination is made of the number of cycles N required to rupture a specimen at a particular stress level σ under a load changing from one magnitude to another (that is, a fluctuating load) by so-called *fatigue tests*. In a commonly used test, *beam test,* complete reversal (tension to compression) of pure bending stress occurs. In other tests, the applied load may be varied so that fluctuating stresses through various ranges result.

Fatigue data are frequently represented in the form of a plot of stress versus the number of cycles with a *semilogarithmic* scale (that is, σ against log N). Figure 3.18 shows two typical *σ-N diagrams* corresponding to rotating-beam tests on a series of identical round steel and aluminum specimens subjected to reversed flexural loads of different magnitude. As may be observed from the figure, when the applied maximum stress is high, a relatively small number of cycles will cause fracture.

For some materials, such as steels, the stress at which the curve levels off is termed the *endurance limit* σ_{es} (Fig. 3.18). Beyond the point (σ_{es}, N_{es}), failure does not occur, even for an infinitely large number of loading cycles.

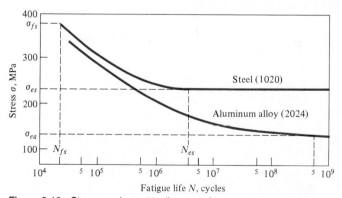

Figure 3.18 Stress-cycle (σ–N) diagrams for two typical materials.

For a structural steel, the endurance limit is nearly one-half of the ultimate strength of the steel. At $N = N_{fs}$ cycles, failure takes place at static *fracture stress* σ_{fs}. When $N < N_{fs}$, the loading is considered to be static.

On the other hand, for nonferrous metals, notably aluminum alloys, the typical σ-N curve in Fig. 3.18 indicates that the stress at failure continues to decrease as the number of cycles is increased. For such materials, the stress corresponding to some arbitrary number of $5(10^8)$ cycles is defined as the *endurance limit* σ_{ea}. It should be noted that many criteria have been suggested for interpreting fatigue data.

PROBLEMS

Secs. 3.1 to 3.4

3.1 A long aluminum alloy wire of specific weight $\gamma = 0.098$ lb/in^3 and yield strength 40 ksi hangs vertically under its own weight. Calculate the greatest length it can have without permanent deformation.

3.2 A spherical balloon changes its diameter from 200 to 201 mm when pressurized. Determine the average circumferential strain.

3.3 A hollow cylinder is subjected to an internal pressure which increases its 200-mm inner diameter by 0.5 mm and its 400-mm outer diameter by 0.3 mm. Calculate (*a*) the maximum normal strain in the circumferential direction and (*b*) the average normal strain in the radial direction.

3.4 Calculate the maximum strain ε_x in the bar seen in Fig. 3.1*a* if the displacement along the member varies as (*a*) $u = (x^2/L)10^{-3}$ and (*b*) $u = L(10^{-3}) \sin(\pi x/2L)$.

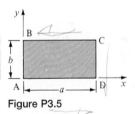

Figure P3.5

3.5 As a result of loading, the thin rectangular plate of Fig. P3.5 deforms into a parallelogram in which sides AB and CD elongate 0.005 mm and rotate 1200 μrad clockwise, while sides AD and BC shorten 0.002 mm and rotate 400 μrad counterclockwise. Calculate the plane strain components. Use $a = 40$ mm and $b = 20$ mm.

3.6 Solve Prob. 3.5, assuming that sides AD and BC elongate 0.001 mm and rotate 1600 μrad clockwise and the other sides remain the same.

3.7 Determine the normal strain in the members AB and CB of the structure shown in Fig. 2.9*a* if point B is displaced leftward 3 mm.

3.8 A thin rectangular plate, $a = 8$ in. and $b = 6$ in. (see Fig. P3.5), is acted upon by a biaxial tensile loading resulting in the uniform strains $\varepsilon_x = 600\ \mu$ and $\varepsilon_y = 400\ \mu$. Calculate the change in length of diagonal AC.

3.9 Redo Prob. 3.8, with the plate in biaxial compression for which $\varepsilon_x = -200\ \mu$ and $\varepsilon_y = -100\ \mu$.

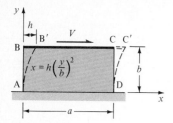

Figure P3.10

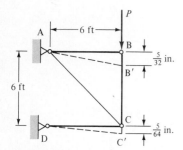

Figure P3.15

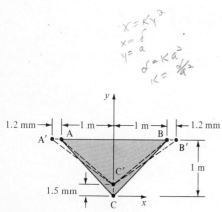

Figure P3.18

3.10 The shear force V deforms plate ABCD into AB'C'D (Fig. P3.10). For $b = 200$ mm and $h = 0.5$ mm, determine the shearing strain in the plate (a) at any point; (b) at the center; and (c) at the origin.

3.11 Redo Prob. 3.10 for the case in which curves AB' = DC' are straight lines.

3.12 A 100-mm by 100-mm square plate is deformed into a 100-mm by 100.2-mm rectangle. Determine the positive shear strain between its diagonals.

3.13 A square plate is subjected to the uniform strains $\varepsilon_x = -500$ μ, $\varepsilon_y = 500$ μ, and $\gamma_{xy} = 0$. Calculate the negative shearing strain between its diagonals.

3.14 When loaded, the plate of Fig. P3.5 deforms into a shape in which diagonal BD elongates 0.2 mm and diagonal AC contracts 0.4 mm while they remain perpendicular and side AD remains horizontal. Calculate the average plane strain components. Take $a = b = 400$ mm.

3.15 The pin-connected structure ABCD is deformed into a shape AB'C'D, as shown by the dashed lines in Fig. P3.15. Calculate the average normal strains in members BC and AC.

3.16 Solve Prob. 3.15, assuming that member BC moves $\frac{3}{16}$ in. down as a rigid body and remains vertical—that is, BB' = CC' = $\frac{3}{16}$ in.

3.17 The handbrakes on a bicycle consist of two blocks of hard rubber attached to the frame of the bike, which press against the wheel during stopping (Fig. P3.17a). Assuming that a force P causes a parabolic deflection ($x = ky^2$) of the rubber when the brakes are applied (Fig. P3.17b), determine the shearing strain in the rubber.

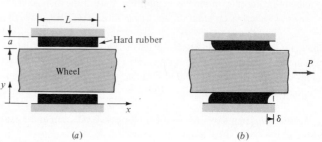

(a)   (b)

Figure P3.17

3.18 The thin, rectangular plate ABC shown in Fig. P3.18 is uniformly deformed into a shape A'B'C'. Calculate (a) the plane stress components ε_x, ε_y, and γ_{xy} and (b) the shearing strain between edges AC and BC.

3.19 Redo Prob. 3.18 for the case in which the plate ABC is uniformly deformed into a shape ABC'.

Secs. 3.5 to 3.10

3.20 What are the yield point and fracture strains for a structural steel bar having the stress-strain curve of Fig. 3.8a? What is the permanent elongation of the bar for a 50-mm gage length?

3.21 A 10-mm by 10-mm square ABCD is drawn on a member prior to loading. After loading, the square becomes the rhombus shown in Fig. P3.21. Determine (a) the modulus of elasticity and (b) Poisson's ratio.

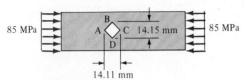

Figure P3.21

3.22 A 2-in.-diameter bar 6 ft long shortens $\frac{3}{64}$ in. under an axial load of 40 kips. If the diameter is increased $0.4(10^{-3})$ in. during loading, calculate (a) Poisson's ratio; (b) the modulus of elasticity; and (c) the shear modulus of elasticity.

3.23 Consider Fig. 3.12 with the block subjected to an axial compression of $P = -400$ kN. Use $a = 60$ mm, $b = 40$ mm, and $L = 200$ mm. If dimensions b and L are changed to 40.02 and 199.7 mm, respectively, calculate (a) Poisson's ratio; (b) the modulus of elasticity; (c) the final value of the dimension a; and (d) the shear modulus of elasticity.

3.24 The data shown in the accompanying table are determined from a tensile test of a mild steel specimen. Plot the data and determine (a) the modulus of elasticity; (b) the yield point; (c) the proportional limit; and (d) the ultimate stress.

Stress, MPa	Strain	Stress, MPa	Strain
35	0.0001	245	0.009
70	0.0003	300	0.025
100	0.0005	340	0.05
135	0.0007	380	0.09
170	0.0008	435	0.15
205	0.0010	450	0.25
240	0.0012	440	0.30
255	0.0025	420	0.36
250	0.0050	325	0.40

3.25 A bar of *any* given material is subjected to uniform *triaxial* stresses. Determine the maximum value of Poisson's ratio.

3.26 The following data are obtained from a tensile test of a 12.7-mm-diameter aluminum specimen having a gage length of 50 mm. After the specimen ruptures, the minimum (neck) diameter is found to be 8.8 mm.

Stress, MPa	Strain	Stress, MPa	Strain
35	0.0005	284	0.0062
70	0.0010	305	0.02
104	0.0014	319	0.05
139	0.0017	326	0.08
173	0.0024	312	0.12
207	0.0030	291	0.15
242	0.0035	256	0.20
259	0.0039		(Fracture)

Plot the engineering stress-strain diagram and determine (*a*) the modulus of elasticity; (*b*) the proportional limit; (*c*) the yield strength at 0.2 percent offset; (*d*) the ultimate strength; (*e*) the percent elongation in 50 mm; (*f*) the percent reduction in area; (*g*) the true ultimate stress; and (*h*) the tangent and secant moduli at a stress level of 310 MPa.

3.27 Calculate the smallest diameter and shortest length that may be selected for a steel control rod of a machine under an axial load of 4 kN if the rod must stretch 2.5 mm. Use $E = 200$ GPa and $\sigma_{\text{all}} = 150$ MPa.

3.28 Determine the axial strain in the block of Fig. 3.12 when subjected to an axial load 4 kips. The block is constrained against *y*- and *z*-directed contractions. Use $a = \frac{3}{4}$ in., $b = \frac{3}{8}$ in., $L = 4$ in., $E = 10 \times 10^6$ psi, and $\nu = \frac{1}{3}$.

3.29 A round steel rod of diameter 1 in. is subjected to axial tensile force P. The decrease in diameter is $0.5(10^{-3})$ in. Compute the largest value of P for $E = 29 \times 10^6$ psi and $\nu = \frac{1}{3}$.

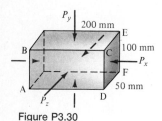

Figure P3.30

3.30 The rectangular concrete block shown in Fig. P3.30 is subjected to loads which have the resultants $P_x = 100$ kN, $P_y = 150$ kN, and $P_z = 50$ kN. Calculate (*a*) the changes in lengths of the block and (*b*) the value of a single force system of compressive forces applied only on the *y* faces that would produce the same *y*-directed deflection as do the initial forces. Use $E = 24$ GPa and $\nu = 0.2$.

3.31 Redo Prob. 3.30 for the *x* faces of the block to be free of stress.

3.32 Show that the changes in the radius *r* and length *L* of a closed-end cylinder (see Fig. 2.11*a*) subjected to internal pressure *p* are

$$\Delta r = \frac{pr^2}{2Et}(2 - \nu) = \frac{\sigma_c r}{2E}(2 - \nu) \qquad (\text{P3.32}a)$$

and

$$\Delta L = \frac{prL}{2Et}(1 - 2v) = \frac{\sigma_a L}{E}(1 - 2v) \qquad \text{(P3.32b)}$$

where t is the thickness of the cylinder.

3.33   For the handbrakes on the bicycle described in Prob. 3.17, express the deflection δ of the hard rubber in terms of P, L, G, a, and b (see Fig. P3.17). Here b and G, respectively, denote the width and the shear modulus of elasticity of a ($a \times b \times L$) rectangular rubber block.

3.34   Verify that the change in the slope of the diagonal line AB, Δ, of a rectangular plate (see Fig. P3.34) subjected to a uniaxial compression stress σ is given by

$$\Delta = \frac{a}{b}\left[\frac{1 + (v\sigma/E)}{1 - (\sigma/E)} - 1\right] \qquad \text{(P3.34)}$$

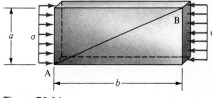

Figure P3.34

where a/b is the initial slope. For $a = 25$ mm, $b = 50$ mm, $v = 0.3$, and $E = 70$ GPa, calculate the value of Δ when $\sigma = 120$ MPa.

3.35   Rework Prob. 3.34, assuming that the plate is subjected to a uniaxial tensile stress σ.

3.36   A $\frac{5}{8}$-in.-diameter bar with a 5-in. gage length is subjected to a gradually increasing tensile load. At the proportional limit, the value of the load is 8 kips, the gage length increases $14.4(10^{-3})$ in., and the diameter decreases $0.6(10^{-3})$ in. Calculate (a) the proportional limit; (b) the modulus of elasticity; (c) Poisson's ratio; and (d) the shear modulus of elasticity.

3.37   A 6-m-long truss member is made of two 50-mm-diameter steel bars ($E = 210$ GPa, $\sigma_{yp} = 230$ MPa, $v = 0.3$). Given a tensile load of 250π kN, calculate the change in (a) the length of the member and (b) the diameter of a bar.

3.38   A 2-in.-diameter solid brass bar ($E = 15 \times 10^6$ psi, $v = 0.3$) is fitted in a hollow bronze tube. Determine the internal diameter of the tube so that its surface and that of the bar are just in contact, with no pressure, when the bar is subjected to an axial compressive load $P = 40$ kips.

3.39   The cast-iron pipe shown in Fig. P3.39 ($E = 70$ GPa, $v = 0.3$), which has length $L = 0.5$ m, outside diameter $D = 130$ mm, and wall thicknesses $t = 15$ mm, is under an axial compressive load $P = 200$ kN. Determine the change in (a) length ΔL; (b) diameter ΔD; and (c) thickness Δt.

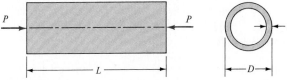

Figure P3.39

3.40 Redo Prob. 3.39 for the case in which the axial load P is tension and the pipe is made of brass ($E = 105$ GPa, $\nu = 0.3$).

3.41 A rectangular aluminum plate ($E = 70$ GPa, $\nu = 0.3$) is subjected to uniformly distributed loading, as shown in Fig. P3.41. Determine the values of w_x and w_y (in kilonewtons per meter) that produce change in length in the x direction of 1.5 mm and in the y direction of 2 mm. Use $a = 2$ m, $b = 3$ m, and $t = 5$ mm.

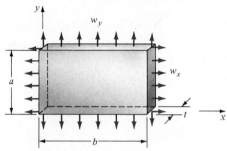

Figure P3.41

3.42 Rework Prob. 3.41, assuming that $a = 4$ m, $b = 2$ m, $t = 6$ mm and that the plate is made of steel ($E = 210$ GPa, $\nu = 0.3$).

3.43 A cylindrical steam boiler made of $\frac{7}{16}$-in. steel plate is 4 ft in inner diameter and 10 ft long. Use $E = 30 \times 10^6$ psi and $\nu = 0.3$. For an internal pressure of 500 psi, calculate (*a*) the change in the inner diameter; (*b*) the change in thickness; and (*c*) the change in length.

3.44 Verify that the change in radius r of a sphere subjected to internal pressure p is given by

$$\Delta r = \frac{pr^2}{2Et}(1 - \nu) \tag{P3.44}$$

3.45 The aluminum rod, 50 mm in diameter and 1.2 m in length, of a hydraulic ram is subjected to the maximum axial loads of ± 200 kN. What are the largest diameter and the largest volume of the rod during service? Use $E = 70$ GPa and $\nu = 0.3$.

3.46 Calculate the smallest diameter and volume of the hydraulic ramrod described in Prob. 3.45.

3.47 A 20-mm-diameter bar is subjected to tensile loading. The increase in length resulting from the load of 50 kN is 0.2 mm for an initial length of 100 mm. Determine (*a*) the conventional and true strains and (*b*) the modulus of elasticity.

3.48 A 1-in.-diameter solid aluminum-alloy bar ($E = 11 \times 10^6$ psi and $\nu = 0.3$) is fitted in a hollow plastic tube of 1.002-in. internal diameter. Determine the maximum axial compressive load that can be applied to the bar for which its surface and that of the tube are just in contact and under no pressure.

3.49 A cast-iron bar ($E = 80$ GPa, $v = 0.3$) of diameter $d = 75$ mm and length $L = 0.5$ m is subjected to an axial compressive load $P = 200$ kN. Determine (a) the increase Δd in diameter; (b) the decrease ΔL in length; and (c) the change in volume ΔV.

3.50 A 200-mm-diameter solid sphere ($E = 70$ GPa, $v = 0.3$) experiences a uniform pressure of 120 MPa. Calculate (a) the decrease in circumference of the sphere and (b) the decrease in volume of the sphere.

3.51 A 2-in.-diameter and 4-in.-long solid cylinder is subjected to uniform axial tensile stresses of $\sigma_x = 7.2$ ksi. Use $E = 30 \times 10^6$ psi and $v = \frac{1}{3}$. Calculate (a) the change in length of the cylinder and (b) the change in volume of the cylinder.

3.52 Redo Prob. 3.51 for the cylinder now subjected to hydrostatic loading with $\sigma_x = \sigma_y = \sigma_z = -7.2$ ksi.

3.53 A steel plate ABCD of thickness $t = 5$ mm is subjected to uniform stresses $\sigma_x = 120$ MPa and $\sigma_y = 90$ MPa (see Fig. P3.5). For $E = 200$ GPa, $v = \frac{1}{3}$, $a = 160$ mm, and $b = 200$ mm, calculate the change in (a) length of edge AB; (b) length of edge AD; (c) length of diagonal BD; and (d) thickness.

3.54 Redo Prob. 3.53, with the plate acted upon by biaxial loading that results in uniform stresses $\sigma_x = 100$ MPa and $\sigma_z = -60$ MPa.

3.55 Calculate the modulus of resilience for two grades of steel (see Table B.1): (a) ASTM-A242 and (b) cold-rolled, stainless steel (302).

3.56 Calculate the modulus of resilience for the following two materials (see Table B.1): (a) aluminum alloy 2014-T6 and (b) annealed yellow brass.

3.57 Using the stress-strain diagram of a structural steel shown in Fig. 3.8a, determine (a) the modulus of resilience and (b) the approximate modulus of toughness.

3.58 From the stress-strain curve of a magnesium alloy shown in Fig. 3.8b, determine (a) the modulus of resilience and (b) the approximate modulus of toughness.

ANALYSIS OF STRESS AND STRAIN

4.1 INTRODUCTION

The stresses and strains thus far treated have been found on sections perpendicular to the coordinates used to describe a member. The principal topics of this chapter will deal with the states of stress and strain at points located on *inclined* or *oblique planes*. The components of these quantities generally also depend upon the position of the point in a loaded body, and the manner of stress variation from point to point will be considered.

The discussions here are limited mainly to *two-dimensional*, or *plane, stress* (Secs. 4.2 through 4.6) and *plane strain* (Secs. 4.7 and 4.8). The formulas derived and the graphical technique introduced in this chapter are helpful in analyzing the transformation of stress and strain at a point under various types of loading. The graphical technique will help the reader to gain a stronger understanding of the stress variation around a point. Section 4.9 uses the transformation laws established to obtain an important relationship between E, G, and ν for linearly elastic materials.

4.2 PLANE STRESS

A two-dimensional state of stress exists when the stresses are independent of one of the coordinate axes, here taken as z. Examples include the stresses arising on inclined sections of an axially loaded rod, a shaft in torsion, and a beam with transversely applied force. This condition of stress also occurs on all planes at a point of a member subjected to more than one load simultaneously.

Definitions. Two-dimensional problems are of two classes: *plane stress* and *plane strain*. The condition that occurs in a thin plate subjected to loading uniformly distributed over the thickness and parallel to the plane of plate typifies the state of plane stress (Fig. 4.1). Because the plate is thin,

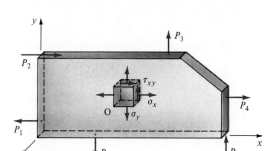

Figure 4.1 Thin plate subjected to plane stress.

the stress distribution may be closely approximated by assuming that two-dimensional stress components do not vary throughout the thickness and that the other components are zero. Another case of plane stress exists on the free surface of a structural or machine component.

The plane stress is therefore specified by

$$\sigma_z = \tau_{xz} = \tau_{yz} = 0 \qquad (a)$$

where the remaining stresses

$$\sigma_x \qquad \sigma_y \qquad \tau_{xy} \qquad (b)$$

may have nonzero values. Substitution of Eqs. (a) and (b) into Eqs. (3.13) yields the following *strain-stress relations* for plane stress:

$$\varepsilon_x = \frac{1}{E}(\sigma_x - \nu\sigma_y)$$

$$\varepsilon_y = \frac{1}{E}(\sigma_y - \nu\sigma_x) \qquad (4.1)$$

$$\gamma_{xy} = \frac{\tau_{xy}}{G}$$

and

$$\gamma_{xz} = \gamma_{yz} = 0 \qquad \varepsilon_z = -\frac{\nu}{E}(\sigma_x + \sigma_y) \qquad (c)$$

Adding the first two of Eqs. (4.1), solving for $\sigma_x + \sigma_y$ from the resulting expression, and introducing it into Eq. (c), we have

$$\varepsilon_z = -\frac{\nu}{1 - \nu}(\varepsilon_x + \varepsilon_y) \qquad (4.2)$$

The above defines the "out-of-plane" principal strain ε_z in terms of the "in-plane" strains ε_x and ε_y.

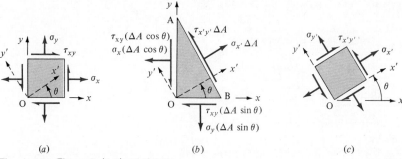

Figure 4.2   Elements in plane stress.

In the case of plane strain, stresses τ_{xz} and τ_{yz} are likewise taken to be zero, but σ_z does not vanish and can be determined from stresses σ_x and σ_y. Also, σ_x, σ_y, and τ_{xy} may have nonzero values. Details for this state of strain will be given in Sec. 4.6.

Our objective is to develop the equations for transformation of stress components σ_x, σ_y, τ_{xy} at any point in a body represented by an *infinitesimal* element (Fig. 4.1). This *stress element* is shown isolated in Fig. 4.2a. The z-directed normal stress, even if nonzero, need not be of concern here. The expressions derived will thus be applicable to all two-dimensional problems. Although the three-dimensional aspect of the stress element should not be forgotten, it is customary, for the sake of convenience, to deal with a *plane representation* of the element. It is assumed that stresses on the mutually parallel faces of the element undergo no change in magnitude, as seen in the figure.

Stresses on Inclined Planes.   Consider an infinitesimal wedge, cut from the element of Fig. 4.2a and represented in Fig. 4.2b. We wish to obtain stresses $\sigma_{x'}$ and $\tau_{x'y'}$, which refer to axes x', y' making an angle θ with axes x, y. In the following derivations, the angle θ locating the x' axis is assumed *positive when measured from the x axis in a counterclockwise direction.* Let side AB be *normal* to the x' axis. It is noted that according to the sign convention (see Sec. 2.3), the stresses are indicated as *positive* values. We observe that if the area of side AB is denoted by ΔA, sides 0A and 0B have areas $A_{0A} = \Delta A \cos \theta$ and $A_{0B} = \Delta A \sin \theta$. Equilibrium of the x'- and y'-directed forces gives

$$\Sigma F_{x'} = 0: \quad \sigma_{x'} \Delta A - \sigma_x (\Delta A \cos \theta) \cos \theta - \tau_{xy} (\Delta A \cos \theta) \sin \theta$$
$$- \sigma_y (\Delta A \sin \theta) \sin \theta - \tau_{xy} (\Delta A \sin \theta) \cos \theta = 0$$

and

$$\Sigma F_{y'} = 0: \quad \tau_{x'y'} \Delta A + \sigma_x (\Delta A \cos \theta) \sin \theta - \tau_{xy} (\Delta A \cos \theta) \cos \theta$$
$$- \sigma_y (\Delta A \sin \theta) \cos \theta + \tau_{xy} (\Delta A \sin \theta) \sin \theta = 0$$

Simplifying and rearranging, we have

$$\sigma_{x'} = \sigma_x \cos^2 \theta + \sigma_y \sin^2 \theta + 2\tau_{xy} \sin \theta \cos \theta \qquad (4.3a)$$

$$\tau_{x'y'} = \tau_{xy} (\cos^2 \theta - \sin^2 \theta) + (\sigma_y - \theta_x) \sin \theta \cos \theta \qquad (4.3b)$$

Stress $\sigma_{y'}$ may readily be obtained by substitution of $\theta + \pi/2$ for θ in the equation for $\sigma_{x'}$ (Fig. 4.2c). This yields the relation

$$\sigma_{y'} = \sigma_x \sin^2 \theta + \sigma_y \cos^2 \theta - 2\tau_{xy} \sin \theta \cos \theta \qquad (4.3c)$$

We can express Eqs. (4.3) in terms of 2θ using the trigonometric identities $2 \cos^2 \theta = 1 + \cos 2\theta$, $2 \sin \theta \cos \theta = \sin 2\theta$, and $2 \sin^2 \theta = 1 - \cos 2\theta$. The *equations of transformation for stress* then appear in the following common form:

$$\sigma_{x'} = \tfrac{1}{2}(\sigma_x + \sigma_y) + \tfrac{1}{2}(\sigma_x - \sigma_y) \cos 2\theta + \tau_{xy} \sin 2\theta \qquad (4.4a)$$

$$\tau_{x'y'} = -\tfrac{1}{2}(\sigma_x - \sigma_y) \sin 2\theta + \tau_{xy} \cos 2\theta \qquad (4.4b)$$

$$\sigma_{y'} = \tfrac{1}{2}(\sigma_x + \sigma_y) - \tfrac{1}{2}(\sigma_x - \sigma_y) \cos 2\theta - \tau_{xy} \sin 2\theta \qquad (4.4c)$$

The foregoing expressions permit the determination of stresses acting on all possible planes AB defined by the angle θ—the state of stress at a point—provided that three components on orthogonal (perpendicular) faces are known. Quantities such as stress (strain and moment of inertia) which are subject to such transformations are tensors of second rank. As we shall see, Mohr's circle (Sec. 4.4) is a *graphical representation* of a tensor transformation.

It is significant to note that addition of Eqs. (4.4a) and (4.4c) leads to

$$\sigma_{x'} + \sigma_{y'} = \sigma_x + \sigma_y = \text{constant} \qquad (d)$$

The sum of the normal stresses on two perpendicular planes is therefore invariant, that is, independent of θ. This assertion is also valid in the case of a three-dimensional state of stress.

The variation of stress with the angle θ is illustrated in the solution of the following problems.

EXAMPLE 4.1

The state of stress at a point in the loaded machine element is shown in Fig. 4.3a. Determine the normal and shearing stresses acting on an inclined plane parallel to (a) line a-a and (b) line b-b. In each case, sketch the results on a properly oriented element.

Solution The x' direction is that of a normal to the inclined plane. We want to obtain the transformation of stress from the xy system of coordinates to the $x'y'$ system. *Care must be exercised* to assure that the stresses and the rotations are designated with their correct signs.

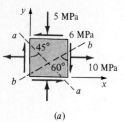

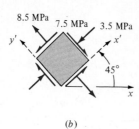

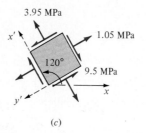

(a)

(b)

(c)

Figure 4.3 Example 4.1.

(a) Applying Eqs. (4.4) for $\theta = 45°$, $\sigma_x = 10$ MPa, $\sigma_y = -5$ MPa, and $\tau_{xy} = -6$ MPa, we obtain

$$\sigma_{x'} = \tfrac{1}{2}(10 - 5) + \tfrac{1}{2}(10 + 5) \cos 90° - 6 \sin 90° = -3.5 \text{ MPa}$$

$$\tau_{x'y'} = -\tfrac{1}{2}(10 + 5) \sin 90° - 6 \cos 90° = -7.5 \text{ MPa}$$

and

$$\sigma_{y'} - \tfrac{1}{2}(10 - 5) - \tfrac{1}{2}(10 + 5) \cos 90° + 6 \sin 90° = 8.5 \text{ MPa}$$

The results are indicated in Fig. 4.3b.

(b) As $\theta = 30 + 90 = 120°$, from Eqs. (4.4), we have

$$\sigma_{x'} = \tfrac{1}{2}(10 - 5) + \tfrac{1}{2}(10 + 5) \cos 240° - 6 \sin 240° = 3.95 \text{ MPa}$$

$$\tau_{x'y'} = -\tfrac{1}{2}(10 + 5) \sin 240° - 6 \cos 240° = 9.5 \text{ MPa}$$

and

$$\sigma_{y'} = \tfrac{1}{2}(10 - 5) - \tfrac{1}{2}(10 + 5) \cos 240° + 6 \sin 240° = 1.05 \text{ MPa}$$

The required sketch is given in Fig. 4.3c.

EXAMPLE 4.2

A two-dimensional condition of stress at a point in a loaded structure is shown in Fig. 4.4a. (a) Write the stress-transformation equations in a form suitable for computer analysis. (b) Compute $\sigma_{x'}$ and $\tau_{x'y'}$ with θ between 0 and 180° in 15° increments for $\sigma_x = 7$ MPa, $\sigma_y = 2$ MPa, and $\tau_{xy} = 5$ MPa. Plot $\sigma_{x'}$ versus θ and $\tau_{x'y'}$ versus θ.

Solution

(a) We express Eqs. (4.4a and b) as follows:

$$\sigma_{x'} = A + B \cos 2\theta + C \sin 2\theta$$
$$\tau_{x'y'} = -B \sin 2\theta + C \cos 2\theta$$

(e)

where

$$A = \tfrac{1}{2}(\sigma_x + \sigma_y) \qquad B = \tfrac{1}{2}(\sigma_x - \sigma_y) \qquad C = \tau_{xy}$$

Note that quantities $\sigma_{x'}$, σ_x, σ_y, τ_{xy}, $\tau_{x'y'}$, and θ may be represented by the following symbols: SXP, SX, SY, TXY, TXYP, and THETA, respectively.

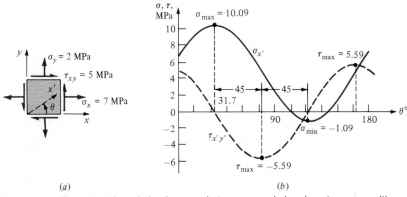

Figure 4.4 Example 4.2: variation in normal stress $\sigma_{x'}$ and shearing stress $\tau_{x'y'}$ with angle θ.

(b) Substitution of the prescribed values into Eqs. (4.4) results in

$$\sigma_{x'} = 4.5 + 2.5 \cos 2\theta + 5 \sin 2\theta$$

$$\tau_{x'y'} = -2.5 \sin 2\theta + 5 \cos 2\theta$$

Here, permitting θ to vary from 0 to 180° in increments of 15° yields the data upon which the curves shown in Fig. 4.4b are based. These *cartesian representations* indicate how the stresses vary around a point. Observe that the direction of maximum (and minimum) shear stress *bisects* the angle between the maximum and minimum normal stresses. Moreover, the normal stress is either a maximum or a minimum on planes $\theta = 31.7°$ and $\theta = 31.7° + 90°$, respectively, for which the shearing stress is zero.

The conclusions drawn from the foregoing are valid for *any* state of stress, as will be seen next. A more convenient approach to the semigraphical transformation of stress will be presented in Sec. 4.4.

4.3 PRINCIPAL STRESSES: MAXIMUM SHEARING STRESS

As has been discussed, the stress at a point varies in magnitude with different orientations of the plane of interest. The largest possible stress and the plane on which such stress occurs are of particular interest, as these are often related to structural failure.

In order to determine the orientation of $x'y'$ corresponding to the extreme values of $\sigma_{x'}$, the condition $d\sigma_{x'}/d\theta = 0$ is applied to Eq. (4.4a) with the result:

$$(\sigma_x - \sigma_y) \sin 2\theta - 2\tau_{xy} \cos 2\theta = 0 \qquad (a)$$

The foregoing gives

$$\tan 2\theta_p = \frac{2\tau_{xy}}{\sigma_x - \sigma_y} \tag{4.5}$$

Here θ_p is used in place of θ to denote the angle defining the plane of maximum or minimum normal stress. Since $\tan 2\theta = \tan(2\theta + \pi)$, two values of $2\theta_p$, differing by 180°, are found to satisfy Eq. (4.5). Hence the two values of θ_p differ by 90°, and the planes of maximum and minimum stresses are mutually perpendicular.

When Eq. (4.4b) is compared with Eq. (a), it becomes clear that $\tau_{x'y'} = 0$ on planes where the extreme values of the normal stress occur. These are called the *principal planes,* and the stresses acting on them—the maximum and minimum normal stresses—are the *principal stresses.* A principal plane is thus a *plane of zero shear.*

We observe from Eqs. (4.5) that

$$\cos 2\theta_p = \pm \frac{\sigma_x - \sigma_y}{r} \qquad \sin 2\theta_p = \pm \frac{2\tau_{xy}}{r} \tag{b}$$

where $r = [(\sigma_x - \sigma_y)^2 + 4\tau_{xy}^2]^{1/2}$. The principal stresses are determined by substituting Eqs. (b) into Eqs. (4.4a):

$$\sigma_{\text{max,min}} = \sigma_{1,2} = \frac{\sigma_x + \sigma_y}{2} \pm \sqrt{\left(\frac{\sigma_x - \sigma_y}{2}\right)^2 + \tau_{xy}^2} \tag{4.6}$$

The algebraically larger stress given above is the *maximum principal stress,* denoted σ_1. The *minimum principal stress* is designated σ_2. Note that Eqs. (4.6) lead to $\sigma_1 + \sigma_2 = \sigma_x + \sigma_y$, as expected.

Two values of θ_p, indicating the σ_1 plane and the σ_2 plane, are represented by θ_p' and θ_p'', respectively. Either of these values may be used to determine the orientation of the principal stress element (Fig. 4.5a and b). Clearly, a particular root of Eq. (4.5) inserted into Eq. (4.4a) will check one of the results obtained from Eqs. (4.6) and will also locate the plane of this principal stress.

We shall now determine the maximum shearing stresses and the planes on which they act, employing an approach similar to that described above.

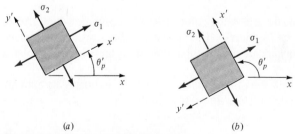

(a)   (b)

Figure 4.5 Planes of principal stresses.

The condition $d\tau_{x'y'}/d\theta = 0$ is applied to Eq. (4.4b) to yield

$$\tan 2\theta_s = -\frac{\sigma_x - \sigma_y}{2\tau_{xy}} \tag{4.7}$$

where θ_s denotes the angles locating the *planes of maximum shear stresses*. As in Eq. (4.5), two directions, mutually perpendicular, are found to satisfy the above expression. These directions may again be designated by attaching to θ_s a prime or a double prime notation. Either direction may be used to sketch the maximum shearing stress element (Fig. 4.6a and b). Furthermore, a comparison of Eqs. (4.5) and (4.7) shows that the planes of maximum shearing stresses are *inclined at 45°* with respect to the planes of principal stresses.

The extreme value of shearing stress is obtained by substituting $2\theta_s$ of Eq. (4.7) into Eq. (4.4b):

$$\tau_{max} = \pm \sqrt{\left(\frac{\sigma_x - \sigma_y}{2}\right)^2 + \tau_{xy}^2} \tag{4.8}$$

From the physical point of view, the algebraic signs are meaningless. For this reason, the largest shearing stress, regardless of algebraic sign, is called the *maximum shearing stress*. If σ_x and σ_y are the principal stresses, τ_{xy} is zero and the foregoing formula reduces to

$$\tau_{max} = \frac{\sigma_1 - \sigma_2}{2} \tag{4.9}$$

Unlike the case of the principal planes on which no shearing stresses occur, the maximum shear stress planes are usually *not* free of normal stresses. The later are determined by introducing the values of $2\theta_s$ from Eq. (4.7) into Eqs. (4.4a) and (4.4c):

$$\sigma' = \frac{\sigma_x + \sigma_y}{2} \tag{4.10}$$

Thus a *normal stress* occurs in addition to a maximum shearing stress unless $\sigma_x + \sigma_y$ is zero.

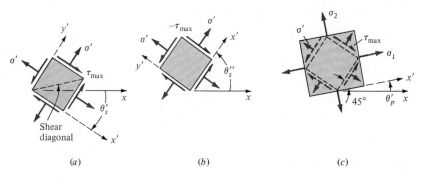

(a)   (b)   (c)

Figure 4.6   Planes of maximum shearing stresses.

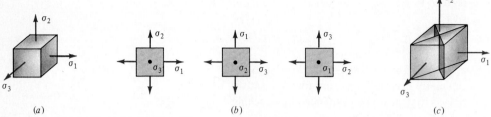

Figure 4.7 Three-dimensional state of stress.

Note that the diagonal of a stressed element that passes between the heads of the arrows denoting the shearing stresses (Fig. 4.6a) is called the *shear diagonal*. It lies in the direction of the maximum principal stress. This assists in predicting the proper direction of the maximum shear stress. Figure 4.6c, where the element associated with the maximum shearing stress is shown by the dashed lines, depicts the results.

In general, *triaxial stresses* σ_1, σ_2, and σ_3 may all act at a point (Fig. 4.7a). For this case $\sigma_1 > \sigma_2 > \sigma_3$, where σ_1, σ_2, and σ_3 represent *algebraic* values. The cubic element representing the point, viewed from three different directions, is sketched in Fig. 4.7b. Observe that the *true maximum shearing stress* is

$$(\tau_{max})_t = \frac{\sigma_1 - \sigma_3}{2} \tag{4.11}$$

and acts on the planes that bisect the planes of maximum and minimum principal stresses (Fig. 4.7c).

EXAMPLE 4.3

A state of plane stress is described in Fig. 4.8a. (a) Write the principal-stress-transformation formulas in a form suitable for computer analysis. (b) Calculate the principal stresses. (c) Calculate the maximum shearing stresses and the associated normal stresses. Sketch the results found in parts (b) and (c) on properly oriented elements.

Solution
(a) Equations (4.6), (4.7), and (4.4a) are written as

$$\sigma_1 = A + (B^2 + C^2)^{1/2}$$

$$\sigma_2 = A - (B^2 + C^2)^{1/2}$$

$$\theta_p = \tfrac{1}{2} \arctan \frac{C}{B} \tag{c}$$

$$\sigma_{x'} = A + B \cos 2\theta + C \sin 2\theta$$

where

$$A = \tfrac{1}{2}(\sigma_x + \sigma_y) \qquad B = \tfrac{1}{2}(\sigma_x - \sigma_y) \qquad C = \tau_{xy}$$

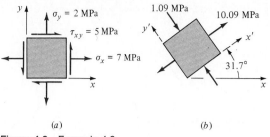

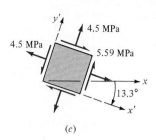

(a)   (b)   (c)

Figure 4.8   Example 4.3.

It is clear that if $\sigma_1 = \sigma_{x'}$, then $\theta'_p = \theta_p$ and $\theta''_p = \theta_p + 90°$, or else $\theta'_p = \theta_p + 90°$ and $\theta''_p = \theta_p$.

(b) Employing Eqs. (4.6), we have

$$\sigma_{1,2} = \frac{7 + 2}{2} \pm \sqrt{\left(\frac{7 - 2}{2}\right)^2 + 5^2} = 4.5 \pm 5.59$$

or

$$\sigma_1 = 10.09 \text{ MPa} \qquad \sigma_2 = -1.09 \text{ MPa}$$

To locate the principal planes, we use Eqs. (4.5):

$$\theta_p = \tfrac{1}{2} \tan^{-1} \frac{2(5)}{7 - 2} = 31.7° \text{ and } 121.7°$$

On which of these planes the principal stresses act is unknown. Substitution of $2\theta = 63.4°$, for example, into Eq. (4.4a) yields $\sigma_{x'} = 10.09$ MPa. This result, aside from providing a check on the previous calculation, means that $\theta'_p = 31.7°$. Consequently, $\theta''_p = 121.7°$. The state of principal stress is shown in Fig. 4.8b.

(c) Equation (4.8) leads to

$$\tau_{\max} = \pm \sqrt{\left(\frac{7 - 2}{2}\right)^2 + 5^2} = \pm 5.59 \text{ MPa}$$

Observe that $(\sigma_1 - \sigma_2)/2$ yields the same result. The planes of maximum shearing stresses are found from Eq. (4.7):

$$\theta_s = \tfrac{1}{2} \tan^{-1} \left(-\frac{7 - 2}{2 \times 5}\right) = -13.3° \text{ and } 76.7°$$

Using Eq. (4.4b) and introducing $2\theta = -26.6°$, $\tau_{x'y'} = 5.59$ MPa. Hence $\theta'_s = -13.3°$ and $\theta''_s = 76.7°$. Applying Eq. (4.10), we have

$$\sigma' = \frac{7 + 2}{2} = 4.5 \text{ MPa}$$

The required sketch is shown in Fig. 4.8c.

Note that the direction of the $\tau_{\max}$s may also be readily predicted by recalling that they act toward the shear diagonal.

One may describe the states of stress at the point in matrix form:

$$\begin{bmatrix} 7 & 5 \\ 5 & 2 \end{bmatrix} \text{MPa}, \qquad \begin{bmatrix} 10.09 & 0 \\ 0 & -1.09 \end{bmatrix} \text{MPa}, \qquad \begin{bmatrix} 4.5 & 5.59 \\ 5.59 & 4.5 \end{bmatrix} \text{MPa}$$

The above representations, associated with the $\theta = 0°$, $\theta = 31.7°$, and $\theta = -13.3°$ planes passing through the point, are equivalent. They confirm the results shown in Fig. 4.4 of Example 4.2.

4.4 MOHR'S CIRCLE FOR PLANE STRESS

The basic equations of stress transformation derived in the last two sections may be interpreted graphically. We now discuss the graphical technique that permits the rapid transformation of stress from one plane to another and also provides an overview of the state of stress at a point (Fig. 4.9a). This method was devised by the German engineer Otto Mohr in 1882.

To develop this representation, we rewrite Eqs. (4.4a) and (4.4b) after setting $\sigma_{x'} = \sigma$ and $\tau_{x'y'} = \tau$:

$$\sigma - \frac{\sigma_x + \sigma_y}{2} = \frac{\sigma_x - \sigma_y}{2} \cos 2\theta + \tau_{xy} \sin 2\theta$$

$$\tau = -\frac{\sigma_x - \sigma_y}{2} \sin 2\theta + \tau_{xy} \cos 2\theta$$

Squaring each equation, adding them, and simplifying, we obtain

$$\left(\sigma - \frac{\sigma_x + \sigma_y}{2}\right)^2 + \tau^2 = \left(\frac{\sigma_x - \sigma_y}{2}\right)^2 + \tau_{xy}^2$$

The foregoing is an equation of a circle in σ, τ coordinates having a radius

$$r = \sqrt{\left(\frac{\sigma_x - \sigma_y}{2}\right)^2 + \tau_{xy}^2}$$

Figure 4.9 (a) Stress element; (b) Mohr's circle of stress; and (c) interpretation of positive shearing stress.

The center is located on the σ axis at $(\sigma_x + \sigma_y)/2$, as shown in Fig. 4.9b. This circle is known as *Mohr's circle of stress*.

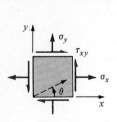

(a)

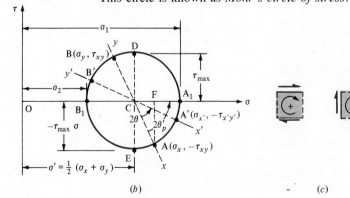

(b)　　　　(c)

Interpretation of Mohr's Circle.   The angles on the circle are measured in the same direction as θ is measured in Fig. 4.9a. An angle of 2θ on the circle corresponds to an angle θ on the stress element. The radius r of the circle is

$$CA = \sqrt{CF^2 + AF^2} \qquad (a)$$

where

$$CF = \frac{\sigma_x - \sigma_y}{2} \qquad AF = \tau_{xy} \qquad (b)$$

and CA is the magnitude of the maximum shearing stress τ_{max}. The state of stress associated with the original x and y planes corresponds to points A and B on the circle, respectively. Points such as A$'$ and B$'$, lying on the diameters other than AB, define states of stress with respect to *any* other set of x' and y' planes rotated relative to the original set through an angle θ.

To demonstrate that the coordinates of point A$'$ on the circle are given by Eqs. (4.4a and b), we let $2\theta'_p$ represent the angle between the radial lines CA and CA$_1$. From Fig. 4.9b, it is apparent that

$$\sigma_{x'} = \frac{\sigma_x + \sigma_y}{2} + CA' \cos (2\theta'_p - 2\theta)$$

$$\tau_{x'y'} = CA' \sin (2\theta'_p - 2\theta)$$

Inasmuch as CA$'$ equals CA, the foregoing become

$$\sigma_{x'} = \frac{\sigma_x + \sigma_y}{2} + CA(\cos 2\theta'_p \cos 2\theta + \sin 2\theta'_p \sin 2\theta)$$

$$\tau_{x'y'} = CA(\sin 2\theta'_p \cos 2\theta - \cos 2\theta'_p \sin 2\theta) \qquad (c)$$

Referring to the figure, note that

$$CA \cos 2\theta'_p = CF \qquad CA \sin 2\theta'_p = AF \qquad (d)$$

When Eqs. (d) and (b) are substituted into Eqs. (c), we obtain Eqs. (4.4a and b). It is shown, therefore, that point A$'$ on Mohr's circle, defined by the angle 2θ, represents the stress components on the x' face of the element, defined by the angle θ.

Clearly, points A$_1$ and B$_1$ on the circle locate the principal stresses and represent the values given by Eqs. (4.5) and (4.6). Similarly, D and E radicate the maximum shearing stresses given by Eqs. (4.7) through (4.9). Mohr's circle shows that the planes of maximum shear stress are always located at 45° from planes of principal stress, as already indicated in Fig. 4.6. The center C of the circle provides the value of the normal stress σ' associated with the τ_{max} planes as expressed by Eq. (4.10).

Construction of Mohr's Circle.   In a Mohr's circle representation, the normal stresses obey the sign convention of Sec. 2.3. However, for the purposes only of *constructing and reading values of stress* from a Mohr's circle, the sign convention for shearing stress is as follows: When the shearing stress on opposite faces of an element would produce shear forces that result in a *clockwise* couple, these stresses are considered *positive* (Fig. 4.9c). Thus the shearing stresses on the y faces of the element in Fig. 4.9a are assumed to be positive (as before), but those on the x faces are now *negative*.

Given σ_x, σ_y, and τ_{xy} with algebraic sign in accordance with the foregoing sign convention, the procedure for determining Mohr's circle is:

1. Set up a rectangular coordinate system, indicating $+\sigma$ and $+\tau$. Both stress scales must be identical.
2. Locate the center C of the circle on the horizontal axis a distance $(\sigma_x + \sigma_y)/2$ from the origin 0.
3. Locate point A by coordinates σ_x, $-\tau_{xy}$. These stresses may relate to any face of the stress element; however, it is usual to specify the stresses on the positive x face.
4. Draw a circle with center at C and with radius CA.
5. Draw line AB through C.

Mohr's circle construction is of fundamental importance because it applies to all tensor quantities. When Mohr's circle is plotted to scale, numerical results can be obtained graphically. However, it is usual to draw only a *rough sketch*; distances and angles are determined with the help of trigonometry.

Various multiaxial states of stress can readily be treated by applying the foregoing procedure. Figure 4.10 shows some examples of Mohr's circles

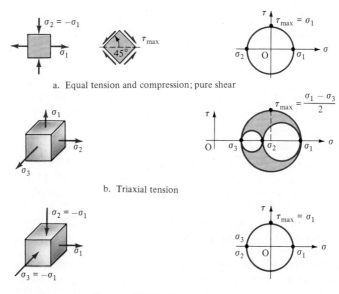

a. Equal tension and compression; pure shear

b. Triaxial tension

c. Tension with lateral pressure

Figure 4.10 Mohr's circle for various states of stress.

for commonly encountered cases. Analysis of material behavior subject to different loading conditions is often facilitated by this type of compilation. Interestingly, for the case of equal tension and compression (Fig. 4.10a), $\sigma_z = 0$ and the z-directed strain does not exist ($\varepsilon_z = 0$; see the third of Eqs. 3.13). Hence the element is in a state of *plane strain as well as plane stress*. An element in this condition can be converted to a condition of pure shear by rotating it 45° as indicated (for details, see Sec. 6.4).

In the case of triaxial tension (Fig. 4.10b), a Mohr's circle is drawn corresponding to each projection of a three-dimensional element (see Fig. 4.7b). The three-circle cluster represents Mohr's circle for triaxial stress. The case of tension with lateral pressure (Fig. 4.10c) is explained similarly.

EXAMPLE 4.4

Redo Example 4.3 in its entirety, employing Mohr's circle.

Solution   Mohr's circle constructed from the given data (Fig. 4.8a) is shown in Fig. 4.11. The center of the circle is at $(7 + 2)/2 = 4.5$ MPa on the σ axis.

(a) The principal stresses are represented by points A_1 and B_1. Referring to the circle, we find the maximum and minimum stresses to be

$$\sigma_{1,2} = OC \pm CA = 4.5 \pm \sqrt{\left(\frac{7-2}{2}\right)^2 + 5^2}$$

from which

$$\sigma_1 = 10.09 \text{ MPa} \qquad \text{and} \qquad \sigma_2 = -1.09 \text{ MPa}$$

The planes of the principal stress are given by

$$2\theta'_p = \tan^{-1}\frac{5}{2.5} = 63.4°$$

and

$$2\theta''_p = 63.4 + 180 = 243.4°$$

Hence

$$\theta'_p = 31.7° \qquad \text{and} \qquad \theta''_p = 121.7°$$

Mohr's circle clearly indicates that θ'_p locates the σ_1 plane. The required sketch is shown in Fig. 4.8b.

(b) The maximum shearing stresses are given by points D and E:

$$\tau_{\max} = \pm\sqrt{\left(\frac{7-2}{2}\right)^2 + 5^2} = \pm 5.59 \text{ MPa}$$

The planes on which these stresses act are defined by

$$\theta''_s = 31.7 + 45 = 76.7°$$

and

$$\theta'_s = 166.7° \text{ (or } -13.3°)$$

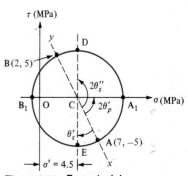

Figure 4.11   Example 4.4.

As Mohr's circle shows, the negative maximum shear stress acts on a plane whose normal x' makes an angle θ_s' with the normal to the x plane. Thus $-\tau_{max}$ acting on two opposite x' faces of the element will be directed so that a counterclockwise couple results. The normal stresses acting on the maximum shear planes are given by OC, $\sigma' = 4.5$ MPa on each face. The result is depicted on a sketch of the rotated element (Fig. 4.8c).

EXAMPLE 4.5

The stresses acting on an element of a wooden structure are shown in Fig. 4.12a. Apply Mohr's circle to determine the normal and shear stresses acting on a grain plane forming an angle of 15° with the horizontal.

Solution   Mohr's circle, Fig. 4.12b, describes the state of stress. Points B_1 and A_1 represent the components on the x and y faces, respectively. The radius of the circle equals $(700 + 300)/2 = 500$ psi. Corresponding to the $\theta = 15 + 90 = 105°$ plane within the element, it is necessary to rotate through 210° counterclockwise on the circle to locate point A'. From the circle,

$$\sigma_{x'} = -200 + 500 \cos 30° = 233 \text{ psi}$$

$$\sigma_{y'} = -200 - 500 \cos 30° = -633 \text{ psi}$$

and

$$\tau_{x'y'} = 500 \sin 30° = 250 \text{ psi}$$

The complete answer is given in Fig. 4.12c. The results may readily be checked by substituting the initial data into Eqs. (4.4).

The foregoing examples demonstrate the usefulness of Mohr's circle as a visual aid. Furthermore, Mohr's circle eliminates the need to remember the formulas of stress transformation. Its applications to various loading conditions will be discussed in the chapters to follow.

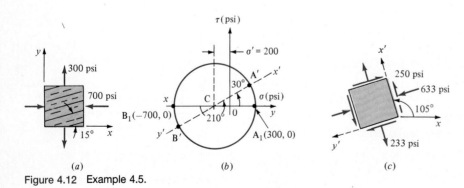

Figure 4.12   Example 4.5.

*4.5 VARIATION OF STRESS THROUGHOUT A MEMBER

As pointed out in Secs. 2.2 and 4.1, the components of stress generally vary from point to point in a loaded body. Such variations of stress, accounted for by the theory of elasticity, are governed by the equations of statics. Fulfillment of these requirements establishes certain relationships referred to as the *differential equations of equilibrium.*

For a two-dimensional case, the stresses acting on an element of sides dx, dy and of unit thickness are shown in Fig. 4.13. As we move from point 0 to 0', the increment of stress, say σ_x, may be expressed by a truncated Taylor's expansion: $\sigma_x + (\partial\sigma_x/\partial x)\,dx$. The partial derivative is used as σ_x varies with x and y. The stresses σ_y, τ_{xy}, and τ_{yx} similarly change.

We now require that the element of Fig. 4.13 satisfy the condition $\Sigma M_0 = 0$:

$$\left(\frac{\partial\sigma_y}{\partial y}\,dx\,dy\right)\frac{dx}{2} - \left(\frac{\partial\sigma_x}{\partial x}\,dx\,dy\right)\frac{dy}{2} + \left(\tau_{xy} + \frac{\partial\tau_{xy}}{\partial x}\,dx\right)dx\,dy$$

$$- \left(\tau_{yx} + \frac{\partial\tau_{yx}}{\partial y}\,dy\right)dx\,dy = 0$$

Neglecting the triple products involving dx and dy, we have

$$\tau_{xy} = \tau_{yx}$$

as already obtained in Sec. 2.3.

The equilibrium of x-directed forces, $\Sigma F_x = 0$, yields

$$\left(\sigma_x + \frac{\partial\sigma_x}{\partial x}\,dx\right)dy - \sigma_x\,dy + \left(\tau_{xy} + \frac{\partial\tau_{xy}}{\partial y}\,dy\right)dx - \tau_{xy}\,dx = 0$$

A similar expression is written for $\Sigma F_y = 0$. Simplifying these relationships, we have

$$\frac{\partial\sigma_x}{\partial x} + \frac{\partial\tau_{xy}}{\partial y} = 0$$

$$\frac{\partial\sigma_y}{\partial y} + \frac{\partial\tau_{xy}}{\partial x} = 0$$

(4.12)

The above differential equations of equilibrium apply for any type of material. These relationships show that the *rate of change* of normal stress must be accompanied by a rate of change in shearing stress. As Eqs. (4.12) contain the three unknown stresses σ_x, σ_y, τ_{xy}, problems in stress analysis are *internally* statically indeterminate. In the mechanics of materials approach, this indeterminacy is eliminated by introducing appropriate assumptions (see Sec. 3.4) and considering the equilibrium of *finite segments* of a member. The equations for the case of three-dimensional stress may be generalized from the foregoing expressions by referring to Eqs. (1.1).

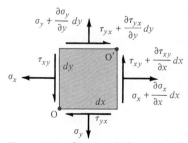

Figure 4.13 Stress variation on an element.

Note that the principles of a complete analysis (see Sec. 1.3) are stated mathematically here and in Chap. 3. In the case of a two-dimensional problem in elasticity, it is required that the following eight quantities be ascertained: σ_x, σ_y, τ_{xy}, ε_x, ε_y, γ_{xy}, u, and v. These components must satisfy eight governing equations throughout the member in addition to the boundary conditions: Eqs. (3.4a), (3.11), and (4.12). Solutions applying the methods of elasticity are not presented in this text.

The objective in introducing the basic equations of elasticity in this book is to acquaint the student with the essentially simple nature of the rigorous foundations of a significant area in the mechanics of solids. Furthermore, we use the formulations of these equations to make critical evaluations of force- (or moment-) deformation relationships, developed in the chapters to follow. Both elasticity and mechanics of materials approaches are of great value, and each method *supplements* the other.

*4.6 PLANE STRAIN

In the case of two-dimensional, or *plane, strain,* as described in Sec. 3.4, each point remains within its plane (xy) following the application of the load. Thus

$$\varepsilon_z = \gamma_{xz} = \gamma_{yz} = 0 \qquad (a)$$

whereas the strains

$$\varepsilon_x \qquad \varepsilon_y \qquad \gamma_{xy} \qquad (b)$$

may have nonzero values. Accordingly, σ_x, σ_y, and τ_{xy} may also have nonzero values. In addition, it follows from the generalized Hooke's law that $\sigma_z = -\gamma(\sigma_x + \sigma_y)$ and $\tau_{xz} = \tau_{yz} = 0$.

The strains occurring at the cross sections of a slender member subjected to lateral loading (for example, a cylinder under pressure) may exemplify an essentially plane-strain distribution. It was shown in Sec. 4.2 that the state of stress at a point can be determined if the stress components on two mutually perpendicular planes are given. A similar operation applies to the state of strain.

Transformation of Plane Strain. Consider the displacement of corners A and D of a linearly deformed and distorted element with dimensions dx, dy and of unit thickness (see Fig. 3.4b). The x and y displacements of A are u and v, respectively. For C, the displacements are $u + du$, $v + dv$, respectively. The variation of displacement is expressed as

$$du = \frac{\partial u}{\partial x}\,dx + \frac{\partial u}{\partial y}\,dy \qquad dv = \frac{\partial v}{\partial x}\,dx + \frac{\partial v}{\partial y}\,dy \qquad (c)$$

The straining of the diagonal AC is shown in Fig. 4.14. It is observed that

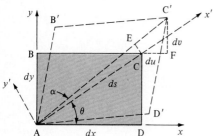

Figure 4.14 Deformations of an element in plane strain.

deformed element A′B′C′D′ has been *translated* so that A coincides with A′. Here CF = du and C′F = dv are the components of the displacement, and α is the *small* angle between AC′ and AC. We shall use $\cos \alpha = 1$ and $\sin \alpha = \tan \alpha = \alpha$ in the derivations which follow.

Let the $x'y'$ coordinate axes be placed as shown in the figure; the components of strain with respect to these axes are $\varepsilon_{x'}$, $\varepsilon_{y'}$, $\gamma_{x'y'}$. Now consider the following relationships:

$$EC' \cos \alpha = EC' = du \cos \theta + dv \sin \theta \qquad (d)$$

By definition, the normal strain equals $\varepsilon_{x'} = EC'/ds$, where ds is the initial length of the diagonal AC. Applying Eq. (d) together with Eqs. (c) and $(3.4a)$, replacing dx/ds by $\cos \theta$ and dy/ds by $\sin \theta$, we have

$$\varepsilon_{x'} = \varepsilon_x \cos^2 \theta + \varepsilon_y \sin^2 \theta + \gamma_{xy} \sin \theta \cos \theta \qquad (4.13a)$$

This is the x'-directed normal strain, which in terms of a double angle becomes

$$\varepsilon_{x'} = \frac{\varepsilon_x + \varepsilon_y}{2} + \frac{\varepsilon_x - \varepsilon_y}{2} \cos 2\theta + \frac{\gamma_{xy}}{2} \sin 2\theta \qquad (4.14a)$$

The normal strain in the y' direction is obtained by substituting $\theta + \pi/2$ for θ in the foregoing expression:

$$\varepsilon_{y'} = \frac{\varepsilon_x + \varepsilon_y}{2} - \frac{\varepsilon_x - \epsilon_y}{2} \cos 2\theta - \frac{\gamma_{xy}}{2} \sin 2\theta \qquad (4.14b)$$

An expression for the shearing strain $\gamma_{x'y'}$ can be determined by first finding the rotation α associated with the x' direction. Referring again to the figure, $\tan \alpha = CE/ds$, where $CE = dv \cos \theta - du \sin \theta - EC' \sin \alpha$. However, $EC' \sin \alpha = \varepsilon_{x'} \, ds \, \alpha = 0$, as we are dealing with a small strain and a small angle. With insertion of Eqs. (c) and $(3.4a)$ into CE, $\alpha = CE/ds$ becomes

$$\alpha = -(\varepsilon_x - \varepsilon_y) \sin \theta \cos \theta + \frac{\partial v}{\partial x} \cos^2 \theta - \frac{\partial u}{\partial y} \sin^2 \theta \qquad (e)$$

The rotation associated with the y' direction is determined by introducing $\theta + \pi/2$ for θ into Eq. (e). Thus

$$\alpha_{\theta + \pi/2} = -(\varepsilon_x - \varepsilon_y) \sin \theta \cos \theta + \frac{\partial v}{\partial x} \sin^2 \theta - \frac{\partial u}{\partial y} \cos^2 \theta$$

The shearing strain $\gamma_{x'y'}$ is equal to the difference between rotations α and $\alpha_{\theta + \pi/2}$. Using Eq. $(3.4a)$, we have

$$\gamma_{x'y'} = 2(\varepsilon_y - \varepsilon_x) \sin \theta \cos \theta + \gamma_{xy}(\cos^2 \theta - \sin^2 \theta) \qquad (4.13b)$$

In terms of the double angle, Eq. (4.13b) becomes

$$\gamma_{x'y'} = - (\varepsilon_x - \varepsilon_y) \sin 2\theta + \gamma_{xy} \cos 2\theta \qquad (4.14c)$$

Equations (4.14) represent the *transformation equations* for the plane *strain* components.

Principal Strains: Maximum Shearing Strain. Comparison of Eqs. (4.4) with Eqs. (4.14) shows an identity of form. It is observed that by replacing σ with ε and τ with $\gamma/2$ the equations for stress are converted to strain relations. Thus the principal strains are obtained from Eqs. (4.6):

$$\varepsilon_{1,2} = \frac{\varepsilon_x + \varepsilon_y}{2} \pm \sqrt{\left(\frac{\varepsilon_x - \varepsilon_y}{2}\right)^2 + \left(\frac{\gamma_{xy}}{2}\right)^2} \qquad (4.15)$$

The directions of the principal strains are given by

$$\tan 2\theta_p = \frac{\gamma_{xy}}{\varepsilon_x - \varepsilon_y} \qquad (4.16)$$

The shearing strains vanish on the principal planes. The maximum shearing strains are found on planes 45° relative to the principal planes and are represented as follows:

$$\gamma_{\max} = \pm 2\sqrt{\left(\frac{\varepsilon_x - \varepsilon_y}{2}\right)^2 + \left(\frac{\gamma_{xy}}{2}\right)^2} \qquad (4.17)$$

On the planes of maximum shear strains, the normal strains are $\varepsilon' = (\varepsilon_x + \varepsilon_y)/2 = (\varepsilon_1 + \varepsilon_2)/2$. The *true* maximum shearing strain of three-dimensional analysis proceeds from Eq. (4.11):

$$(\gamma_{\max})_t = \varepsilon_1 - \varepsilon_3 \qquad (4.18)$$

Here ε_1 and ε_3 are the algebraically largest and smallest principal strains, respectively.

It is important to note that when x, y, and z are the directions of principal stress, $\tau_{xy} = \tau_{yz} = \tau_{xz} = 0$, and from Eqs. (3.13) for the shearing strains, $\gamma_{xy} = \gamma_{yz} = \gamma_{xz} = 0$. This indicates that x, y, and z are also the axes of principal strains. Therefore, for isotropic materials, the principal axes of stress and principal axes of strain *coincide*. This means that in determining the location of the principal axes corresponding to a given state of stress, one may employ either Eq. (4.5) for stress or Eq. (4.16) for strain.

Mohr's Circle for Plane Strain. It is apparent that Mohr's circle for plane strain is constructed in the same manner for plane stress (compare Figs. 4.15 and 4.9). The normal strains—ε—are plotted on the horizontal axis, and the shearing strains divided by 2—$\gamma/2$—are plotted on the vertical axis. The positive axes are taken in the usual manner, upward and to the right. Also, the convention for the shear strain, used only in *constructing and reading values* from Mohr's circle, is consistent with the convention

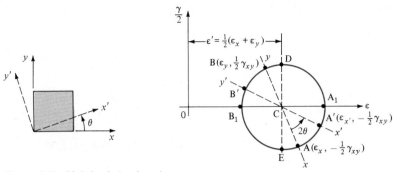

Figure 4.15 Mohr's circle of strain.

employed for shearing stress in Sec. 4.4. That is, when the shearing strain is *positive,* point A representing the *x* axis is plotted a distance $\gamma/2$ *below* the ε line, and the *y* axis point B a distance *above* the ε line; this is reversed when the shearing strain is negative.

All quantities corresponding to points A', A_1, B_1, D, and E can readily be determined with the aid of the circle, as illustrated in the examples following.

EXAMPLE 4.6

The strain components at a point in a machine member are given by $\varepsilon_x = 900\ \mu$, $\varepsilon_y = -100\ \mu$, and $\gamma_{xy} = 600\ \mu$. Using Mohr's circle, determine the principal strains and the maximum shearing strains. Show the results on a properly oriented deformed element.

Solution A sketch of Mohr's circle is shown in Fig. 4.16a, constructed by obtaining the position of point C at $(\varepsilon_x + \varepsilon_y)/2 = 400\ \mu$ on the horizontal axis and of point A at $(\varepsilon_x, -\gamma_{xy}/2) = (900\ \mu, -300\ \mu)$ from the origin 0.

The principal strains are represented by points A_1 and B_1. Therefore, referring to the figure, we have

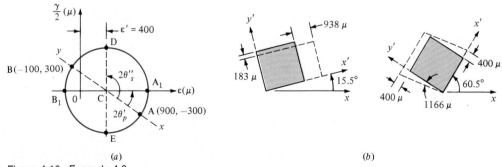

(a)   (b)

Figure 4.16 Example 4.6.

$$\varepsilon_{1,2} = 400 \pm \sqrt{\left(\frac{900 + 100}{2}\right)^2 + 300^2}$$

from which

$$\varepsilon_1 = 983 \ \mu \qquad \varepsilon_2 = -183 \ \mu$$

As a check, note that $\varepsilon_x + \varepsilon_y = \varepsilon_1 + \varepsilon_2 = 800 \ \mu$. The planes of principal strains are

$$2\theta_p' = \tan^{-1}\frac{300}{500} = 31° \qquad \text{and} \qquad 2\theta_p'' = 31 + 180 = 211°$$

or

$$\theta_p' = 15.5° \qquad \text{and} \qquad \theta_p'' = 105.5° \qquad\qquad (f)$$

From Mohr's circle, θ_p' locates the ε_1 direction. The results may be verified by introducing the two values of θ_p into Eq. (4.14a).

The maximum shearing strains are given by points D and E:

$$\gamma_{\text{max}} = \pm 2 \sqrt{\left(\frac{900 + 100}{2}\right)^2 + 300^2} = \pm 1166 \ \mu$$

Alternatively, $\gamma = \varepsilon_1 - \varepsilon_2 = 983 + 183 = 1166 \ \mu$. We observe from the circle that the axes of maximum shearing strain make an angle of 45° with respect to the principal axes. The normal strains associated with the axes of γ_{max} are represented by OC, $\varepsilon' = 400 \ \mu$. Mohr's circle shows that the maximum negative shear strain axis is represented by point D, which is plotted *above* the ε line. Thus $-\gamma_{\text{max}}$ occurs at an element rotated through $\theta_s'' = \theta_p' + 45° = 60.5°$. Note that substitution of θ_s'' and the given data into Eq. (4.14c) also yields the same result.

The required sketches are given in Fig. 4.16b, where the initial and deformed elements are indicated by the solid and dashed lines, respectively.

If a state of plane stress exists in the member under consideration, the third principal strain for $\nu = 0.3$ is from Eq. (4.2):

$$\varepsilon_z = \varepsilon_3 = -\frac{0.3}{1 - 0.3}(900 - 100) = -343 \ \mu$$

Then the true maximum shear strain equals

$$(\gamma_{\text{max}})_t = \varepsilon_1 - \varepsilon_3 = 983 + 343 = 1326 \ \mu$$

In this case, the maximum in-plane shear strain does *not* represent the largest shearing strain.

EXAMPLE 4.7

For the state of strain described in Example 4.6, determine the principal stresses and the maximum shear stress—and their directions. The material properties are $E = 200$ GPa and $\nu = 0.3$.

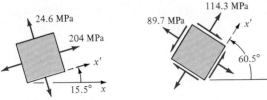

Figure 4.17   Example 4.7.

Solution   From the preceding example, we have

$$\varepsilon_1 = 983\ \mu \qquad \varepsilon_2 = -183\ \mu \qquad \gamma_{max} = 1166\ \mu \qquad (g)$$

The first two of Eqs. (3.12) together with (g) result in

$$\sigma_1 = \frac{200 \times 10^3}{1 - 0.09}\,[983 + 0.3(-183)] = 204\ \text{MPa}$$

$$\sigma_2 = \frac{200 \times 10^3}{1 - 0.09}\,[-183 + 0.3(983)] = 24.6\ \text{MPa}$$

Applying the last of Eqs. (3.12), we obtain

$$\tau_{max} = \frac{200 \times 10^3}{2(1 + 0.3)}\,(1166) = 89.7\ \text{MPa}$$

From Eq. (4.9), we have

$$\sigma' = \tfrac{1}{2}(204 + 24.6) = 114.3\ \text{MPa}$$

As a check, we note that $(\sigma_1 - \sigma_2)/2$ yields the identical value for τ_{max}.

The directions of σ_1 and σ_2 are given by Eq. (f) of Example 4.6, and the complete answer is presented in Fig. 4.17.

*4.7 MEASUREMENT OF STRAIN

Various mechanical, electrical, and optical systems have been developed for measuring the normal strain on the *free surface* of a member where a state of *plane stress* exists.* An extensively used, convenient, and accurate method employs electrical strain gages. In this section we briefly discuss a typical *bonded strain gage* and its special combinations.

Taking the outward normal to the surface as the z direction, we have $\sigma_z = \tau_{yz} = \tau_{xz} = 0$. Inasmuch as this stress condition offers no restraint to out-of-plane elastic deformation, a normal strain develops in addition to the in-plane strain components (ε_x, ε_y, γ_{xy}). It follows from the generalized

*See, for example, M. Hetenyi, *Handbook of Experimental Stress Analysis*, Wiley, New York, 1957, and J. M. Dally and W. F. Riley, *Experimental Stress Analysis*, McGraw-Hill, New York, 1978.

Hook's law that $\gamma_{xz} = \gamma_{yz} = 0$; thus the strain ε_z is a principal strain, as already specified in Sec. 4.2. This out-of-plane normal strain is significant in the determination of the true maximum shear strain (as shown in Examples 4.6 and 4.9). Observe that Eqs. (4.14) were derived for a state of plane strain. However, the principal strain ε_z is obtained in terms of ε_x and ε_y, applying Eq. (4.2). Thus the derivations of Sec. 4.6 remain valid.

Strain Gage. This device consists of a grid of fine wire or foil filament cemented between two sheets of treated paper foil or plastic backing (Fig. 4.18a). The backing serves to insulate the grid from the metal surface on which it is to be bonded. Generally, 0.03-mm-diameter wire or 0.003-mm foil filament is used. As the surface is strained, the grid is lengthened or shortened, which changes the electrical resistance of the gage. A bridge circuit, connected to the gage by means of wires, is then used to translate variations in electrical resistance into strains. An instrument used for this purpose is the *Wheatstone bridge*.

Strain Rosette. Consider three strain gages with angles θ_a, θ_b, and θ_c with respect to the reference x axis (Fig. 4.18b). The a-, b-, and c-directed normal strains are from Eq. (4.13a):

$$\varepsilon_a = \varepsilon_x \cos^2 \theta_a + \varepsilon_y \sin^2 \theta_a + \gamma_{xy} \sin \theta_a \cos \theta_a$$

$$\varepsilon_b = \varepsilon_x \cos^2 \theta_b + \varepsilon_y \sin^2 \theta_b + \gamma_{xy} \sin \theta_b \cos \theta_b \qquad (4.19)$$

$$\varepsilon_c = \varepsilon_x \cos^2 \theta_c + \varepsilon_y \sin^2 \theta_c + \gamma_{xy} \sin \theta_c \cos \theta_c$$

When the values of ε_a, ε_b, ε_c are measured for prescribed θ_a, θ_b, and θ_c, the values of ε_x, ε_y, and γ_{xy} can be found by simultaneous solution of Eqs. (4.19). The arrangement of gages employed for this type of measurement is known as a *strain rosette*.

A rosette usually consists of three specifically arranged gages whose axes are either 45 or 60° apart, as illustrated in the following examples.

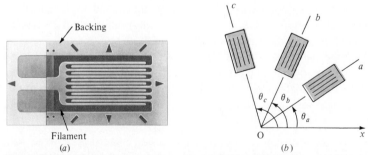

Backing

Filament

(a)

(b)

Figure 4.18 (a) Strain gage (*photo courtesy of Micro-Measurements Division, Measurements Group, Inc., Raleigh, N.C.*) and (b) strain rosette.

EXAMPLE 4.8

Using a 45° rosette, the following strains are measured at a point on the free surface of a stressed member:

$$\varepsilon_a = 900 \; \mu \qquad \varepsilon_b = 700 \; \mu \qquad \varepsilon_c = -100 \; \mu$$

These correspond to $\theta_a = 0°$, $\theta_b = 45°$, and $\theta_c = 90°$ (Fig. 4.18b). Determine the strain components ε_x, ε_y, and γ_{xy}.

Solution   For the situation described, Eqs. (4.19) become

$$\varepsilon_a = \varepsilon_x \qquad \varepsilon_c = \varepsilon_y \qquad \varepsilon_b = \tfrac{1}{2}(\varepsilon_x + \varepsilon_y + \gamma_{xy})$$

or   (4.20)

$$\varepsilon_x = \varepsilon_a \qquad \varepsilon_y = \varepsilon_c \qquad \gamma_{xy} = 2\varepsilon_b - (\varepsilon_a + \varepsilon_c)$$

Upon substitution of numerical values, $\varepsilon_x = 900 \; \mu$, $\varepsilon_y = -100 \; \mu$, and $\gamma_{xy} = 600 \; \mu$. The principal stresses and the maximum shearing stresses for these data were found in Example 4.6.

EXAMPLE 4.9

At a point on the free surface of a structure being tested, the 60° rosette readings indicate that

$$\varepsilon_a = 70 \; \mu \qquad \varepsilon_b = 850 \; \mu \qquad \varepsilon_c = 250 \; \mu$$

for $\theta_a = 0°$, $\theta_b = 60°$, and $\theta_c = 120°$ (Fig. 4.18b). Calculate (a) the in-plane principal strains and the in-plane maximum shearing strains and (b) the true maximum shearing strain ($\nu = 0.3$). Show the results obtained in part (a) on properly oriented deformed elements.

Solution   In this case, Eqs. (4.19) reduce to

$$\varepsilon_a = \varepsilon_x$$

$$\varepsilon_b = \tfrac{1}{2}(\varepsilon_x + \varepsilon_y) - \tfrac{1}{4}(\varepsilon_x - \varepsilon_y) + \frac{\sqrt{3}}{4} \gamma_{xy}$$

$$\varepsilon_c = \tfrac{1}{2}(\varepsilon_x + \varepsilon_y) - \tfrac{1}{4}(\varepsilon_x - \varepsilon_y) - \frac{\sqrt{3}}{4} \gamma_{xy}$$

which, when solved simultaneously, yield

$$\varepsilon_x = \varepsilon_a$$

$$\varepsilon_y = \tfrac{1}{3}[2(\varepsilon_b + \varepsilon_c) - \varepsilon_a] \qquad\qquad (4.21)$$

$$\gamma_{xy} = \frac{2}{\sqrt{3}}(\varepsilon_b - \varepsilon_c)$$

Introducing the given data into the equations above, $\varepsilon_x = 70\ \mu$, $\varepsilon_y = 710\ \mu$, and $\gamma_{xy} = 693\ \mu$, rounded off to three significant digits.

(a) Equations (4.14) and (4.15) are therefore

$$\varepsilon_{1,2} = \frac{70 + 710}{2} \pm \sqrt{\left(\frac{70 - 710}{2}\right)^2 + \left(\frac{693}{2}\right)^2}$$

$$\gamma_{max} = \pm 2\sqrt{\left(\frac{70 - 710}{2}\right)^2 + \left(\frac{693}{2}\right)^2}$$

from which

$$\varepsilon_1 = 862\ \mu \qquad \varepsilon_2 = -82\ \mu \qquad \gamma_{max} = \pm 943\ \mu$$

Normal strains corresponding to γ_{max} are

$$\varepsilon' = \tfrac{1}{2}(70 + 710) = 390\ \mu$$

Orientations of the principal axes are, from Eq. (4.16),

$$2\theta_p = \tan^{-1}\frac{693}{70 - 710} \qquad \text{or} \qquad \theta_p = -23.6° \text{ and } 66.4°$$

We use $\theta_p = 66.4°$ in Eq. (4.14) and find that $\varepsilon_{x'} \approx \varepsilon_1$. Similarly, for $\theta_s = 66.4 + 45 = 111.4°$, Eq. (4.15) yields $\gamma_{x'y'} \approx -\gamma_{max}$. Thus $\theta'_p = 66.4°$ and $\theta''_s = 111.4°$. The results are given in Fig. 4.19a.

(b) Using Eq. (4.2), the out-of-plane principal strain is

$$\varepsilon_z = \varepsilon_3 = -\frac{0.3}{1 - 0.3}(70 + 710) = -334\ \mu$$

The true maximum shearing strain equals

$$(\gamma_{max})_t = 862 + 334 = 1196\ \mu$$

The plane-strain components found can also be used to construct a Mohr's circle, as already illustrated in Example 4.6. The Mohr's circle solution of the problem is presented in Fig. 4.19b.

Figure 4.19 Example 4.9.

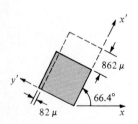

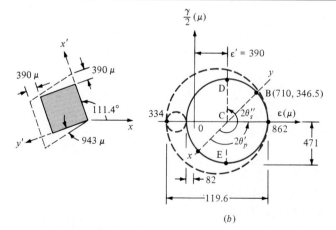

(a)

(b)

*4.8 RELATIONS INVOLVING *E*, *v*, AND *G*

We are now in a position to develop the fundamental relationship, Eq. (3.14), connecting the modulus of elasticity E, shear modulus of elasticity G, and Poisson's ratio v. Consider an element subjected to pure shear. Mohr's circle for this case shows that $\sigma_1 = \tau_{max}$ and $\sigma_2 = -\tau_{max}$ on the planes making 45° with the shear planes (see Fig. 4.10*a*).

Thus application of the first of Eqs. (3.13) for $\varepsilon_x = \varepsilon_1$, $\sigma_x = \sigma_1$, $\sigma_y = -\sigma_2$, and $\tau_{xy} = 0$ results in

$$\varepsilon_1 = \frac{\sigma_1}{E} - v\frac{\sigma_2}{E} = \frac{\tau_{max}}{E}(1 + v) \qquad (a)$$

On the other hand, for the state of pure shear strain, it is observed from Mohr's circle that ε_1 is $\gamma_{max}/2$. Hooke's law connects the shear strain and shear stress, $\gamma_{max} = \tau_{max}/G$. Hence

$$\varepsilon_1 = \frac{\tau_{max}}{2G} \qquad (b)$$

Finally, equating the alternative expressions for ε_1 in Eqs. (*a*) and (*b*), we obtain

$$G = \frac{E}{2(1 + v)} \qquad (3.14)$$

It can be shown that for *any* choice of orientation of coordinate axes *x* and *y*, the same result—Eq. (3.14)—is obtained. Therefore, as stated in Sec. 3.8, the two- and three-dimensional stress-strain relations for an isotropic and elastic material can be written in terms of two constants.

PROBLEMS

Secs. 4.1 and 4.2

4.1 A 60-mm by 40-mm plate of 5-mm thickness is subjected to uniformly distributed biaxial tensile forces (Fig. P4.1). What normal and shearing stresses exist along

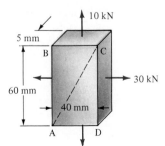

Figure P4.1

diagonal AC? As is done in the derivations given in Sec. 4.2, use an approach based upon the equilibrium equations applied to the wedge-shaped half ABC of the plate.

4.2 Redo Prob. 4.1 using Eqs. (4.4).

4.3 through 4.5 The states of stress at three points in a loaded body are represented in Figs. P4.3 through P4.5. Determine for each point the normal and shearing stresses acting on the indicated inclined plane. As is done in the derivations given in Sec. 4.2, use an approach based upon the equilibrium equations applied to the wedge-shaped element shown.

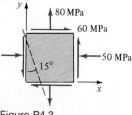

Figure P4.3

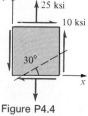

Figure P4.4

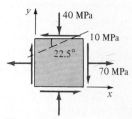

Figure P4.5

4.6 through 4.8 Redo Probs. 4.3 through 4.5 using Eqs. (4.4).

4.9 The stresses at a point (see Fig. 4.2a) are $\sigma_x = \sigma_y = 0$ and $\tau_{xy} = 15$ ksi (pure shear). Determine the stresses on all sides of an element rotated through an angle $\theta = 22.5°$ and show their sense on the element.

4.10 At a point in a loaded member, the stresses are as shown in Fig. P4.10. The normal stress at the point on the indicated plane is 7 ksi (tension). What is the magnitude of the shearing stress τ?

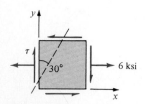

Figure P4.10

4.11 A triangular plate is subjected to stresses as shown in Fig. P4.11. Determine σ_x, σ_y, and τ_{xy} and sketch the results on a properly oriented element.

4.12 Calculate the normal and shearing stresses acting on the plane indicated in Fig. P4.10 for $\tau = 4$ ksi.

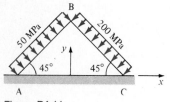

Figure P4.11

4.13 At a critical point A in the loaded member in Fig. P4.13, the stresses on the inclined plane are $\sigma = 28$ MPa and $\tau = 10$ MPa, and the normal stress on the y plane is zero. Calculate the normal and shear stresses on the x plane through the point. Show the results on a properly oriented element.

4.14 At a particular point in a loaded machine part (use Fig. 4.2a), the stresses are $\sigma_x = 6$ ksi, $\sigma_y = -4$ ksi, and $\tau_{xy} = 0$. Determine the normal and shear stresses on the plane whose normals are at angles of -30 and $120°$ with the x axis. Sketch the results on properly oriented elements.

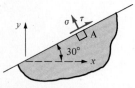

Figure P4.13

4.15   The stresses at a point in the enclosure plate of a boiler are as shown in the element of Fig. P4.15. Calculate the normal and shear stresses at the point on the indicated inclined plane. Show the results on a properly oriented element.

4.16   Solve Prob. 4.15, with the state of the tensile stresses reversed so as to act in compression.

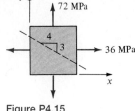

Figure P4.15

4.17   The stresses acting uniformly at the edges of a thick, rectangular plate are shown in Fig. P4.17. Determine the stress components on planes parallel and perpendicular to *a-a*. Show the results on a properly oriented element.

4.18   Figure P4.18 represents the state of stress at a point in a structural member. Calculate the normal and shear stresses at the point on the indicated inclined plane. Sketch the results on a properly oriented element.

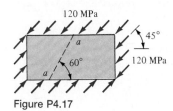

Figure P4.17

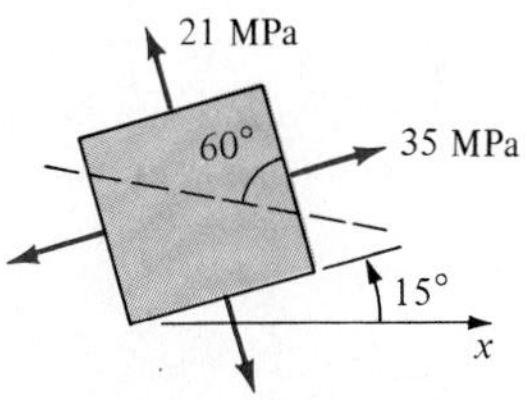

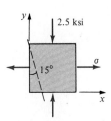

Figure P4.18

4.19   At a point in a loaded member, the stresses are as shown in Fig. P4.19. Determine the allowable value of σ if the normal and shearing stresses acting simultaneously in the indicated inclined plane are limited to 5 and 3 ksi, respectively.

4.20   Calculate the normal and shear stresses acting on the inclined plane in Fig. P4.19 for $\sigma = 2$ ksi.

Sec. 4.3

4.21   Calculate and show the stresses on planes of maximum shearing stresses for an element subjected to principal stresses: (*a*) $\sigma_1 = 6$ ksi and $\sigma_2 = 2$ ksi; (*b*) $\sigma_1 = 4$ ksi, $\sigma_2 = 3$ ksi, and $\sigma_3 = -1$ ksi.

4.22   A closed cylindrical tank fabricated of 10-mm-thick plate is subjected to an internal pressure of 6 MPa. (*a*) Determine the maximum diameter if the maximum shear stress is limited to 30 MPa. (*b*) For the diameter found in part (*a*) determine the limiting value of tensile stress.

4.23   A welded plate is subjected to the uniform biaxial tension shown in Fig. P4.23. Calculate the maximum stress σ, if the weld has (*a*) an allowable shear stress of 25 MPa and (*b*) an allowable normal stress of 60 MPa.

Figure P4.19

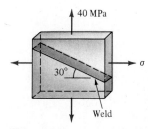

Figure P4.23

4.24   Obtain the principal stresses and the related orientations for the following stresses:

$$(a) \begin{bmatrix} 3 & 4 \\ 4 & 2 \end{bmatrix} \text{MPa}, \qquad (b) \begin{bmatrix} 15 & 1 \\ 1 & 6 \end{bmatrix} \text{ksi}, \qquad (c) \begin{bmatrix} 14 & -2 \\ -2 & 6 \end{bmatrix} \text{MPa}$$

Show the results on properly oriented elements.

4.25   A closed cylindrical vessel, constructed of a thin plate 1.2 m in diameter, is subjected to an external pressure of 1 MPa. Calculate (a) the wall thickness if the maximum allowable shear stress is set at 20 MPa and (b) the corresponding maximum principal stress.

4.26   For the state of stress given in Fig. P4.3, determine the magnitude and orientation of the principal stresses. Show the results on a properly oriented element.

4.27   Determine the maximum shearing stresses and the associated normal stresses for the state of stress of Fig. P4.5. Sketch the results on a properly oriented element.

4.28 through 4.30   The stresses at three points in a loaded member are represented in Figs. P4.28 through P4.30. For each point, calculate and sketch (a) the principal stresses and (b) the maximum shearing stresses with the associated normal stresses.

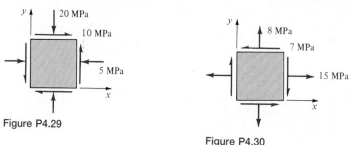

Figure P4.29

Figure P4.30

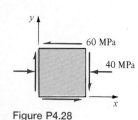

Figure P4.28

4.31   Given the stresses acting uniformly at the edges of a block (Fig. P4.31), calculate (a) the stresses σ_x, σ_y, τ_{xy} and (b) the maximum shearing stresses with the associated normal stresses. Sketch the results on properly oriented elements.

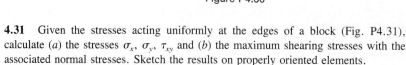

Figure P4.31

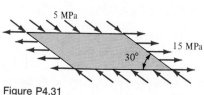

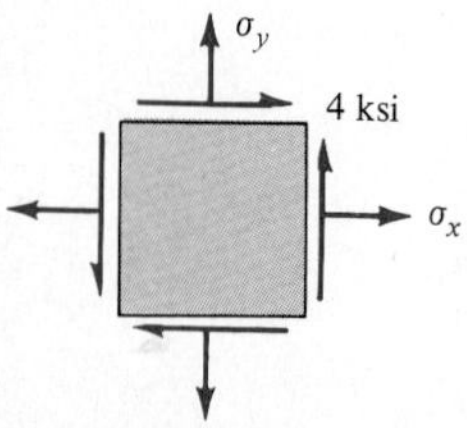

Figure P4.32

4.32   The shearing stress at a point in a loaded member is $\tau_{xy} = 4$ ksi. (See Fig. P4.32.) The principal stresses at this point are $\sigma_1 = 5$ ksi and $\sigma_2 = -8$ ksi. Determine the σ_x and σ_y and indicate the principal and maximum shear stresses on an appropriate sketch.

4.33 The state of stress at a point is shown in Fig. P4.33. For that point determine (a) the magnitude of the shear stress τ if the maximum principal stress is not to exceed 70 MPa and (b) the corresponding maximum shearing stresses and the planes at which they act.

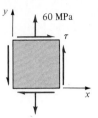

Figure P4.33

4.34 The side wall of the cylindrical steel pressure vessel has butt-welded seams. (See Fig. P4.34.) The allowable tensile strength of the joint is 90 percent of that of steel. Determine the maximum value of the seam angle ϕ if the tension in the steel is to be limiting.

Figure P4.34

·4.35 A cylindrical vessel of internal diameter 200 mm and wall thickness 2 mm has a welded helical seam angle of $\phi = 60°$ (Fig. P4.34). If the allowable tensile stress in the weld is 100 MPa, determine (a) the maximum value of internal pressure p and (b) the corresponding shear stress in the weld.

4.36 A structural member is subjected to two different loadings, each separately producing stresses at point A, as indicated in Fig. P4.36. Calculate, and show on a sketch, the principal planes and the principal stresses under the effect of the combined loading.

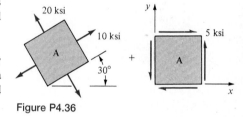

Figure P4.36

4.37 Redo Prob. 4.36 for the case shown in Fig. P4.37.

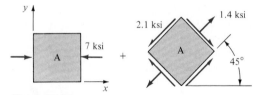

Figure P4.37

4.38 Redo Prob. 4.36 for the case shown in Fig. P4.38, where σ and θ are known constants.

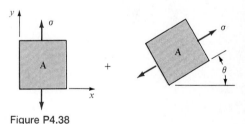

Figure P4.38

4.39 The state of stress on the horizontal and vertical planes at a point is only incompletely known, as shown in Fig. P4.39. However, at the point σ_2 and τ_{max} are prescribed as -7 and 8 ksi, respectively. Determine stresses σ, τ, and σ_1. Show the results on properly oriented elements.

4.40 Taking the σ_z of Prob. 4.30 to be 0, determine the true maximum shear stresses and the associated normal stresses.

Figure P4.39

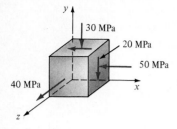

Figure P4.41

4.41 Consider a point in a loaded body subjected to the stresses shown in Fig. P4.41. Determine (a) the principal stresses and (b) the maximum shear stresses. Sketch the results on properly oriented elements.

4.42 The state of stress at a point A in a structure is shown in Fig. P4.42. Determine the normal stress σ and the angle θ.

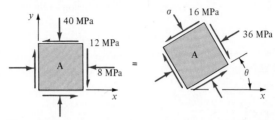

Figure P4.42

4.43 Redo Prob. 4.42 for the case shown in Fig. P4.43.

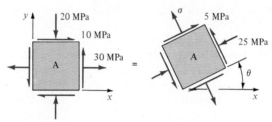

Figure P4.43

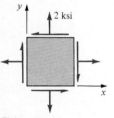

Figure P4.44

4.44 At a particular point in a loaded member (see Fig. P4.44) there exists on the horizontal plane a normal stress $\sigma_y = 2$ ksi and a negative shear stress. One of the principal stresses at the point is 1 ksi (tensile), and the maximum shearing stress has a magnitude of 5 ksi. Calculate (a) the unknown stresses on the horizontal and vertical planes and (b) the unknown principal stress. Show the principal stresses and maximum shear stresses on a sketch of a properly oriented element.

Secs. 4.4 and 4.5

4.45 Using Mohr's circle, solve Prob. 4.3.

4.46 Using Mohr's circle, solve Prob. 4.4.

4.47 Using Mohr's circle, solve Prob. 4.5.

4.48 Using Mohr's circle, solve Prob. 4.10.

4.49 Using Mohr's circle, solve Prob. 4.15.

4.50 Using Mohr's circle, solve Prob. 4.18.

4.51   Using Mohr's circle, solve Prob. 4.21.

4.52   Using Mohr's circle, solve Prob. 4.23.

4.53   Using Mohr's circle, solve Prob. 4.25.

4.54   Using Mohr's circle, solve Prob. 4.28.

4.55   Using Mohr's circle, solve Prob. 4.33.

4.56   Using Mohr's circle, solve Prob. 4.35.

4.57   Using Mohr's circle, solve Prob. 4.37.

4.58   Using Mohr's circle, solve Prob. 4.38.

4.59   Using Mohr's circle, solve Prob. 4.39.

4.60   Using Mohr's circle, solve Prob. 4.41.

4.61   Using Mohr's circle, solve Prob. 4.42.

4.62   Using Mohr's circle, solve Prob. 4.44.

Sec. 4.6

4.63 through 4.67   The state of strain at specific points is given in the table below. Determine the state of strain associated with the specified angle θ. Use Eqs. (4.14).

Problem	ε_x	ε_y	γ_{xy}	θ
4.63	$-400\ \mu$	$800\ \mu$	$1500\ \mu$	$30°$
4.64	$500\ \mu$	$1200\ \mu$	$-1000\ \mu$	$-45°$
4.65	$200\ \mu$	$410\ \mu$	$100\ \mu$	$40°$
4.66	$-300\ \mu$	$420\ \mu$	$200\ \mu$	$10°$
4.67	$-720\ \mu$	$-200\ \mu$	$240\ \mu$	$-30°$

4.68   Using Mohr's circle, redo Prob. 4.63.

4.69   Using Mohr's circle, redo Prob. 4.64.

4.70   Using Mohr's circle, redo Prob. 4.65.

4.71   Using Mohr's circle, redo Prob. 4.66.

4.72   Using Mohr's circle, redo Prob. 4.67.

4.73 through 4.77   The state of strain at a point in a thin plate (Fig. 4.1) is given. Calculate (a) the in-plane principal strains and the maximum in-plane shear strain and (b) the true maximum shearing strain ($v = 0.3$). Sketch the results found in part (a) on properly oriented deformed elements.

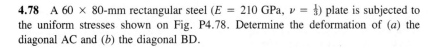

Problem	ε_x	ε_y	γ_{xy}
4.73	400 μ	100 μ	200 μ
4.74	-900 μ	400 μ	-200 μ
4.75	-720 μ	0	300 μ
4.76	200 μ	600 μ	600 μ
4.77	500 μ	-100 μ	150 μ

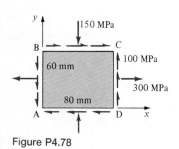

Figure P4.78

4.78   A 60×80-mm rectangular steel ($E = 210$ GPa, $\nu = \frac{1}{3}$) plate is subjected to the uniform stresses shown on Fig. P4.78. Determine the deformation of (a) the diagonal AC and (b) the diagonal BD.

4.79   Show that for plane strain, $\varepsilon_x + \varepsilon_y = \varepsilon_{x'} + \varepsilon_{y'} = \varepsilon_1 + \varepsilon_2$.

Secs. 4.7 and 4.8

4.80   During the static test of an aircraft panel, a 45° rosette measures the following normal strains on the free surface: $\varepsilon_a = -400$ μ, $\varepsilon_b = -500$ μ, $\varepsilon_c = 200$ μ. Calculate the principal strains. Show the results on a properly oriented deformed element.

4.81   Using a 60° rosette, we find the following strains at a critical point on the frame of a stressed beam: $\varepsilon_a = -200$ μ, $\varepsilon_b = -350$ μ, and $\varepsilon_c = -500$ μ. Determine (a) the maximum in-plane shear strains and the accompanying normal strains and (b) the true maximum shear strain. Use $\nu = 0.3$. Sketch the results found in part (a) on a properly oriented distorted element.

4.82   Verify that, for a 45° rosette, the principal strains are expressed as follows:

$$\varepsilon_{1,2} = \frac{\varepsilon_a + \varepsilon_c}{2} \pm \frac{1}{2}[2\varepsilon_a(\varepsilon_a - 2\varepsilon_b) + 2\varepsilon_c(\varepsilon_c - 2\varepsilon_b) + 4\varepsilon_b^2]^{1/2} \quad \text{(P4.82)}$$

4.83   At a point on the free surface of a steel member ($E = 200$ GPa, $\nu = 0.3$) subjected to plane stress, a 60° rosette measures the strains $\varepsilon_a = 1200$ μ, $\varepsilon_b = -650$ μ, and $\varepsilon_c = 500$ μ. Determine (a) the principal strains and their directions and (b) the corresponding principal stresses and the maximum shear stresses. Sketch the results found in part (b) on a properly oriented element.

4.84   Redo Prob. 4.83 for $\varepsilon_a = 400$ μ, $\varepsilon_b = 500$ μ, and $\varepsilon_c = -700$ μ.

4.85   At a point on the surface of a stressed structure (see Fig. 4.18b), the strain readings are $\varepsilon_a = -200$ μ, $\varepsilon_b = -500$ μ, and $\varepsilon_c = -900$ μ for $\theta_a = 0°$, $\theta_b = 120°$, and $\theta_c = 240°$. Calculate (a) the in-plane principal strains and (b) the in-plane maximum shear strains. Show the results on properly oriented deformed elements.

4.86   The strain readings at a point on the free surface of a member subjected to plane stress (see Fig. 4.18b) are $\varepsilon_a = 400$ μ, $\varepsilon_b = 350$ μ, and $\varepsilon_c = 800$ μ for $\theta_a = 0°$, $\theta_b = 45°$, and $\theta_c = 135°$. Calculate (a) the maximum in-plane shearing strains and (b) the true maximum shear strain ($\nu = \frac{1}{3}$). Sketch the results obtained in part (a) on a properly oriented distorted element.

5

AXIALLY LOADED MEMBERS

5.1 INTRODUCTION

Normal stresses in bars under simple tension or compression were discussed in Chap. 2. Our main concern here will be the stresses and associated deformations due to more complex axial loadings. The method of determining the deformations of *axially loaded members* is based upon the previously considered procedures and expressions. We shall treat both statically determinate and indeterminate structures, including those composed of different materials, subjected to external forces or to temperature changes. A representation of stress states on differently oriented planes in a bar in tension will be given in Sec. 5.6, and the last two sections of this chapter will describe the effect of localized stress and strain.

Clearly, it is often necessary to determine not only the stresses in an axially loaded member but the deflections as well. The latter, for example, may have to be kept within limits so that certain clearances are maintained. Moreover, deformation must be considered in the analysis of statically indeterminate systems, although we are interested only in the forces (Sec. 5.3).

The formulas to be developed in this chapter apply to materials having a linearly elastic stress-strain relationship. The widely used method of superposition (Sec. 5.4) is also important for torsion and bending problems. We shall describe the plastic behavior of axially loaded bars in Secs. 12.2 through 12.4.

5.2 DEFLECTION OF AXIALLY LOADED MEMBERS

This section deals with the elongation or contraction of slender members under axial loading. The axial stress in these cases is assumed not to exceed the proportional limit of the linearly elastic range of the material. The definitions of normal stress and normal strain, and the relationship between the two given by Hooke's law, are therefore employed.

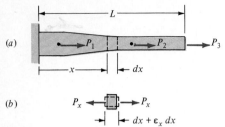

Figure 5.1 An axially loaded bar.

Consider the *deformation* or *deflection* of a bar of varying cross section (Fig. 5.1*a*) subjected at several points to axial loads of various magnitudes. The free-body diagram of an arbitrary element cut from the bar is depicted in Fig. 5.1*b*. Both the internal axial force P_x acting at the cut section and the cross-sectional area A_x of the element depend, in general, upon axial distance x. The infinitesimal deformation $d\delta$ that occurs within this element upon application of the forces is equal to the strain ε_x multiplied by the length dx—that is, $d\delta = \varepsilon_x \, dx$. The magnitude of the axial stress at the cut section is found from $\sigma_x = P_x/A_x$. These results are combined with $\varepsilon_x = \sigma_x/E$ and are integrated over the length L of the bar to give the following equation for the deformation of the bar:

$$\delta = \int_0^L \frac{P_x \, dx}{A_x E} \tag{5.1}$$

The *positive* sign indicates *elongation*, which is associated with a tensile force. A negative sign would represent contraction. Since the units of P, A, and E cancel out of the equation, the deformation δ has units of length L. Note that for *tapered bars*, the above formula yields results of acceptable accuracy provided the angle between the sides of the rod is not larger than $20°$.

If the bar is of *uniform* cross-sectional area and is loaded at its ends, Eq. (5.1) becomes

$$\delta = \frac{PL}{AE} \tag{5.2}$$

where the product AE is known as *axial rigidity* of the bar. For an axially loaded rod composed of portions having *constant* values of A, E, and P, the deformation of the entire bar is expressed as

$$\delta = \sum_{i=1}^n \frac{P_i L_i}{A_i E_i} \tag{5.3}$$

where the subscript i is an index denoting the various parts of the bar and n is the total number of parts.

The preceding expressions show that the deflection of the prismatic rod is directly proportional to the applied load and length and is inversely proportional to cross-sectional area and elastic modulus. The *flexibility*—that is, the deflection of the bar owing to a unit value of the load—is, from Eq. (5.2), $f = L/AE$. Similarly, the *stiffness*—that is, the force required to produce a unit deflection of the bar—is $k = AE/L$, and thus $f = k^{-1}$. These deformation characteristics play a significant role in the analysis of structures.*

---

*For details, see J. M. Gere and S. P. Timoshenko, *Mechanics of Materials,* 3d ed., PWS-Kent, Boston, Mass., 1990.

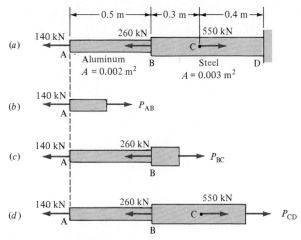

Figure 5.2 Example 5.1.

EXAMPLE 5.1

Steel and aluminum rods are rigidly attached at point B of a bar, as shown in
Fig. 5.2a. Determine the deformation of the composite bar due to the axial loads
applied where indicated. Let $E_s = 200$ GPa and $E_a = 70$ GPa.

Solution To investigate the variation of P along the length, we pass sections
through each of the component parts. For each section, the free-body diagram of
the portion of the rod is located to the left of the section (Fig. 5.2b–d). These
free bodies are in equilibrium if the internal forces acting are as follows: $P_{AB} =$
140 kN, $P_{BC} = 400$ kN, and $P_{CD} = -150$ kN. Applying Eq. (5.3) and substituting numerical values, we obtain

$$\delta = \sum \frac{PL}{AE} = \frac{140 \times 10^3(0.5)}{0.002(70 \times 10^9)} + \frac{400 \times 10^3(0.3)}{0.003(200 \times 10^9)} - \frac{150 \times 10^3(0.4)}{0.003(200 \times 10^9)}$$

$$= 0.6 \times 10^{-3} \text{ m} = 0.6 \text{ mm}$$

The foregoing represents the displacement of a point A with respect to point D,
or the total elongation of the bar.

EXAMPLE 5.2

Determine the horizontal displacement δ_H and vertical displacement δ_V of joint
B of the two-bar truss shown in Fig. 5.3a. The cross-sectional areas are $A_{AB} =$
0.004 m² and $A_{BC} = 0.002$ m². Use $L_{AB} = 1.5$ m, $L_{BC} = 2.5$ m, $P = 40$ kN,
and $E_{AB} = E_{BC} = 70$ GPa.

Solution Analyze the truss using the principles of Sec. 1.3.

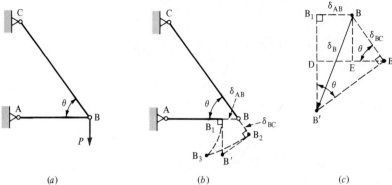

(a)   (b)   (c)

Figure 5.3 Example 5.2.

Statics. The axial forces in the members are found from the equilibrium of forces acting at joint B (recall Example 2.2). These are

$$F_{AB} = -P \cot \theta \qquad F_{BC} = P \csc \theta$$

Introducing the data given, we obtain

$$F_{AB} = 40(\tfrac{3}{4}) = -30 \text{ kN} \qquad F_{BC} = 40(\tfrac{5}{4}) = 50 \text{ kN}$$

Deformations. The contraction of bar AB and the elongation of bar BC are provided by Eq. (5.2):

$$\delta_{AB} = -\frac{PL_{AB} \cot \theta}{A_{AB} E_{AB}} \qquad \delta_{BC} = \frac{PL_{BC} \csc \theta}{A_{BC} E_{BC}} \qquad (5.4)$$

After substituting the numerical values, we are led to

$$\delta_{AB} = \frac{-30 \times 10^3 (1.5)}{0.004(70 \times 10^9)} = -0.16 \times 10^{-3} \text{ m} = -0.16 \text{ mm}$$

$$\delta_{BC} = \frac{50 \times 10^3 (2.5)}{0.002(70 \times 10^9)} = 0.893 \times 10^{-3} \text{ m} = 0.893 \text{ mm}$$

Geometry. The conditions of geometric fit require that the rods AB and BC remain straight and remain fastened together at B subsequent to their deformation. To ascertain the displacement of joint B, assume for the moment that bars AB and BC are disconnected at B and allowed to change lengths to AB_1 and CB_2, respectively, as shown greatly exaggerated in Fig. 5.3b. We can bring the ends together by rotating them about A and C to meet at B_3. Because the displacement of point B is very small, the arcs generated can be replaced by straight lines drawn perpendicular to AB and BC. The two perpendicular lines intersect at point B', the final location of joint B.

The displacements δ_{AB} and δ_{BC} are drawn to a larger scale in Fig. 5.3c. Referring to this displacement diagram, we observe that the horizontal component of δ_B equals

$$\delta_H = -\frac{PL_{AB} \cot \theta}{A_{AB}E_{AB}} \qquad (5.5a)$$

and its vertical component B_1B' is the sum of $B_1D = BE = \delta_{BC} \sin \theta$ and $DB' = (\delta_{BC} \cos \theta + \delta_{AB}) \cot \theta$. It follows that

$$\delta_V = \delta_{BC} \csc \theta + \delta_{AB} \cot \theta \qquad (5.5b)$$

Inserting the data into Eqs. (5.5), we obtain

$$\delta_H = 0.16 \text{ mm} \leftarrow$$
$$\delta_V = 0.893(\tfrac{5}{4}) + 0.16(\tfrac{3}{4}) = 1.24 \text{ mm} \downarrow$$

Having calculated the values of δ_H and δ_V of joint B, we can readily obtain the resultant displacement:

$$\delta_B = \sqrt{\delta_H^2 + \delta_V^2} = \sqrt{(0.16)^2 + (1.24)^2} = 1.25 \text{ mm}$$

The preceding approach can be used for two-bar trusses of *any* geo-metrical arrangement. The conclusion to be drawn from the foregoing dis-cussion is that, for *small displacements,* the elongation or contraction of any bar of the truss is equal to the component of the relative displacement of its ends taken in the *original direction* of the bar. In Chap. 11, we shall rework Example 5.2, employing a more convenient method.

EXAMPLE 5.3

Determine the elongation due to its own weight of the solid truncated cone shown in Fig. 5.4a. The weight per unit volume of the material is γ; the elastic modulus is E.

Solution In this case P_x and A_x are variables. They are conveniently expressed in terms of x if the origin is taken at the vertex 0 of the cone. From similar triangles, $h/a = (h + L)/b$, or

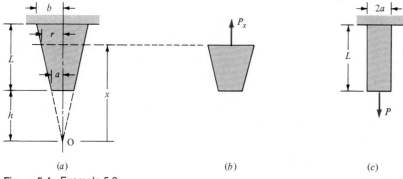

(a)   (b)   (c)

Figure 5.4 Example 5.3.

$$h = \frac{aL}{b - a} \tag{a}$$

where a, b, and L are known dimensions of the member. In a like manner, the radius at level x is obtained by

$$r = \frac{a}{h} x \tag{b}$$

Hence

$$A_x = \pi r^2 = \pi x^2 \left(\frac{a}{h}\right)^2 \tag{c}$$

The internal force P_x can be found from the free-body diagram depicted in Fig. 5.4b, which shows this force to be equal to the weight of the isolated portion:

$$P_x = \tfrac{1}{3}\pi\gamma \left(r^2 x - a^2 h\right) \tag{d}$$

Substituting Eqs. (b) through (d) into Eq. (5.1) and performing the indicated integration gives

$$\delta = \int_h^{h+L} \frac{P_x \, dx}{A_x E} = \frac{\gamma}{3E} \int_h^{h+L} \left(x - \frac{h^3}{x^2}\right) dx$$

$$= \frac{\gamma}{3E} \left.\left|\frac{x^2}{2} + \frac{h^3}{x}\right.\right|_h^{h+L} = \frac{\gamma L^2(L + 3h)}{6E(h + L)}$$

Finally, we substitute Eq. (a) into the above and obtain the following displacement of the free end:

$$\delta = \frac{\gamma L^2(b + 2a)}{6bE} \tag{5.6}$$

Note that when $a = b$, the foregoing becomes

$$\delta = \frac{WL}{2AE} \tag{e}$$

where W is equal to the weight of the *cylindrical rod* and A is its cross-sectional area. If an axial tensile load P acts at the lower end in addition to the rod's own weight W (Fig. 5.4c), then

$$\delta = \frac{PL}{AE} + \frac{WL}{2AE} = \frac{[P + (W/2)]L}{AE} \tag{f}$$

This is the total end displacement owing to the two loads and is obtained by superposition of the results given by Eqs. (5.2) and (e).

5.3 STATICALLY INDETERMINATE STRUCTURES

So far it has always been possible to determine the axial forces in the bars of the structure by applying the equations of equilibrium. That is, the members are invariably statically determinate. There are certain problems of considerable interest, however, in which the equilibrium equations are insufficient to compute the reactions and the internal forces. For such *statically indeterminate structures,* all unknown forces can be obtained only if the deformations of the members are taken into account.

The three principles of mechanics of materials (recall Sec. 1.3) are necessary and sufficient in the analysis of statically indeterminate members. The solutions of the following problems serve to illustrate the procedure in determining the axial forces as well as the deformations.

EXAMPLE 5.4

As shown in Fig. 5.5a, a rigid horizontal bar is supported by a hinge at A and by two steel cables BD and CE, which are of equal length, $L = 0.8$ m, and cross-sectional area, $A = 140$ mm^2. Calculate the stress in each cable due to a force of 40 kN, applied as shown in the figure. Assume that the yielding stress is 250 MPa and that $E = 200$ GPa.

Solution The free-body diagram of the bar given in Fig. 5.5b shows four unknown forces. Note that the rotated position of the bar after loading is indicated by the dashed lines in the figure.

Statics. Applying equations of equilibrium to Fig. 5.5b, we have

$$\rightarrow \Sigma F_x = 0: \quad R_{Ax} = 0$$

$$\uparrow \Sigma F_y = 0: \quad -R_{Ay} + F_{BD} + F_{CE} - 40 = 0 \quad (a)$$

$$\circlearrowright \Sigma M_A = 0: \quad F_{BD} + 2F_{CE} - 40 \,(1.4) = 0$$

It is observed that the system is statically indeterminate to the first degree.

Deformations. The elongations of members BD and EC are given by

$$\delta_B = \frac{F_{BD}L}{AE} \downarrow \qquad \delta_C = \frac{F_{CE}L}{AE} \downarrow \qquad (b)$$

Geometry. The condition of geometric compatibility is determined by referring to Fig. 5.5b. From similar triangles,

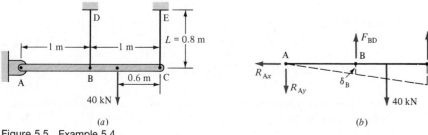

(a) (b)

Figure 5.5   Example 5.4.

$$\delta_C = 2\delta_B \qquad (c)$$

This, together with Eqs. (b), leads to

$$F_{CE} = 2F_{BD} \qquad (d)$$

Solving Eqs. (a) and (d), F_{BD} = 11.2 kN, F_{CE} = 22.4 kN, and R_{Ay} = 6.4 kN. Having found the axial forces, we find the deformations of the cables, from Eqs. (b):

$$\delta_B = \frac{(11.2 \times 10^3)(0.8)}{140 \times 10^{-6}(200 \times 10^9)} = 0.32 \times 10^{-3} \text{ m} = 0.32 \text{ mm} \downarrow$$

$$\delta_C = 0.64 \text{ mm} \downarrow$$

The stresses in the members are

$$\sigma_{BD} = \frac{11.2 \times 10^3}{140 \times 10^{-6}} = 80 \text{ MPa} \qquad \sigma_{CE} = 160 \text{ MPa}$$

Since the stresses are below the yielding strength of the material, the solution is acceptable. If either of these stresses were above the yield point, the results would not be valid and a redesign would be required.

EXAMPLE 5.5

A copper rod of length L_c, cross-sectional area A_c, and modulus of elasticity E_c is inserted into an aluminum tube as shown in Fig. 5.6. The aluminum tube has a cross-sectional area of A_a, modulus of elasticity E_a, and length L_a. Determine (a) the deformations of each member due to axial load P, which is large enough to close the small gap Δ and (b) the axial strain ε if no gap exists.

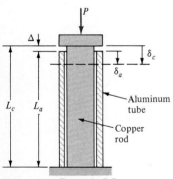

Figure 5.6   Example 5.5.

Solution

(a) Consider the elastic deformations, following application of load, of the aluminum tube and the copper rod, δ_a and δ_c, respectively.

Statics.   A force P_a will be developed in the aluminum and a force P_c in the copper. These are related by the following equations of equilibrium:

$$P_a + P_c = P \qquad (e)$$

Clearly, the above is not sufficient to obtain the two unknown forces P_a and P_c.

Deformations.   Applying Eq. (5.2), we find that the contractions of the members are

$$\delta_a = \frac{P_a L_a}{A_a E_a} \qquad \delta_c = \frac{P_c L_c}{A_c E_c} \qquad (f)$$

Geometry.   Since deformation of the copper rod is equal to that of aluminum tube plus the initial gap, $\delta_c = \delta_a + \Delta$, or

$$\frac{P_c L_c}{A_c E_c} = \frac{P_a L_a}{A_a E_a} + \Delta \qquad (g)$$

Solving Eqs. (a) and (g) simultaneously, we have

$$P_a = \frac{PL_c - A_c E_c \Delta}{L_c + (A_c E_c L_a / A_a E_a)}$$

$$P_c = \frac{PL_a + A_a E_a \Delta}{L_a + (A_a E_a L_c / A_c E_c)} \qquad (5.7)$$

Expressions (f) may then be used to determine the deformations of the aluminum tube and the copper rod.

(b) In this case, we have $\Delta = 0$ and hence $L_a = L_c$. Then Eqs. (5.7) become

$$P_a = \frac{(A_a E_a)P}{A_a E_a + A_c E_c} \qquad P_c = \frac{(A_c E_c)P}{A_a E_a + A_c E_c} \qquad (5.8)$$

indicating that the forces in the members are proportional to the axial rigidities.

Compressive stresses σ_a in the aluminum and σ_c in the copper are found upon division of P_a by A_a and P_c by A_c, respectively. Finally, using Hooke's law together with Eqs. (5.8), we obtain the compressive strain:

$$\varepsilon = \frac{P}{A_a E_a + A_c E_c} \qquad (5.9)$$

From the foregoing, we observe that the strain is equal to the applied load divided by the sum of the axial rigidities of the members. The deformation of the aluminum tube is obtained either from Eq. (f) with Eq. (5.8) or by multiplying the strain ε by the length L_a. The deformation δ_c can be found in a like manner.

EXAMPLE 5.6

A stepped steel bar AB (Fig. 5.7a) is supported at each end by thrust bearings and loaded by a force P at point A. Determine the support reactions. The modulus of elasticity of the material is E.

Solution The reactions at A and B produced by P are denoted R_A and R_B, respectively.

Statics. The only equation of equilibrium for the rod is

$$R_A + R_B = P \qquad (h)$$

which does not suffice to determine the reactions.

Deformations. From the free-body diagram shown in Fig. 5.7b, we have $P_{AC} = -R_A$ and $P_{BC} = R_B$. Using force-deformation relations (5.2), we obtain

$$\delta_a = -\frac{R_A a}{A_a E} \qquad \delta_b = \frac{R_B b}{A_b E} \qquad (i)$$

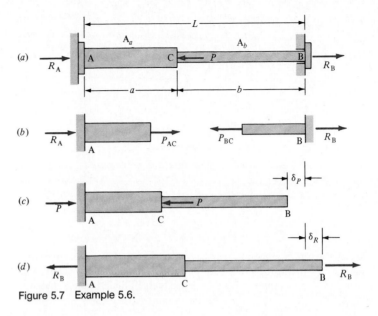

Figure 5.7   Example 5.6.

Geometry. From geometrical considerations, the total deformation of the bar must be zero, and thus

$$\delta_a + \delta_b = 0$$

Introducing Eqs. (i) into the above, we get

$$\frac{R_A a}{A_a E} = \frac{R_B b}{A_b E} \qquad (j)$$

Solving Eqs. (h) and (j) simultaneously yields

$$R_A = \frac{P}{1 + (aA_b/bA_a)} \qquad R_B = \frac{P}{1 + (bA_a/aA_b)} \qquad (5.10)$$

The axial stresses in AC and BC are found from R_A/A_a and R_B/A_b, respectively. Either of Eqs. (i), together with Eqs. (5.10), may be used to obtain the displacement of point C.

5.4   APPLICATION OF THE METHOD OF SUPERPOSITION

For many statically indeterminate situations, the principle of superposition offers an effective approach to solution. Reconsider, for example, the bar of Fig. 5.7a, replaced by the bars shown in Fig. 5.7c and d. At point B, the bar now experiences the displacements δ_P and δ_R due, respectively, to P and R_B. Superposition of these displacements, in order that there be no movement of the right end of the bar, yields

$$\delta_P + \delta_R = 0$$

Applying Eq. (5.2) and taking the elongations to be positive, we have

$$-\frac{Pa}{A_a E} + \left(\frac{R_B a}{A_a E} + \frac{R_B b}{A_b E}\right) = 0 \qquad\qquad (a)$$

from which

$$R_B = \frac{P}{1 + (bA_a/aA_b)}$$

This result is the same as that given by Eq. (5.10). The remaining reaction can be obtained from the condition of statics: $R_A + R_B = P$.

The *method of superposition* employed above may be summarized as follows:

1. One of the unknown reactions is designated as *redundant* and released from the member by removing the support.
2. The remaining member, which is rendered statically determinate, is loaded by the actual load (P) and the redundant (R_B) itself. Note that the redundant is considered to be an *unknown load*.
3. The expressions for the displacements due to these loads are obtained and substituted into the equation of geometric compatibility to calculate the redundant reaction. The other unknown reaction is found by applying statics.

In problems involving a large degree of indeterminacy, the principle of superposition may be used to good advantage for simplification of analysis. (We shall see this in Sec. 9.6.) Since this approach employs the redundant reactions as the unknowns, it is often referred to as the *force method*. Another name is the *flexibility method*, because flexibilities (as discussed in Sec. 5.2) appear in the compatibility requirement, Eq. (a). Clearly, once the redundant reaction is determined, the calculations of stress and deformation proceed in the usual manner.

5.5   THERMAL DEFORMATION AND STRESS

Consider the consequences of increasing or decreasing the *uniform* temperature of an unconstrained isotropic body. The resultant expansion or contraction occurs in such a way as to cause a cubic element of the solid to remain cubic while undergoing changes of length on each of its sides. Normal strains occur in all directions, unaccompanied by normal stresses. In addition, there are neither shearing strains nor shearing stresses.

The strain owing to a 1° temperature change is denoted by α and is called the *coefficient of thermal expansion*. *Thermal strain* caused by a uniform increase in temperature ΔT is therefore

$$\varepsilon_t = \alpha \, \Delta T \tag{5.11}$$

The coefficient of expansion is approximately constant over a moderate range of temperature change. It represents a quantity per degree Celsius (1/°C), or per degree Fahrenheit (1/°F). Values of this property for common materials are included in Table B.1, App. B.

The *free* expansion or contraction owing to a change in temperature is readily found by applying Eq. (5.11). In an elastic body, *thermal deformation* caused by a uniform temperature change is

$$\delta_t = \alpha(\Delta T)L \tag{5.12}$$

where the L is the dimension of the body. The thermal strain and deformation are *positive* if the temperature *increases* and negative if it decreases.

Stresses due to the restriction of thermal expansion or contraction of a body are called *thermal stresses*. In a statically determinate structure, a uniform temperature change will not cause any stresses, as deformations are permitted to occur freely. On the other hand, a temperature change in a structure that is supported in a statically *indeterminate* manner will induce stresses in the member.

The loads and stresses produced when temperature deformation is prevented can be calculated using the methods described in Secs. 5.3 and 5.4. The following examples demonstrate the approach.

EXAMPLE 5.7

A steel rod of length L and uniform cross-sectional area A (Fig. 5.8a) is secured between two walls. Calculate the stresses for a temperature increase of ΔT for two cases: (a) the walls are immovable and (b) the walls move apart a total distance δ_w. Compare the values obtained in parts (a) and (b) for $L = 5$ ft, $E = 30 \times 10^6$ psi, $\alpha = 6.5 \times 10^{-6}/°F$, $\Delta T = 80°F$, and $\delta_w = 0.01$ in.

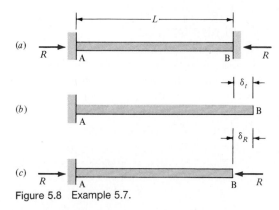

Figure 5.8 Example 5.7.

Solution   The supports exert equal and opposite axial forces R on the rod following a temperature increase in order to prevent elongation (Fig. 5.8a). Thus the condition that the sum of the axial forces acting on the bar equal zero is fulfilled and the reaction R is unknown.

(a) If end B is released (Fig. 5.8b and c), its rightward displacement caused by the temperature change alone is δ_t, and the leftward displacement owing to R is δ_R. The condition of geometric fit, together with Eqs. (5.2) and (5.12), yields

$$\delta_t - \delta_R = 0$$

$$\alpha(\Delta T)L - \frac{RL}{AE} = 0$$

from which

$$R = AE\alpha(\Delta T)$$

The compressive stress and strain in the bar are therefore

$$\sigma = \frac{R}{A} = E\alpha(\Delta T) \qquad \varepsilon = \frac{\sigma}{E} = \alpha\,\Delta T \qquad\qquad (5.13)$$

We observe that temperature change can produce stresses in a statically indeterminate structure even in the absence of any external loading. It is also noted that the stress is *independent* of the length of the bar.

(b) In this case, the bar can expand an amount δ_w, and hence the geometric compatibility of deformations becomes

$$\alpha(\Delta T)L - \frac{RL}{AE} = \delta_w$$

or

$$R = AE\left(\alpha\Delta T - \frac{\delta_w}{L}\right)$$

The compressive stress is then

$$\sigma = E\left(\alpha\Delta T - \frac{\delta_w}{L}\right) \qquad\qquad (5.14)$$

Substitution of the given numerical data into Eqs. (5.13) and (5.14) gives 15.6 and 10.6 ksi, respectively. Comparing these results, we conclude that the thermal stress can be reduced considerably by providing a small amount of support displacement.

EXAMPLE 5.8

As shown in Fig. 5.9a, a 30-mm-diameter bronze cylinder is secured between a rigid cap and slab tightening two 20-mm-diameter steel bolts. At 20°C, no deformation and stress exist in the assembly. Determine the stress in the bronze and steel at 70°C. Use $E_s = 200$ GPa, $E_b = 83$ GPa, $\alpha_s = 11.7 \times 10^{-6}/°C$, and $\alpha_b = 18.9 \times 10^{-6}/°C$.

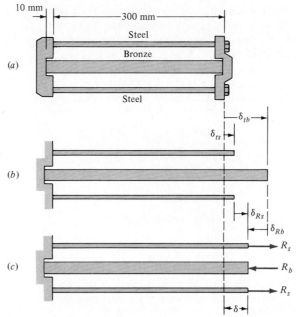

Figure 5.9 Example 5.8.

Solution After the temperature is raised by $\Delta T = 70 - 20 = 50°C$, the cap and slab exerts equal and opposite forces R_b in the bronze cylinder and R_s in each steel bolt. With the cap detached (Fig. 5.9b), thermal elongations δ_{tb} and δ_{ts} take place in the bronze cylinder and steel bolts, respectively. It is necessary, however, that the net deflection of these two members be the same. Thus the cylinder stretches the bolts, and the bolts in turn compress the cylinder until a final equilibrium position is reached (Fig. 5.9c).

Statics. The only equation of equilibrium for the system leads to

$$R_b = 2R_s \tag{a}$$

containing both unknown reactions.

Deformations. Equations (5.12) and (5.2), together with the given data, yield

$$\delta_{tb} = 18.9 \times 10^{-6}(50)(0.31) = 0.293 \times 10^{-3}\,\text{m}$$
$$\delta_{ts} = 11.7 \times 10^{-6}(50)(0.3) = 0.176 \times 10^{-3}\,\text{m} \tag{b}$$

and

$$\delta_{Rb} = \frac{R_b(0.31)}{\pi(0.015^2)(83 \times 10^9)} = 5.28(10^{-9})R_b$$

$$\delta_{Rs} = \frac{R_s(0.3)}{\pi(0.01^2)(200 \times 10^9)} = 4.77(10^{-9})R_s \tag{c}$$

Geometry. Referring to the figure, we see that the deformations are related by

$$\delta = \delta_{tb} - \delta_{ts} = \delta_{Rb} + \delta_{Rs}$$

Substituting Eqs. (b) and (c) into the above and solving the resulting expression and Eq. (a) yields

$$R_b = 15.26 \text{ kN} \qquad R_s = 7.63 \text{ kN}$$

Thus

$$\sigma_b = \frac{15.26 \times 10^3}{\pi(0.015^2)} = 21.59 \text{ MPa} \qquad \sigma_s = \frac{7.63 \times 10^3}{\pi(0.01^2)} = 24.29 \text{ MPa}$$

are the compressive stress in the cylinder and the tensile stress in each bolt, respectively.

5.6 STRESSES ON INCLINED PLANES

Consider a bar in *simple tension,* as shown in Fig. 5.10a, and assume that the axial stress ($\sigma_x = P/A$) is uniformly distributed on cross sections ($A = ab$) along the length. The normal and shearing stresses on an *inclined plane* c-c whose normal x' makes an angle θ with the axial direction are designated $\sigma_{x'}$ and $\tau_{x'y'}$, respectively. Application of Eqs. (4.4a and b) to the free-body diagram of Fig. 5.10b yields

$$\sigma_{x'} = \tfrac{1}{2}\sigma_x (1 + \cos 2\theta) = \frac{P}{2A}(1 + \cos 2\theta) \qquad (5.15a)$$

$$\tau_{x'y'} = -\tfrac{1}{2}\sigma_x \sin 2\theta = -\frac{P}{2A}\sin 2\theta \qquad (5.15b)$$

These stresses are uniformly distributed over the area $A/\cos\theta$ of the oblique plane.

Equations (5.15) show that $\sigma_{x'}$ is a maximum (σ_1) when $\theta = 0°$, that $\tau_{x'y'}$ is a maximum (τ_{max}) when $\theta = 45°$, and that $\tau_{max} = \tfrac{1}{2}\sigma_1$. The maximum stresses are then

$$\sigma_1 = \frac{P}{A} \qquad \tau_{max} = \frac{P}{2A} \qquad (5.16)$$

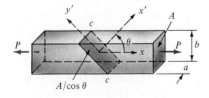

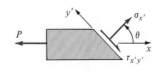

(a)   (b)   Figure 5.10   Prismatic bar in tension.

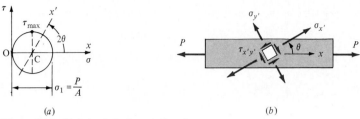

Figure 5.11 Mohr's circle for axial loading.

Thus, for a brittle material which is weaker in tension than in shear, failure occurs on the transverse plane, whereas for a ductile material which is weaker in shear than in tension, failure tends to be along the 45° plane. Mohr's circle provides a convenient way of checking the foregoing results, as illustrated in Fig. 5.11a.

Observe that the normal stress is either maximum or minimum on planes ($\theta = 0°$, $\theta = 90°$, $\theta = 180°$) for which the shearing stress vanishes. Clearly, when $\theta > 90°$, the sign of $\tau_{x'y'}$ in Eq. (5.15b) changes; the shearing stress *changes sense*. However, the *magnitude* of the shearing stress for any angle θ obtained from Eq. (5.15b) is equal to that for $\theta + 90°$, as expected (see Sec. 2.3). The stresses on a sketch of a rotated element in a bar in tension are portrayed in Fig. 5.11b.

Note that Eqs. (5.15) can be used as well for *axial compression* by assigning P a negative value. The sense of each stress direction is then reversed in Fig. 5.11b.

*5.7 STRESS CONCENTRATIONS

In cases in which the cross section of an axially loaded member varies gradually, reasonably accurate results can be obtained if one applies Eq. (2.3) derived on the basis of constant section. However, where abrupt changes of the shape exist, this equation cannot predict the high value of stress which actually exists near the discontinuity. Such *local*, or *peak*, stress is called *stress concentration*. This condition occurs in stress raisers (for example, holes, notches, fillets, and threads) which are starting points of material failure by fracture. While the stress and accompanying deformation near a discontinuity can, in some cases, be analyzed by applying the theory of elasticity, it is more usual to rely upon experimental techniques and/or the finite-element method. It is important to note that stress concentration is the primary cause of fatigue failures and static failures in brittle materials.

Two common examples of stress concentration near a discontinuity in axially loaded flat bars are illustrated in Fig. 5.12. In each situation, the ratio of the maximum stress to the *average*, or *nominal*, stress on the section is called the *stress concentration factor* and is denoted K. The *maximum* normal stress is therefore

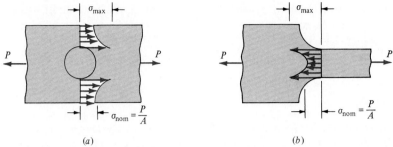

Figure 5.12 Stress-distribution under axial loading: (a) bar with a circular hole and (b) bar with fillets.

$$\sigma_{max} = K\sigma_{nom} = K\frac{P}{A} \qquad (5.17)$$

where A is usually taken as the *net area* at the *reduced* section. Figure 5.13 shows the corresponding stress concentration factors. Both curves in the figure indicate the advisability of streamlining junctures and transitions of portions that make up a structural member. This is accomplished by removing the material adjacent to a stress raiser. The technical literature containing stress concentration factors in the form of graphs, tables, and formulas is voluminous.*

Note that Eq. (5.17) is valid as long as σ_{max} does *not* exceed the proportional limit of the material since the Ks in Fig. 5.13 are based upon a linear stress-strain relationship. Factors of stress concentration for repeated loading are usually lower than the values for static loading given in this figure. Behavior of the bar seen in Fig. 5.12a when stressed beyond yield point will be discussed in Sec. 12.3. Those seeking more thorough treatment

*See, for example, R. E. Peterson, *Stress Concentration Design Factors*, Wiley, New York, 1974; and W. C. Young, *Roark's Formulas for Stress and Strain*, 6th ed., McGraw-Hill, New York, 1989.

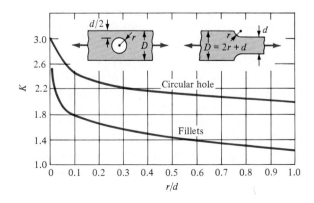

Figure 5.13 Stress concentration factors for flat bars under axial loading. (*Source: M. M. Frocht, "Photoelastic Studies in Stress Concentration," Mechanical Engineering, August 1936, pp. 485–489.*)

are referred to texts dealing with machine design. Stress concentrations due to discontinuities in various members subjected to torsion and bending will be considered in Secs. 6.7 and 7.10.

EXAMPLE 5.9

Calculate the maximum stress in a flat bar of thickness t with shoulder fillets (Fig. 5.12b). Let $t = \frac{1}{2}$ in., $r = \frac{3}{8}$ in., $D = 2\frac{3}{4}$ in., $d = 2$ in., and $P = 10$ kips.

Solution From the data, $r/d = \frac{3}{16} = 0.1875$. From Fig. 5.13, $K = 1.67$. The nominal stress on the net section is

$$\sigma_{\text{nom}} = \frac{P}{A_{\text{net}}} = \frac{10 \times 10^3}{(2\frac{3}{4} - \frac{3}{8})(\frac{1}{2})} = 8.42 \text{ ksi}$$

We thus have $\sigma_{\max} = 1.67(8.42) = 14.06$ ksi. This result shows that a large local increase in stress occurs at the fillets.

5.8 SAINT-VENANT'S PRINCIPLE

It is apparent that the strain must be a maximum in the vicinity of concentrated loads and reactions. Therefore, the corresponding stresses must also be a maximum there. Saint-Venant (1797–1886) observed that if the actual distribution of forces is replaced by a statically equivalent system, the distribution of stress throughout the member is changed only near the regions of force application. The foregoing statement is referred to as *Saint-Venant's principle* and is useful in engineering design.

According to this principle, for example, the nonuniform distribution of force at the ends of a tensile specimen which results from end clamping may be replaced by a uniform distribution to determine the stresses some distance away from the ends. For practical purposes, the local effect of a concentrated load (or reaction) on a bar can be assumed to disappear at a section *equal to or greater than* the *largest* cross-sectional dimension of the bar away from the load (or end).

Saint-Venant's principle applies to other types of loadings as well—for example, to beams, shafts, plates, and shells. Consider, for instance, the substitution of the vertical and horizontal forces and an end couple at the fixed end of a cantilever beam for the actual distribution of force exerted by the wall. If we require the stress in a region *one depth* of the beam *away* from the wall, the stress variation at the fixed end need not be of concern as it does not lead to considerable stress variation in the region of interest.

PROBLEMS

Secs. 5.1 and 5.2

5.1   Calculate the elongation of a 600-ft steel wire, fixed at its upper end and carrying its own weight only. Use $E = 29 \times 10^6$ psi and specific weight $\gamma = 0.284$ lb/in.^{3}.

5.2   As seen in Fig. P5.2, a steel rod ABC of cross-sectional area $A = 100$ mm^2 and a brass rod CD of the same cross-sectional area are joined at point C to form the 5.5-m rod ABCD. Let $E_s = 200$ GPa and $E_b = 105$ GPa. Calculate the displacement relative to A of (a) point B and (b) point D.

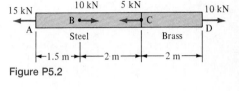

Figure P5.2

5.3   Redo Prob. 5.2, with the left end of the bar fixed and the 10-kN force at D directed leftward.

5.4   A steel bar of $\frac{9}{16}$-in. by 2-in. cross section and 6 ft long ($E = 30 \times 10^6$ psi) is to support an axial tensile load. Given that the allowable elongation is $\frac{1}{16}$ in. and normal stress is 25 ksi, calculate the largest permissible value of the load.

5.5   A tapered plate of uniform thickness t is subjected to an axial load P, as shown in Fig. P5.5. (a) Determine the elongation δ in terms of P, L, a, t, and E. (b) Calculate the value of δ for $P = 50$ kN, $a = 25$ mm, $L = 5$ m, $t = 10$ mm, and $E = 80$ GPa.

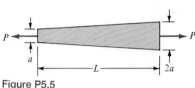

Figure P5.5

5.6   An axially loaded member 15 mm thick and 5 m long is subjected to the forces shown in Fig. P5.6. Assuming a modulus of elasticity of $E = 210$ GPa, determine the total deformation in the axial direction.

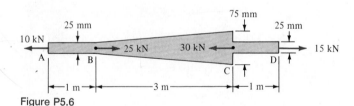

Figure P5.6

5.7   Rework Prob. 5.6, with the member fixed at its left end and the 30-kN force at C directed rightward.

5.8   A foundation pile, which has been driven to a depth L in the earth, supports a vertical load Q, as shown in Fig. P5.8. Assume that this load is resisted by a frictional force whose intensity varies parabolically $f(x) = kx^2$. Here k, a constant, is to be found from the equilibrium of vertical forces. Determine the total shortening δ of the pile in terms of Q, L, A, and E.

5.9   Rework the preceding problem, with the frictional force varying linearly [that is, $f(x) = kx$]. Calculate the total shortening of the pile by taking $Q = 80$ kips, $L = 30$ ft, $A = 0.75$ in.^{2}, and $E = 1.45 \times 10^6$ psi.

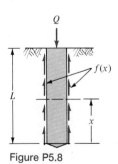

Figure P5.8

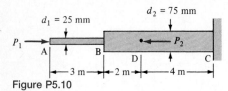

Figure P5.10

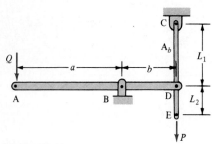

Figure P5.13

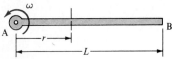

Figure P5.16

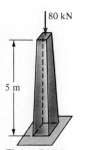

Figure P5.17

5.10 As seen in Fig. P5.10, a stepped steel bar ABC ($E = 200$ GPa) of circular cross sections $d_1 = 25$ mm and $d_2 = 75$ mm is subjected to loads P_1 and P_2. Calculate, for $P_2 = 100$ kN, (a) the value of P_1 such that A does not move and (b) the corresponding displacement of D.

5.11 Calculate the magnitude of compressive force P_1 necessary to produce a total shortening $\delta = 5$ mm in the steel bar of Fig. P5.10 for $P_2 = 0$.

5.12 Referring again to Fig. P5.10, what is the deflection δ at the free end of a steel bar ($E = 200$ GPa) due to loads $P_1 = 5$ kN and $P_2 = 15$ kN?

5.13 As seen in Fig. P5.13, a brass rod CE ($E = 105$ GPa) has a cross-sectional area of $A_b = 200$ mm². What is the vertical displacement caused by loads Q and P (a) at the end A of rigid member AD? (b) at point E? Let $Q = 15$ kN, $P = 5$ kN, $a = 2b$, and $L_1 = 2L_2 = 0.4$ m.

5.14 Rework Prob. 5.13, with load P directed upward at end E of member CE.

5.15 Referring to Fig. P5.13, derive the expression for load Q in terms of a, b, L_1, L_2, A_b, and P for which vertical displacement of point E is zero.

5.16 Bar AB (Fig. P5.16) rotates about a fixed vertical axis at A with constant angular velocity ω rad/s. If the weight per unit volume of the bar is γ, acceleration of gravity is g, and the modulus of elasticity is E, what is the total elongation? Note that $P = m(\omega^2 r)$.

5.17 As shown in Fig. P5.17, a concrete column 5 m high supports a 80-kN load. Its dimensions at the top and bottom are 0.3 m by 0.3 m and 0.4 m by 0.4 m, respectively. Determine the total axial deflection δ of the column. Let $E = 20$ GPa.

5.18 The composite bar seen in Fig. P5.18 is made of an aluminum of diameter d_1 and a steel of diameter d_2. The bar is subjected to tensile force P. For $a = 9$ ft, $b = 6$ ft, $d_2 = \frac{3}{4}$ in., and $P = 5$ kips, calculate the diameter d_1 for a total elongation $\delta = \frac{1}{16}$ in.

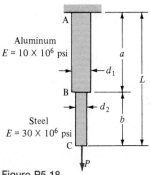

Figure P5.18

5.19 Determine the lengths *a* and *b* for the aluminum and steel portions of the composite rod shown in Fig. P5.18 for which both materials deflect an equal amount. Use $d_1 = d_2 = 1$ in., $P = 20$ kips, and $L = 3.6$ ft. Calculate the total deformation of the rod.

5.20 An assembly consists of a vertical composite bar and a horizontal rigid member, loaded as shown in Fig. P5.20. For $E_b = 105$ GPa, $E_s = 210$ GPa, and $b = 1.5a$, determine the ratio of the cross-sectional area of steel to brass A_s/A_b so that the axial deformations of parts CB and BD are equal.

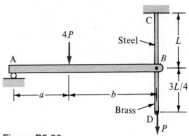

Figure P5.20

5.21 Redo Example 5.3, with an additional axial tensile load of $\gamma \pi a^2 L$ applied on the rod at the free end.

5.22 As shown in Fig. P5.22, a rigid rod ABC is suspended by two wires. The wire AD is made of steel having $A_s = 0.14$ in.^{2} and $E_s = 30 \times 10^6$ psi. For the aluminum-alloy wire CE, $A_a = 0.28$ in.^{2}, and $E_a = 10 \times 10^6$ psi. Determine the displacement of point B caused by $P = 6$ kips.

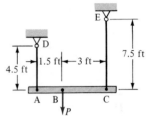

Figure P5.22

5.23 Two horizontal rigid members AB and CD are connected by bars as shown in Fig. P5.23. The bars have length *L*, modulus of elasticity *E*, and cross-sectional areas *A* and 2*A*. Show that the displacement between the two forces *P* acting at points E and F is given by

$$\delta = \frac{PL}{AE}\left[1 - 2\left(\frac{x}{a}\right) + \frac{3}{2}\left(\frac{x}{a}\right)^2\right] \qquad (P5.23)$$

Calculate the values of δ, using $E = 200$ GPa, $A = 1000$ mm^2, $a = 1$ m, $L = 4$ m, and $P = 100$ kN, for (a) $x = 0$; (b) $x = a/2$; and (c) $x = a$.

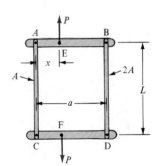

Figure P5.23

Secs. 5.3 to 5.5

5.24 As shown in Fig. P5.24, an axially loaded bar AB is attached to rigid supports at its ends. The bar has cross-sectional area A_1 from A to C and $A_2 = 2A_1$ from C to D. If $P_C = 3P_B$ and the elastic modulus is *E*, determine (a) the reactions at A and D and plot the axial force diagram and (b) the displacement δ_C of point C.

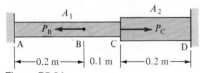

Figure P5.24

5.25 Redo Prob. 5.24 for a bar that has a constant cross-sectional area $A_1 = A_2 = 250$ mm^2, $E = 70$ GPa, and $P_C = 3P_B = 150$ kN.

5.26 A 6-ft reinforced-concrete column 8 in. in diameter is designed to carry an axial compressive load of 80 kips. The cross-sectional area of the axial reinforcing steel bars is $A_s = 0.8$ in.^{2}, $E_s = 30 \times 10^6$ psi, and $E_c = 3.5 \times 10^6$ psi. Calculate the stress in each material.

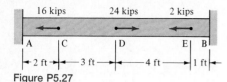

Figure P5.27

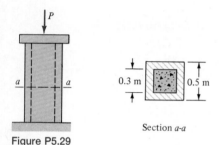

Figure P5.29

5.27 Referring to Fig. P5.27, calculate the reactions at A and B for the $\frac{3}{4}$-in.-diameter rod supported by rigid walls at its ends. Plot the axial load diagram.

5.28 Two round bars are supported between rigid walls, as shown in Fig. P5.28. The material properties are $\alpha_a = 24 \times 10^{-6}/°C$, $E_a = 70$ GPa, $\alpha_s = 12 \times 10^{-6}/°C$, and $E_s = 200$ GPa. If the temperature is increased from 20 to 50°C, determine the stress in each bar.

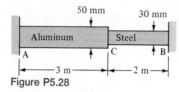

Figure P5.28

5.29 As shown in Fig. P5.29, a square cast-iron column shell is filled with concrete, and a load $P = 400$ kN is applied through a rigid cap. Use $E_{ci} = 70$ GPa and $E_c = 25$ GPa. Determine stresses in each material.

5.30 In the assembly of the bronze tube ($E_b = 83$ GPa, $A_b = 900$ mm²) and steel bolt ($E_s = 200$ GPa, $A_s = 500$ mm²) shown in Fig. P5.30, the bolt is single-threaded, with a 2-mm pitch. If the nut is tightened one-half of a full turn, calculate the normal stresses produced in the bolt and in the tube.

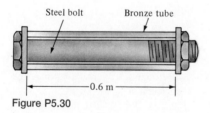

Figure P5.30

5.31 As seen in Fig. P5.31, a rigid member AB is suspended by aluminum bars AD and BE, each having a 400-mm² cross-sectional area and an E of 70 GPa. If member AB is to remain horizontal under a load of $P = 15$ kN applied at point C, determine (a) the tensile stress in bar BE, (b) the vertical displacement of C, and (c) the value of x.

5.32 Rigid member AB in the assembly of Fig. P5.31 is to remain horizontal under a load of $P = 100$ kN applied at $x = 150$ mm. Calculate the smallest cross-sectional area of steel bars AD and BE if $\sigma_{all} = 150$ MPa and $E = 200$ GPa.

5.33 Determine the stresses in the bolt-and-tube assembly of Fig. P5.30 after a temperature rise of 80°C. Let $\alpha_b = 18 \times 10^{-6}/°C$ and $\alpha_s = 12 \times 10^{-6}/°C$. (The constants A_b, E_b, A_s, and E_s are given in Prob. 5.30.)

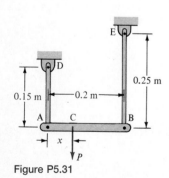

Figure P5.31

5.34 An axial eccentric force $Q = 12$ kips is applied to the $\frac{1}{2}$-in.-thick, 1-in.-wide, and 2-in.-long brass ($E = 15 \times 10^6$ psi) and steel ($E = 30 \times 10^6$ psi) blocks shown in Fig. P5.34. If the rigid end plate AB is to remain horizontal, determine (a) the value of x and (b) the normal stress in each material.

5.35 At 20°C there is a gap $\Delta = 0.4$ mm between the ends of the aluminum and magnesium bars shown in Fig. P5.35. Use $E_a = 70$ GPa, $\alpha_a = 23 \times 10^{-6}/°C$, $A_a = 500$ mm², $E_m = 45$ GPa, $\alpha_m = 26 \times 10^{-6}/°C$, and $A_m = 1200$ mm². Determine (a) the compressive stress in each rod when the temperature rises to 120°C and (b) the change in length of the magnesium bar.

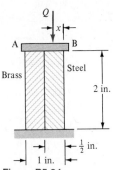

Figure P5.34

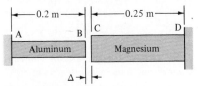

Figure P5.35

5.36 Redo Prob. 5.35 for a gap between the ends of the bars equal to zero.

5.37 A 9-ft-long steel bar ($E_s = 30 \times 10^6$ psi, $\alpha_s = 6.5 \times 10^{-6}/°F$) of $1\frac{5}{8}$-in. diameter and a 6-ft-long brass bar ($E_b = 15 \times 10^6$ psi, $\alpha_b = 11 \times 10^{-6}/°F$) of the same diameter are joined end to end. If the bars are subjected to axial compressive loads of 20 kips at the free ends and simultaneously to a temperature increase of 90°F, calculate the total change in length of the composite bar.

5.38 A stepped brass bar 150 mm long is inserted into a steel link with rigid ends, as shown in Fig. P5.38. Initially, no axial force exists in the bar. If the temperature increases 40°C, determine the maximum normal stress produced in the bar. Use $E_b = 105$ GPa, $\alpha_b = 20 \times 10^{-6}/°C$, $E_s = 200$ GPa, $\alpha_s = 12 \times 10^{-6}/°C$, and the cross-sectional areas $A_1 = 500$ mm², $A_2 = 400$ mm², $A_3 = A_4 = 450$ mm².

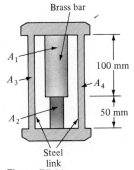

Figure P5.38

5.39 Three identical wires support a load P hung from a rigid rod ABC at point D, as shown in Fig. P5.39. Determine the tensile forces T_A, T_B, and T_C. Let $x = L/2$.

5.40 According to Fig. P5.39, three wires, two of aluminum (AE and BG) and one of steel (CF), support a load P hung at the center ($x = L$) of rigid rod ABC. Determine the largest value of P if $(\sigma_s)_{all} = 18$ ksi and $(\sigma_a)_{all} = 12$ ksi. Assume that diameter $d = \frac{1}{8}$ in., $E_s = 30 \times 10^6$ psi, and $E_a = 10 \times 10^6$ psi.

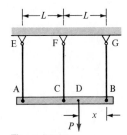

Figure P5.39

5.41 A 3-m-long bronze bar of 40 mm diameter is fixed at one end. At room temperature (20°C) an 0.5-mm gap exists between the other end and a rigid wall. Material properties are $E = 70$ GPa, $\alpha = 18 \times 10^{-6}/°C$, and $\gamma = 0.3$. Calculate, for a temperature rise of 40°C, (a) the axial strain in the bar and (b) the diameter of the bar.

5.42 In the assembly of Fig. P5.42, brass rod AE ($E_b = 105$ GPa) and aluminum rod CF ($E_a = 70$ GPa) each have a cross-sectional area of 500 mm². Determine the displacement of end D of the rigid member ABCD that would be caused by a 40-kN load.

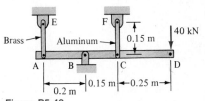

Figure P5.42

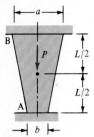

Figure P5.44

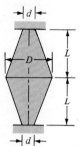

Figure P5.45

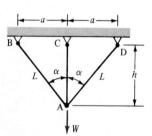

Figure P5.47

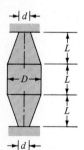

Figure P5.50

5.43 A steel ring ($E = 210$ GPa, $\alpha = 12 \times 10^{-6}/°C$) of internal radius 499.8 mm is to be shrink-fit over a rigid cylinder of a radius $r = 500$ mm. (a) How much must the temperature of the ring be increased above normal for this to occur? (b) Determine the stress in the ring when the temperature returns to normal.

5.44 A tapered member of constant thickness t is securely fastened between two rigid walls, as shown in Fig. P5.44. Determine the reactions at the walls due to a load of P at the center. Express the answer in terms of P, a, and b.

5.45 As shown in Fig. P5.45, an elastic conical compression member of diameters $D = L$ and $d = L/4$ and length $2L$ just touches the two surfaces at temperature T_1. Determine the maximum stress in the bar in terms of E, α, T_1, and T_2 if the temperature is raised to T_2.

5.46 Determine the reactions at A and B for the stepped bar and loading shown in Fig. 5.7 if the walls move 0.4 mm closer together during a temperature drop of 50°C. Let $P = 100$ kN, $a = 0.2$ m, $A_a = 1500$ mm^2, $E_a = 70$ GPa, $\alpha_a = 23 \times 10^{-6}/°C$, $b = 0.3$ m, $A_b = 500$ mm^2, $E_b = 200$ GPa, and $\alpha_b = 18 \times 10^{-6}/°C$.

5.47 The assembly shown in Fig. P5.47 consists of three bars, each with identical cross-sectional area A and material properties E, α. Determine the axial stress in the bar AC owing to a temperature increase ΔT in *all* members. Use $W = 0$, $a = 0.6L$, and $h = 0.8L$.

5.48 Redo Prob. 5.47 for a temperature increase ΔT in bar AC *only*. Let $W = 0$, $a = 0.6L$, and $h = 0.8L$.

5.49 As shown in Fig. P5.47, the assembly of three steel bars, each of equal axial rigidity AE, supports a vertical load W. (a) Verify that the forces in the bars AB = AD and AC are, respectively,

$$F_{AB} = F_{AD} = \frac{W \cos^2 \alpha}{1 + 2 \cos^3 \alpha} \qquad F_{AC} = \frac{W}{1 + 2 \cos^3 \alpha} \qquad (P5.49)$$

(b) For $W = 100$ kN, $a = 0.6L$, $h = 0.8L$, determine the value of the cross-sectional area A so that the axial stress does not exceed 160 MPa.

5.50 As shown in Fig. P5.50, the compression cylinder of diameters $d = L/4$ and $D = L$ just touches the fixed surfaces at a temperature T_1. Determine the maximum stress in the member in terms of E, α, T_1, and T_2 if the temperature is raised to T_2.

Secs. 5.6 to 5.8

5.51 What are the maximum normal and shearing stresses in a circular bar of $1\frac{5}{8}$-in. diameter subjected to an axial compression load of $P = 20$ kips?

5.52 Determine the maximum axial load P that can be applied to the rectangular wooden bar of Fig. 5.10a without exceeding a shearing stress of $\tau_{x'y'} = 1.7$ ksi or a normal stress of $\sigma_{x'} = 4.2$ ksi on the inclined plane c-c parallel to its grain. Use $\theta = 60°$, $a = \frac{3}{4}$ in., and $b = 2$ in.

5.53   Solve the preceding problem for $\sigma_{x'} = 2$ ksi and $\tau_{x'y'} = 1.3$ ksi.

5.54   A cylinder of 50-mm inner radius and 5-mm wall thickness has a welded spiral seam at an angle of 40° with the axial (x) direction. The cylinder is subjected to an axial compressive load of 10 kN applied through rigid end plates. Determine the normal σ and shear τ stresses acting simultaneously in the plane of welding.

5.55   Redo Prob. 5.54 for a seam angle of 55°.

5.56   Calculate the normal and shearing stresses on a plane through the bar of Fig. 5.10a that makes an angle of 30° with the direction of force $P = 100$ kN. Use $a = 12$ mm and $b = 25$ mm.

5.57   The stresses on an inclined plane c-c in a bar in tension (see Fig. 5.10a) are $\sigma_{x'} = 16$ MPa and $\tau_{x'y'} = 8$ MPa. Determine the axial load P and the angle θ for $a = 30$ mm and $b = 40$ mm.

5.58   Solve Prob. 5.56, with the 30° angle changed to 20°.

5.59   Determine the maximum stress caused by an axial load of 150 kN applied to a 200-mm-wide by 10-mm-thick plate with a hole of 25-mm diameter through its center.

5.60   The machine component of Fig. P5.60 is 8 mm thick and is made of mild steel. If $\sigma_{\text{all}} = 120$ MPa, calculate the maximum value of the axial load P.

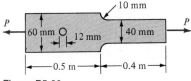
Figure P5.60

5.61   An axial load P produces a total elongation of $\delta = 0.5$ mm in the 15-mm-thick bar ($E = 200$ GPa) of Fig. P5.60. Neglecting the effect of the stress concentrations on axial deformation, calculate the maximum stress in the bar.

Calculate the maximum stress in the 10-mm-thick steel bar shown in Fig. P5.60 if (a) $P = 25$ kN and (b) $P = 40$ kN and the hole is not present.

5.63   Determine the fillet radius r and width d of the steel plate in tension, seen in Fig. P5.63. Use a maximum allowable stress of 19.5 ksi and an allowable nominal stress in the reduced section of 12 ksi.

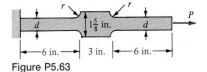
Figure P5.63

5.64   For the $\frac{3}{4}$-in.-thick steel bar ($\sigma_{yp} = 36$ ksi) of Fig. P5.63, given the ratio of $r/d = 0.5$, calculate the maximum axial load that can be applied without causing permanent deformation.

5.65   Due to the axial load P, a total elongation of $\delta = 8 \times 10^{-3}$ in. occurs in the $\frac{3}{16}$-in.-thick plate of Fig. P5.63. Given $r = \frac{5}{32}$ in., $d = 1\frac{5}{16}$ in., and $E = 30 \times 10^6$ psi, and omitting the effect of the stress concentrations on the elongation, calculate the maximum stress.

CHAPTER

6

TORSION

6.1 INTRODUCTION

In addition to exploring the general concepts of solid mechanics, previous chapters dealt with the behavior of axially loaded bars. Consideration is now given to stresses and deformations in prismatic bars subject to torsion. Usually these twisted members are slender and circular in cross section—for example, shafts. The principal practical application of shafts is the transmission of power as from a steam turbine to an electric generator or from an engine to the propeller of an outboard motor. In other situations, a twisted rod can serve as a spring, offering rotational stiffness, as does an automobile torsion-bar spring system.

   The equations describing the stress and deformation of a twisted circular bar will be derived by applying the principles of Sec. 1.3. All relationships derived in this chapter, with the exception of those in Sec. 6.9, are based upon the assumptions introduced in the next section. The torsion of slender bars with other than circular cross sections is, of course, also encountered in some engineering applications. Thin-walled tubes of *any shape* will be discussed in Sec. 6.9.

   In the treatment presented here, the bars are considered free of *all* constraints. Thus the study of complex stress patterns at the supports or at the locations of torque-applying devices is avoided. According to Saint-Venant's principle, the actual stress distribution closely approximates that given by the torsion formula everywhere, except near the restraints and geometric discontinuities in the shaft.

6.2 BEHAVIOR OF A TWISTED CIRCULAR SHAFT

A torsion problem may conveniently be treated by first examining the *geometric behavior* of a twisted circular bar; from this the deformation pattern is constructed. The stress-strain relationships related to the *material behavior* are then introduced. Finally, the conditions of *equilibrium* are applied, as will be demonstrated in the next section.

Figure 6.1*a* shows a circular bar subject to equal and opposite twisting couples or torques *T*, applied at the ends, about the longitudinal (*x*) axis coinciding with the center of the cross section. A bar loaded in this manner is said to be in *pure torsion*. Considerations of symmetry of geometry and loading dictate the following information regarding the deformations. Imagine the bar cut into two equal parts across the middle. For static equilibrium of each half bar, the midsections must be subjected to the same internal torques *T* as the end sections. Hence they must deform identically. We conclude therefore that *cross sections* of the bar *rotate identically* as rigid figures about the *x* axis; the longitudinal line AB deforms into the helix AB′, and the angle BOB′ represents the total angle of twist ϕ (phi).

From the foregoing, we obtain the following basic assumptions associated with the small deformation of a slender circular bar in pure torsion:

1. Plane cross sections perpendicular to the axis of the bar *remain plane* after the application of a torque; points in a given plane remain in that plane after twisting. Furthermore, expansion or contraction of a cross section does not occur nor does a shortening or lengthening of the bar. Thus all *normal* strains are zero.

2. Subsequent to twisting, cross sections are undistorted in their individual planes; the *shearing strain γ varies linearly* from zero at the center to a maximum on the outer surface. This supposition is illustrated by considering the turning of the right end through a small angle of twist $d\phi$ relative to the left end (that is, $m = m'$) of an element of length dx in Fig. 6.1*a*, shown isolated in Fig. 6.1*b*. Note that the straight line *mn*, initially parallel to the axis of the bar, distorts into line *mn′*, which is approximately straight, and the magnitude of $\gamma_{\max}$ is given by angle *nmn′*. We observe that a plane such as mnO_1O_2 moves to $mn'O_1O_2$ subsequent to twisting. In other

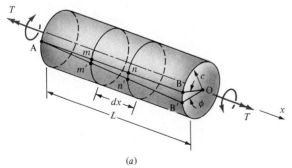

(*a*)

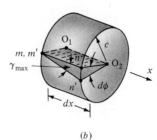

(*b*)

Figure 6.1   Variation of strain in a circular shaft in torsion.

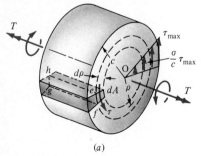

words, if a radius O_1m is regarded as fixed in direction, a similar radius originally at O_2n rotates to become O_2n' and *remains straight*.

The preceding assumptions hold for both elastic and inelastic material behavior (see Sec. 12.5). In the *elastic case,* the following also applies:

3. The material is homogeneous and obeys Hooke's law; the shear stress is *proportional* to the shear strain. Thus the magnitude of the torque is such that the shear stresses in the bar remain below the yielding strength or proportional limit of the material. Hence the magnitude of angle γ_{max} must be less than the yield angle.

Note that in treating noncircular prismatic bars, initially plane sections (Fig. 6.2*a*) experience out-of plane deformation or warping (Fig. 6.2*b*). Thus assumptions 1 and 2 are no longer appropriate, and it is necessary to apply the theory of elasticity, computer-oriented numerical approaches, or experimental methods.

Figure 6.2 Deformation of a rectangular bar in torsion: (*a*) plane cross section and (*b*) warped cross section.

6.3 THE TORSION FORMULA

The relationship between applied torque and resulting stresses will now be treated. In accordance with assumptions 2 and 3 of the preceding section, the *shearing stresses vary linearly* with the distance from the axis of a circular bar. They are oriented in the tangential direction and *lie in the cross section,* as shown in Fig. 6.3*a*. Note that, unlike the case of an axially loaded bar, the stress is not constant. The largest shear stress occurs at points most remote from the center O and is denoted τ_{max}. As can be seen from the figure, at any arbitrary distance ρ (rho) from O, $\tau = (\rho/c)\tau_{max}$. This stress distribution leaves the *external* cylindrical *surface* of the bar free of stress, as it should.

We next require that the stresses acting at a section satisfy the conditions of equilibrium. That is, the resultant of the rotationally symmetric stress distribution depicted in Fig. 6.3*a* must be statically equivalent to the torque

Figure 6.3 Distribution of stress in a circular shaft in torsion.

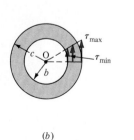

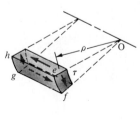

(*a*)   (*b*)   (*c*)

applied. The shear force acting on an element of area dA is $\tau\, dA$, and the moment of this force about the central axis is $(\tau\, dA)\rho$. The applied torque T is equal to the sum over the entire area A of the cross section of such elemental moments:

$$T = \int_A \left(\frac{\rho}{c}\, \tau_{max}\, dA \right) \rho = \frac{\tau_{max}}{c} \int_A \rho^2\, dA \qquad (a)$$

where $\int \rho^2\, dA$ is called the *polar moment of inertia* J of the cross-sectional area (see App. A). Referring to Fig. 6.3a, for a circle of radius c and diameter d, $dA = 2\pi\rho\, d\rho$. Thus

$$J = \int_A \rho^2\, dA = \frac{\pi c^4}{2} = \frac{\pi d^4}{32} \qquad (6.1)$$

Expressed in terms of J for a circular area, Eq. (a) may be written as

$$\tau_{max} = \frac{Tc}{J} \qquad (6.2)$$

This is known as the *torsion formula* and is applicable to circular shafts. It was derived by C. A. Coulomb, a French engineer, in about 1775. This expression shows that the maximum shearing stress is proportional to the applied torque T and inversely proportional to the polar moment of inertia of the cross section. The shearing stress at any point a distance ρ from the center of a section is

$$\tau = \frac{T\rho}{J} \qquad (6.3)$$

In the torsion formula, if SI units are used, T is measured in newton-meters (N·m), c or ρ in meters (m), and J in meters^4 (m^4); and the resulting shearing stress τ is expressed in pascals (Pa). Similarly, in U.S. customary units, T is measured in pound-inches (lb·in.), c or ρ in inches (in.), and J in inches^4 (in.^{4}), making the units of stress τ pounds per square inch (psi). Equation (6.2), when written as $\tau = T/(J/c)$, is analogous to $\sigma = P/A$ for the stress in axially loaded bars.

The torsion formula is based upon Hooke's law. It thus applies only when stresses do not exceed the shearing proportional limit. Nevertheless, Eq. (6.2) is sometimes used to calculate a fictitious *shearing stress at rupture* τ_r, with the experimentally obtained ultimate torque T_u (which we shall deal with in Sec. 12.5) substituted for T:

$$\tau_r = \frac{T_u c}{J} \qquad (6.4)$$

Also known as the *modulus of rupture in torsion*, this stress is used to compare the static resistance of various materials under torsional loading.

The torsion analysis of a *hollow circular tube* or cylinder is based upon the same assumptions as used for a solid bar and hence Eqs. (6.2) and (6.3) may still be employed. In this case, the radial distance ρ is limited to the range b to c (Fig. 6.3b), and the polar moment of inertia is

$$J = \frac{\pi}{2}(c^4 - b^4) \tag{6.5}$$

where b and c represent the inner and outer radii of the circular tube, respectively. For thin-walled circular members, Eq. (6.1) can be approximated by $J \approx c^2 \int dA \approx 2\pi c^3 t$, where t is the thickness of the tube (see Example 6.9). If the ratio of the thickness to the inner radius is less than approximately $\frac{1}{10}$, the cylinder is generally classified as *thin-walled*, and the variation of stress with the radius can be disregarded.

The torsion formula was derived for a uniform bar in pure torsion. It may also be employed, however, for a bar of variable cross section or a bar subjected to torques at locations other than the ends. In these situations, the shearing stress in a cross section of interest is determined by applying Eq. (6.2), with J and T representing the polar moment of inertia and the internal torque at that section, respectively. Similar to the case of an axially loaded bar, the value of the internal torque is determined by employing the method of sections.

The directions of the shearing stresses acting in the plane of the cross section, defined. by Eqs. (6.2) and (6.3), coincide with the direction of the torque T (Fig. 6.3a). These stresses must be accompanied by equal shearing stresses taking place on the mutually perpendicular or axial planes of the bar (recall Sec. 2.3). Thus the state of stress is as illustrated in Fig. 6.3c, where τ is the only stress component acting on the element *efgh* removed from the bar. That is, the element is in *pure shear*; τ designates the shearing stresses in the *axial and tangential* directions. Hence, if a material is weaker in shear axially than laterally (for example, wood), the failure in a twisted bar occurs longitudinally along axial planes.

EXAMPLE 6.1

A stepped circular shaft is in equilibrium under the torques applied to the pulleys fastened to it, as shown in Fig. 6.4a. Calculate the maximum shearing stress in the shaft.

Solution We begin by idealizing the situation (Fig. 6.4b). The only equation of statics available for the shaft, $\Sigma T = 0$, yields

$$T_{\mathrm{B}} = 36 + 14 = 50 \text{ kN} \cdot \text{m}$$

Using T_{AB} to denote the torque in segment AB, we imagine a transverse plane passing anywhere between A and B. A free body of a part of the shaft is shown in Fig. 6.4c. Similarly, in Fig. 6.4d, the plane passes through the segment BC,

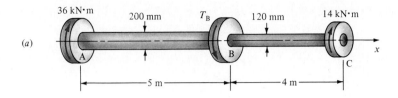

(a)

(b)

(c)

(d)

Figure 6.4   Examples 6.1 and 6.2.

and T_{BC} is the resisting torque on this section. These free bodies are in equilibrium if the internal torques acting on the segments are $T_{AB} = 36$ kN·m and $T_{BC} = -14$ kN·m.

As the location of the maximum shearing stress is not apparent, we must determine the stress at both sections by applying the torsion formula. For segment AB of the shaft,

$$\tau_{max} = \frac{Tc}{J} = \frac{2T}{\pi c^3} = \frac{2(36 \times 10^3)}{\pi(0.1^3)} = 22.9 \text{ MPa}$$

while for the segment BC,

$$\tau_{max} = \frac{2(14 \times 10^3)}{\pi(0.06^3)} = 41.3 \text{ MPa}$$

The maximum shearing stress is thus 41.3 MPa.

6.4   STRESSES ON INCLINED PLANES

Up to now, our treatment has been limited to shearing stresses on planes parallel or perpendicular to the axis of a bar, defined by the torsion formula. This state of stress is shown acting on a surface element of the shaft in Fig. 6.5a. An alternative representation of the stresses at the same point may be given on an infinitesimal wedge whose outer normal (x') on its oblique face makes an angle θ with the axial axis (x).

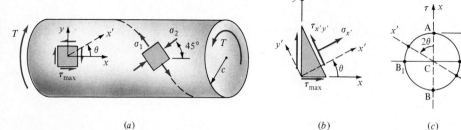

(a) (b) (c)

Figure 6.5 (a) Stress acting on a surface element; (b) free body diagram of wedge cut from element; and (c) Mohr's circle for torsional loading.

Consider the wedge cut from the shaft element (Fig. 6.5b). Here we analyze the stresses $\sigma_{x'}$ and $\tau_{x'y'}$ which must act on the *inclined plane* to keep the isolated body in equilibrium, as discussed in Sec. 4.2. The equations of equilibrium of forces in the x' and y' directions, Eqs. (4.4a) and (4.4b), yield

$$\sigma_{x'} = -\tau_{max} \sin 2\theta \qquad (6.6a)$$

$$\tau_{x'y'} = -\tau_{max} \cos 2\theta \qquad (6.6b)$$

The foregoing expressions indicate that $\sigma_{min} = \sigma_2 = -\tau_{max}$ and $\tau_{x'y'} = 0$ when $\theta = 45°$ and $\sigma_{max} = \sigma_1 = \tau_{max}$ and $\tau_{x'y'} = 0$ when $\theta = 135°$. Thus the *maximum normal stress* equals Tc/J and the minimum normal stress equals $-Tc/J$, acting in the directions shown in Fig. 6.5a. Furthermore, no shearing stress is developed which is greater than τ_{max}.

Figure 6.5c, a *Mohr's circle for torsional loading,* provides a convenient manner of checking the foregoing results. Note that points A(0, τ_{max}) and B(0, τ_{min}) are located on the τ axis. Points A_1 and B_1 define the principal planes and the principal stresses.

It becomes evident, therefore, that for a material such as cast iron, which is weaker in tension than in shear, failure occurs in tension along a helix indicated by the dashed lines in Fig. 6.5a. Ordinary chalk behaves in this way. Shafts made of materials weak in shearing strength (for example, mild steel) break along a line perpendicular to the axis. Experiments show that a very thin walled hollow shaft buckles or wrinkles in the direction of maximum compression, while in the direction of maximum tension, tearing occurs.

6.5 ANGLE OF TWIST

In the preceding two sections, concern was with torsion stresses. We now turn to deformation of twisted circular members. Here we assume that the entire bar remains elastic. For most structural materials, the amount of twist is small and hence the member behaves as described in Sec. 6.2. But in the case of materials such as rubber, where twisting is large, the basic assumptions must be reexamined.

Let us consider the deformation of a slice of a circular bar subjected to torsion (Fig. 6.1b). According to assumption 2 of Sec. 6.2, radius $O_2 n$ remains straight and rotates through an angle $d\phi$ to a new position $O_2 n'$. The shearing strain may be approximated by taking tan $\gamma_{max} = \gamma_{max}$ for *small angles*: $\gamma_{max} dx = c\, d\phi$. Thus,

$$\gamma_{max} = c\frac{d\phi}{dx} \tag{6.7}$$

where $d\phi/dx$ represents the *angle of twist per unit length*, or the rate of twist, a constant along the length of the bar under pure torsion.

As a consequence of assumption 3 of Sec. 6.2, $\gamma_{max} = \tau_{max}/G$ or, recalling the torsion formula, $\gamma_{max} = Tc/GJ$. Introducing the latter relationship into Eq. (6.7) and solving for $d\phi$, we obtain

$$d\phi = \frac{T\,dx}{GJ} \tag{6.8}$$

Equation (6.8) shows that the angle of twist per unit length varies directly with torque T and inversely with the product GJ, called the *torsional rigidity* of the bar. This is in agreement with experimental evidence, which justifies the assumptions used in the derivation.

The *total angle of twist* ϕ represents the angle through which one end cross section of a bar rotates with respect to another. The general expression for ϕ for a shaft of a linearly elastic material is

$$\phi = \int_0^L \frac{T_x\,dx}{GJ_x} \tag{6.9}$$

where ϕ is measured in radians. The *sense* of ϕ is the same as that of the torque T_x. Equation (6.9) is suitable when the quantities T_x or J_x vary along the length L of the shaft. The foregoing applies to both solid and hollow circular shafts. Observe the close similarity between Eq. (6.9) and Eq. (5.1) for the deformation of axially loaded bars. If the bar is homogeneous, has a *uniform* polar moment of inertia, and is twisted at its ends, Eq. (6.9) reduces to

$$\phi = \frac{TL}{GJ} \tag{6.10}$$

In general, the total angle of twist for shafts consisting of several segments, each having *constant* values for G, J, and T, can be expressed as

$$\phi = \sum_{i=1}^{n} \frac{T_i L_i}{G_i J_i} \tag{6.11}$$

Here the subscript i is an index identifying the various parts of the shaft and n is the total number of segments.

Examination of Eq. (6.10) suggests a method for determining the shear

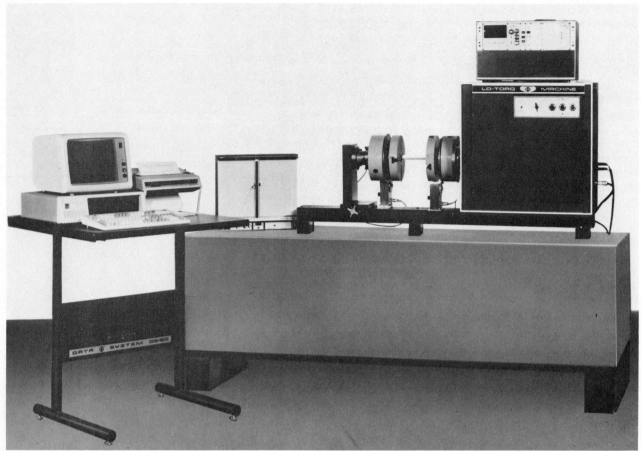

Figure 6.6 Typical 10,000 in.·lb Lo-Torq bench-type torsion testing machine, shown with direct digital readout of torque values and the DS-50 Data System. *(Photo courtesy of Tinius Olsen Testing Machine Co., Inc., Willow Grove, Pa.)*

modulus of elasticity G for a given material. A cylindrical specimen of the material, of known diameter and length, is placed in a torque-testing machine (Fig. 6.6). As the specimen is twisted, the increasing value of the applied torque T and the corresponding values of the angle of twist ϕ between the two ends of the specimen are recorded as a *torque-twist curve*. The slope of this diagram (T/ϕ) in the linearly elastic region represents the quantity GJ/L, from which the magnitude of G can readily be calculated.

EXAMPLE 6.2

Calculate the relative angle of rotation between pulleys A and C of a stepped shaft in equilibrium (Fig. 6.4a). Let $G = 80$ GPa.

Solution From Example 6.1, the torques transmitted by each segment of the shaft are known to be $T_{AB} = 36$ kN·m and $T_{BC} = -14$ kN·m. The total angular distortion is found by applying Eq. (6.11). Hence

$$\phi_{AC} = \sum \frac{TL}{GJ} = \frac{10^3}{(80 \times 10^9)(\pi/2)} \left[\frac{36 \times 5}{(0.1)^4} - \frac{14 \times 4}{(0.06)^4} \right]$$

from which $\phi_{AC} = -0.02$ rad $= -1.15°$. The negative result means that the rotation of C with respect to A is in the clockwise direction.

EXAMPLE 6.3

Hollow and solid cylindrical shafts (Fig. 6.7) are constructed of the same material. Both have the same length and cross-sectional area, and both are subjected to pure torsion. What is the ratio of the largest torque that can be applied to the shafts (a) if the allowable stress is τ_{all}, and (b) if the allowable angle of twist is ϕ_{all}?

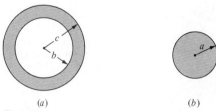

(a) (b)

Figure 6.7 (a) A hollow shaft and (b) a solid shaft of the *same* cross-sectional area.

Solution As the areas of both shafts are the same, we have $\pi(c^2 - b^2) = \pi a^2$ or $a^2 = c^2 - b^2$.

(a) The maximum permissible torque transmitted by the hollow shaft, from $T = J\tau_{all}/c$, is

$$T_h = \frac{\pi}{2c}(c^4 - b^4)\tau_{all}$$

Similarly, for the solid shaft, we have

$$T_s = \frac{\pi}{2}a^3\tau_{all}$$

The ratio of the torques is therefore

$$\frac{T_h}{T_s} = \frac{c^4 - b^4}{ca^3} = \frac{c^4 - b^4}{c(c^2 - b^2)^{3/2}} \tag{6.12a}$$

If, for example, $c = 1.25b$, the foregoing quotient leads to

$$\frac{T_h}{T_s} = 2.73 \tag{a}$$

(b) In this case, the maximum permissible torque transmitted by the hollow shaft, from $T = GJ\phi_{all}/L$, is found to be

$$T_h = \frac{\pi}{2L}(c^4 - b^4)G\phi_{all}$$

and for the solid shaft,

$$T_s = \frac{\pi}{2L}a^4G\phi_{all}$$

Hence

$$\frac{T_h}{T_s} = \frac{c^4 - b^4}{a^4} = \frac{c^2 + b^2}{c^2 - b^2} \tag{6.12b}$$

For $c = 1.25b$, the above yields

$$\frac{T_h}{T_s} = 4.56 \tag{b}$$

Results (a) and (b) indicate that the hollow shaft can transmit greater torque than the solid shaft. It appears that a given amount of material is used more *efficiently* in torsion when the shaft has a *hollow* cross section. However, fabrication constraints may make it impractical to produce a hollow section of exact dimensions. For this reason, a solid shaft is often used, except in applications where weight is critical.

6.6 STATICALLY INDETERMINATE SHAFTS

The methods used in the analysis of problems dealing with statically indeterminate bars subject to torsion are the same as those employed for statically indeterminate axially loaded members. The only difference is that the condition of compatibility relates to the angles of twist rather than of axial deformations.

The solution to a statically indeterminate torsion problem is obtained by following the basic steps outlined in Sec. 1.3. This consists of establishing equilibrium between internal and applied torques and satisfying the torque–angular deformation relationships—Eq. (6.10). With the internal torques found, the stress and the angle of twist are calculated by using Eqs. (6.2) and (6.10), respectively.

Application of these principles and equations is illustrated in the examples that follow.

EXAMPLE 6.4

A round shaft of elastic modulus G is fixed to rigid walls at both ends, as illustrated in Fig. 6.8a. The left segment of the bar has a polar moment of inertia J_a;

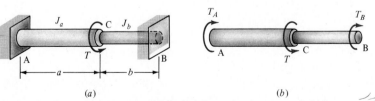

(a)   *(b)*

Figure 6.8 Example 6.4: statically indeterminate circular stepped shaft.

the polar moment of inertia of the right part is J_b. Determine the reactions resulting from the application of a torque T at the point of discontinuity of the section. Use $J_a = 1.5\pi$ in.^{4}, $J_b = 0.5\pi$ in.^{4}, $a = 9$ in., $b = 6$ in., and $T = 12$ kip·in.

Solution   We apply the procedure stated in Sec. 1.3 to the free-body diagram of the shaft (Fig. 6.8*b*), where T_A and T_B denote the reactive torques exerted by the supports.

Statics.   The only condition of equilibrium available concerns the sum of the moments about the axis of the shaft:

$$T_A + T_B = T \tag{a}$$

The problem is thus statically indeterminate.

Deformations.   The angle of twist at C is expressed in terms of the left and right segments of the shaft, respectively:

$$\phi_a = \frac{T_A a}{GJ_a} \qquad \phi_b = \frac{T_B b}{GJ_b} \tag{b}$$

Geometry.   Inasmuch as each part must undergo the identical twist,

$$\phi_a = \phi_b$$

or

$$\frac{T_A a}{J_a} = \frac{T_B b}{J_b} \tag{c}$$

Expressions (*a*) and (*c*) can be solved simultaneously to yield

$$T_A = \frac{T}{1 + (aJ_b/bJ_a)}$$

$$T_B = \frac{T}{1 + (bJ_a/aJ_b)} \tag{6.13}$$

If the shaft is of constant cross section, the foregoing becomes

$$T_A = \frac{Tb}{L} \qquad T_B = \frac{Ta}{L} \tag{6.14}$$

Upon substitution of the given data into Eqs. (6.13), we obtain

$$T_A = \frac{12 \times 10^3}{1 + (9 \times 0.5\pi/6 \times 1.5\ \pi)} = 8 \text{ kip·in.}$$

$$T_B = \frac{12 \times 10^3}{1 + (6 \times 1.5\pi/9 \times 0.5\pi)} = 4 \text{ kip·in.}$$

The angle of rotation at C and shearing stresses in each segment may now be obtained from Eqs. (*b*) and (6.2), respectively.

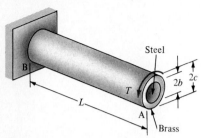

Figure 6.9 Example 6.5: statically indeterminate composite bar in torsion.

EXAMPLE 6.5

A composite circular shaft composed of steel and brass is bonded securely together at the interface b and loaded by torque T, as illustrated in Fig. 6.9. The shear moduli of elasticity are G_b (outer shaft) and G_s (inner shaft). Derive expressions for the maximum shear stress in the brass τ_b and steel τ_s. What is the rotation ϕ_A of the end A with respect to the fixed end B?

Solution Let T_s and T_b represent the resisting torques in the core and in the tube, respectively. We now apply the principles of analysis to Fig. 6.9.

Statics. Equilibrium conditions for the loading require that

$$T_b + T_s = T \tag{d}$$

The shaft is statically indeterminate, as T_b and T_s cannot be determined from Eq. (d) alone.

Deformations. Applying Eq. (6.10), we obtain the expression for the angle of twist ϕ_b of the tube and ϕ_s of the core at the free end:

$$\phi_b = \frac{T_b L}{G_b J_b} \qquad \phi_s = \frac{T_s L}{G_s J_s} \tag{e}$$

where the polar moments of inertia are $J_b = (c^4 - b^4)/2$ and $J_s = c^4/2$.

Geometry. Continuity of the distortions in the composite shaft at the free end A requires that

$$\phi_A = \phi_b = \phi_s \tag{f}$$

Combining Eqs. (d), (e), and (f), we obtain

$$T_b = \frac{(G_b J_b)T}{G_b J_b + G_s J_s} \qquad T_s = \frac{(G_s J_s)T}{G_b J_b + G_s J_s} \tag{g}$$

showing that the torques in the members are proportional to their rigidities. Insertion of Eqs. (g) into Eq. (6.2) gives the maximum shear stress in the brass and in the steel:

$$\tau_b = \frac{(cG_b)T}{G_b J_b + G_s J_s} \qquad \tau_s = \frac{(bG_s)T}{G_b J_b + G_s J_s} \tag{6.15}$$

The angle of twist of the free end, from Eq. (f) together with Eqs. (g) and (e), is

$$\phi_A = \frac{TL}{G_b J_b + G_s J_s} \tag{6.16}$$

Thus, the angle of twist per unit length, ϕ_A/L, is equal to the applied torque divided by the sum of the torsional rigidities of the two parts.

*6.7 STRESS CONCENTRATIONS

If the cross section of the shaft is not constant, but changes gradually, the torsion formula which was derived for a uniform shaft still yields fairly accurate results. However, in situations in which the cross-sectional shape of a shaft changes abruptly, high concentrations of stress will occur near the discontinuity. These local stresses can be determined by employing an approach similar to that described in Sec. 5.7. Thus the largest value of the shearing stress at the fillets in a stepped shaft is expressed as

$$\tau_{max} = K\tau_{nom} = K\frac{Tc}{J} \qquad (6.17)$$

where the nominal or average stress is obtained for the *smaller* diameter shaft. The *stress-concentration factor K* corresponding to given r/d and D/d ratios is read from Fig. 6.10 which indicates the importance of rounding corners where a transition in shaft diameter occurs.

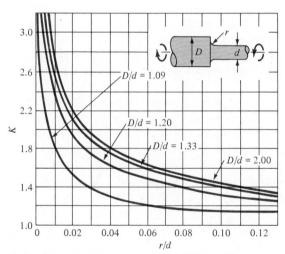

Figure 6.10 Stress-concentration factors for fillets in circular shafts. *(Source: L. S. Jacobsen, "Torsional-Stress Concentrations in Shafts of Circular Cross-Section and Variable Diameter," Trans. ASME, vol. 47, 1925, pp. 619–638.)*

Of great importance also is the keyway and small oil hole found in shafts (Fig. 6.11). For both sections, stress-concentration effects are particularly notable at points A. It is not surprising, therefore, that most torsional fatigue failures have their origin at these locations.

Figure 6.11 Shafts with a keyway and a hole.

EXAMPLE 6.6

Determine the largest torque that can be sustained by a stepped shaft if the allowable shear stress is 15 ksi. Let $D = 4$ in., $d = 2$ in., and $r = \frac{1}{8}$ in.

Solution The geometric proportions are

$$\frac{D}{d} = 2 \qquad \frac{r}{d} = \frac{1}{16} = 0.0625$$

and therefore from Fig. 6.10, $K = 1.62$. Then, from Eq. (6.17), substituting $J/c = \pi c^3/2$ and the given numerical values, we have

$$T = \frac{\pi c^3}{2K} \tau_{max} = \frac{\pi(1)^3(15 \times 10^3)}{2 \times 1.62} = 14.54 \text{ kip·in.}$$

the required magnitude of the torque.

*6.8 DESIGN OF CIRCULAR SHAFTS IN TORSION

One of the most common applications of a circular shaft is for the transmission of power at a specified speed of rotation. In the design of such shafts, the material and the dimensions of the cross section are selected so as not to exceed the allowable shearing stress and/or a limiting angle of twist when the shaft is rotating. Thus a designer needs to know the torque acting on the power-transmitting shaft.

The SI unit often used in industry to measure the power transmitted by a shaft is the kilowatt (kW). A relationship connecting the power and the torque acting through the shaft is established as follows. One watt does the work of one newton-meter per second. If the velocity f is expressed in cycles per second, then the angle through which the shaft rotates is $2\pi f$ rad/s. Hence, for a constant torque T, the work done per unit time is $2\pi fT$. This is equal to the power delivered: $2\pi fT = \text{kW}(1000)$. The torque is therefore

$$T = \frac{159 \text{ kW}}{f} \qquad \text{(N·m)} \tag{6.18a}$$

where f is the frequency in revolutions per second (rps) or hertz (Hz) of the shaft.

In U.S. engineering practice, power is usually expressed in horsepower (hp), a unit equal to 6600 in.·lb/s, torque in pound-inches (lb·in.), and frequency in revolution per minute (rpm), denoted by the letter N. Hence, $2\pi NT = \text{hp}(6600)(60)$ and the expression for torque is

$$T = \frac{63,000 \text{ hp}}{N} \qquad \text{(lb·in.)} \tag{6.18b}$$

Equations (6.18) convert the kilowatts or horsepower supplied to the shaft into a constant torque exerted on it during rotation.

Once the torque to be transmitted is determined, the design of circular shafts to meet strength requirements can be accomplished by the use of the process outlined in Sec. 1.9:

1. *Evaluate the mode of possible failure:* Assume that, as is usually the case, shear stress is closely associated with failure. Note, however, that in some materials, the maximum tensile and compression stresses occurring on planes at 45° to the shaft axis may cause the failure.

2. *Determine the relationship between the applied torque and stress:* A significant value of the shear stress is given by $\tau_{max} = Tc/J$. In accurate analysis of shafts, the formulas describing the failure criteria are frequently used, as will be illustrated in Sec. 8.10.

3. *Determine the maximum usable value of stress:* The maximum usable value of τ_{max} without failure is the yield shear stress τ_{yp} or ultimate shear stress τ_u.

4. *Select the factor of safety:* A factor of safety f_s is applied to τ_{max} to calculate the allowable stress $\tau_{all} = \tau_{yp}/f_s$ or $\tau_{all} = \tau_u/f_s$. The *required parameter* J/c of the shaft is therefore

$$\frac{J}{c} = \frac{T}{\tau_{all}} \tag{6.19}$$

For a given stress specification, Eq. (6.19) can be used to design both solid and hollow circular shafts subjected to torques only.

EXAMPLE 6.7

A gray cast-iron shaft having an ultimate strength of 210 MPa is to transmit a torque of 500 N·m with a factor of safety of $f_s = 1.2$. Determine the required diameter d of the shaft.

Solution The maximum tensile stress σ_u occurs on a diagonal plane and is equal in magnitude to the maximum shear stress τ_{max} at a point on the surface of the shaft (Fig. 6.5a). Thus Eq. (6.19) becomes

$$\frac{J}{d/2} = \frac{\pi d^3}{16} = \frac{Tf_s}{\sigma_u} \tag{a}$$

Introducing the given data, we get

$$\frac{\pi d^3}{16} = \frac{(500)1.2}{210(10^6)}$$

from which $d = 24.4 \times 10^{-3}$ m = 24.4 mm. Hence a 25-mm shaft should be used.

EXAMPLE 6.8

A steel propeller shaft ($G = 80$ GPa) is to transmit 400 kW at 900 rpm without exceeding a yield shear stress of 250 MPa or a twisting through more than $3°$ in a length of 2 m. Using a factor of safety of 1.4, calculate the diameter of the shaft.

Solution The frequency in cycles per second is $f = \frac{900}{60} = 15$. The torque is given by Eq. (6.18a):

$$T = \frac{159(400)}{15} = 4.24 \text{ kN·m}$$

Using Eq. (6.19), we find that the shaft size needed to satisfy the stress specification is

$$\frac{\pi}{2} c^3 = \frac{4240(1.4)}{250(10^6)}$$

This yields $c = 24.7 \times 10^{-3}$ m $= 24.7$ mm.

The size of the shaft required to fulfill the distortion specification is obtained from Eq. (6.10):

$$\frac{\phi_{\text{all}}}{L} = \frac{T}{GJ} \tag{6.20}$$

Upon substitution of the numerical values, we have

$$\frac{3°(\pi/180°)}{2} = \frac{(4240)2}{(80 \times 10^9)\pi c^4}$$

The foregoing gives $c = 33.7 \times 10^{-3}$ m $= 33.7$ mm.

Thus the minimum allowable diameter of the shaft must be 67.4 mm. For practical purposes, a 70-mm shaft would probably be used.

*6.9 THIN-WALLED HOLLOW MEMBERS

In this section, we shall discuss the torsion of *thin-walled hollow members* or *tubes* of any cross-sectional form. Exact solutions in these cases are rather involved, requiring fulfillment of all principles stated in Sec. 1.3 and application of the methods of the theory of elasticity. However, a good *approximation* of the distribution of shear stresses in the twisted tube can be determined by a simple computation on the basis of force equilibrium only.

Consider a hollow cylindrical member of arbitrary shape and varying wall thickness t subjected to torque T. A small segment of length dx of such a shaft is shown in Fig. 6.12a. The shearing stresses τ produced by the torques can be assumed to be *uniform* across the small thickness of the tube, although they may vary around the cross section.

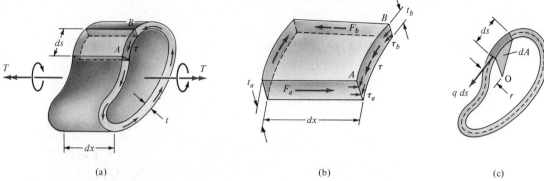

Figure 6.12   Stresses in a thin-walled tube.

A free-body diagram of an element isolated from the tube segment is sketched in enlarged scale in Fig. 6.12b. The shearing stresses on mutually perpendicular planes must be equal at a corner of an element (see Sec. 2.3). Accordingly, τ_a and τ_b denote the stresses at corners A and B, as shown in the figure. The stress resultants F_a and F_b on the longitudinal planes can be expressed by

$$F_a = \tau_a t_a \, dx \qquad F_b = \tau_b t_b \, dx$$

where t_a and t_b are the thicknesses of the tube at points A and B. The equilibrium of the longitudinal forces yields $F_a = F_b$, or

$$\tau_a t_a = \tau_b t_b \qquad\qquad (a)$$

This means that the product of the longitudinal stress and of the wall thickness at two *arbitrary* points must be equal. It follows that

$$q = \tau t = \text{constant} \qquad\qquad (6.21)$$

where q possesses the dimensions of force per unit length, representing the resisting force per unit length along the tube perimeter. For this reason, q is known as the *shear flow*.

We now evaluate the shear flow in terms of the applied torque T by considering the element of length ds in the cross section of the tube (Fig. 6.12c). The moment of the force $q \cdot ds$ acting on this element about any point O is $(q \, ds)r$, where r is the perpendicular distance from O to the line of action of the force. The total torque is the sum of all such moments:

$$T = \oint rq \, ds = q \oint r \, ds \qquad\qquad (b)$$

in which integration is performed along the *centerline* of the perimeter (shown as a dashed line in the figure). Note that the shaded area of triangle dA in Fig. 6.12c is $r \, ds/2$. Then, letting A_m represent the *area enclosed by the*

mean perimeter, Eq. (*b*) becomes

$$T = 2A_m q \quad \text{or} \quad q = \frac{T}{2A_m} \qquad (6.22)$$

Combining Eqs. (6.21) and (6.22), we obtain the shearing stress at *any* point of a tube:

$$\tau = \frac{q}{t} = \frac{T}{2A_m t} \qquad (6.23)$$

where t is the wall thickness at that point. Application of Eq. (6.23) is limited to linearly elastic behavior of thin-walled tubes displaying no abrupt changes in thickness. In the inelastic range, Eq. (6.23) is valid only if thickness t is constant. One should be alert to the possibility that if the tube under torsion has *very thin* walls, buckling of the walls may occur.

To develop a relationship for the rotation angle, consider the shearing strain energy $U = \int (\tau^2/2G)\, dV$ for the member (see Sec. 3.9) and the work done by the applied torque per unit distance $W = \frac{1}{2}T(\phi/L)$. Here $\tau = T/2tA_m$ and for a unit distance the volume $dV = t\, ds$. Thus, setting $W = U$, we find that the *angle of twist ϕ of a thin-walled shaft* is

$$\phi = \frac{TL}{4A_m^2 G} \oint \frac{ds}{t} \qquad (6.24)$$

where L and G represent the length and the shear modulus of elasticity of the tube, respectively. Equations (6.23) and (6.24) are sometimes called *Bredt's formulas*.

The validity and the application of the preceding relationships are illustrated below in the solution of the torsion problems involving two typical tubes.

EXAMPLE 6.9

As shown in Fig. 6.13, a thin, circular tube of constant thickness t and mean radius r is twisted by torques T acting at the ends. Determine the approximate and exact values of the shearing stress and the angle of twist.

Solution   Referring to Fig. 6.13, we find the area enclosed by the mean perimeter of length $\phi\, ds = 2\pi r$ is $A_m = \pi r^2$. We note that $c = r + t/2$ and $b = r - t/2$. Hence the *approximate* shear-stress and the angle-of-twist formulas, Eqs. (6.23) and (6.24), yield

$$\tau_a = \frac{T}{2\pi r^2 t} \qquad (6.25)$$

$$\phi_a = \frac{TL}{2\pi G r^3 t} \qquad (6.26)$$

Figure 6.13   Example 6.9.

The polar moment of inertia is expressed as

$$J = \frac{\pi}{2}(c^4 - b^4) = \frac{\pi r t}{2}(4r^2 + t^2)$$

The *exact* maximum shearing stress and angle of twist are

$$\tau_e = \frac{Tc}{J} = \frac{T(2r + t)}{\pi r t(4r^2 + t^2)} \qquad (6.27)$$

$$\phi_e = \frac{TL}{GJ} = \frac{2TL}{\pi G r t(4r^2 + t^2)} \qquad (6.28)$$

To gage the magnitude of the deviation between the stress and the angle of twist components, consider the ratios

$$\frac{\tau_a}{\tau_e} = \frac{4\alpha^2 + 1}{2\alpha(2\alpha + 1)} \qquad \frac{\phi_a}{\phi_e} = \frac{4\alpha^2 + 1}{4\alpha^2}$$

where $\alpha = r/t$. If, for example $\alpha = 10$, the above quotients are 0.955 and 1.003, respectively. For a thin-walled tube, $r \geq 10t$, and it is clear that the approximate formulas provide results of sufficient accuracy.

EXAMPLE 6.10

A thin-walled brass tube possesses the rectangular box section shown in Fig. 6.14. The dimensions are $t_1 = t_3 = 6$ mm, $t_4 = 2t_2 = 4$ mm, $a = 194$ mm, and $b = 97$ mm. Let $G = 39$ GPa. (*a*) Calculate the shear stress in each wall when the tube is subjected to a torque of 5 kN·m. (*b*) What torque will cause an allowable shearing stress of 50 MPa? For this torque, calculate the angle of twist in an 0.5-m length of the tube.

Solution   The area bounded by the centerline is equal to $A_m = 0.194 \times 0.097 = 18.818 \times 10^{-3}$ m^2.
(*a*) The shear flow in the tube section, from Eq. (6.22), is

$$q = \frac{T}{2A_m} = \frac{5 \times 10^3}{2(18.818 \times 10^{-3})} = 132.85 \text{ kN·m}$$

Then, substituting successively $t_1 = t_3 = 6$ mm, $t_2 = 2$ mm, and $t_4 = 4$ mm into Eq. (6.23), we have

$$\tau_{AC} = \tau_{BD} = \frac{q}{t_1} = \frac{132.85 \times 10^3}{6 \times 10^{-3}} = 22.1 \text{ MPa}$$

$$\tau_{AB} = \frac{q}{t_2} = 66.4 \text{ MPa}$$

and

$$\tau_{CD} = \frac{q}{t_4} = 33.2 \text{ MPa}$$

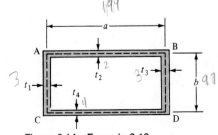

Figure 6.14   Example 6.10.

Clearly, the stress in a given wall depends only upon the thickness for a given torque.

(b) The allowable stress occurs in the thinnest wall, AB. The shear flow is thus

$$q = \tau_{AB} t_2 = 50 \times 10^6 (2 \times 10^{-3}) = 100 \text{ kN/m}$$

The largest permissible torque is obtained from Eq. (6.22):

$$T = 2qA_m = 2(100 \times 10^3)(18.818 \times 10^{-3}) = 3.76 \text{ kN·m}$$

The angle of twist, according to Eq. (6.24), is

$$\phi = \frac{3.76 \times 10^3 (0.5)}{4(18.818 \times 10^{-3})^2 (39 \times 10^9)} \left(\frac{2 \times 97}{6} + \frac{194}{2} + \frac{194}{4} \right)$$

from which $\phi = 6.05 \times 10^{-3}$ rad $= 0.35°$.

Note that the stress concentration at the corners of a rectangular (or any other form) tube is neglected in the approximate solution.

PROBLEMS

Secs. 6.1 to 6.4

6.1   A hollow steel shaft of 60-mm outer diameter is subjected to pure torque, $T = 2$ kN·m. If the shearing stress is not to exceed 80 MPa, what is the required inner diameter?

6.2   A 2-in. diameter shaft is to be replaced by a hollow circular tube of the same material, resisting the same maximum shear stress and the same torque. Determine the outer diameter D of the tube if its wall thickness is $D/25$.

6.3   The circular shaft is subjected to the torques shown in Fig. P6.3. What is the largest shear stress in the member and where does it occur?

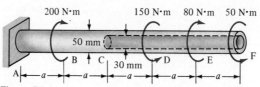

Figure P6.3

6.4   A solid shaft of diameter d and a hollow shaft of outer diameter D and thickness $D/4$ are to transmit the same torsional loading at the same maximum shear stress. Compare the weights of equal lengths of these two shafts.

6.5 Four pulleys, attached to a solid stepped shaft, transmit the torques shown in Fig. P6.5. Calculate the maximum shear stress for each segment of the shaft.

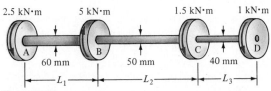

Figure P6.5

6.6 Redo Prob. 6.5, with a hole of 20-mm diameter drilled axially through the shaft to form a tube.

6.7 The torques T_B = 30 kip·in. and T_C = 12 kip·in. are exerted on the stepped shaft shown in Fig. P6.7. Calculate the maximum shear stress for d_1 = 2 in. and d_2 = $1\frac{3}{4}$ in.

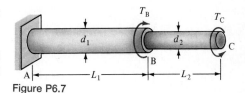

Figure P6.7

6.8 Determine the values of the torques T_B and T_C so that each segment of the shaft shown in Fig. P6.7 is stressed to a permissible shear strength of 14 ksi. Use d_1 = $2\frac{3}{8}$ in. and d_2 = 2 in.

6.9 Rework Prob. 6.7, assuming that the torque at B is 12 kip·in. and the torque at C is 20 kip·in.

6.10 A 50-mm-diameter solid axle is made of cast iron having an ultimate strength of 170 MPa in tension, 650 MPa in compression, and 240 MPa in shear. Calculate the largest torque that may be applied to the axle.

6.11 A hollow shaft is made by rolling a 5-mm-thick plate into a cylindrical shape and welding the edges along the helical seams oriented θ = 60° to the axis of the member (Fig. P6.11). If the permissible tensile and shear stresses in the weld are 100 and 55 MPa, respectively, what is the maximum torque that can be applied to the shaft?

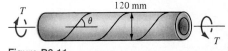

Figure P6.11

6.12 Redo Prob. 6.11, assuming that θ = 40°.

6.13 A solid steel bar (G = 12 × 10⁶ psi) of 2-in. diameter is subjected to a torque T = 20 kip·in. Determine the maximum shear strain and the maximum normal strain in the member.

6.14 A tube of outer diameter 100 mm and inner diameter 80 mm experiences a torque T = 10 kN·m (Fig. P6.11). At this loading, if a strain gage positioned at θ = 30° measures ε = −500 μ, determine the shear modulus of elasticity G of the material.

6.15 Rework Prob. 6.14 for θ = 45°.

Sec. 6.5

6.16   Consider the aluminum shaft AF loaded as shown in Fig. P6.3. Let $a = 100$ mm and $G = 28$ GPa. Determine the angle of twist (*a*) at C, and (*b*) at F.

6.17   A brass rod AB ($G = 6 \times 10^6$ psi) is bonded to an aluminum rod BC ($G = 4 \times 10^6$ psi), as shown in Fig. P6.7. What is the angle of twist at C? Let $T_B = 2T_C = 40$ kip·in., $d_1 = 2d_2 = 4$ in., and $L_1 = 2L_2 = 16$ in.

6.18   For the steel shaft loaded as shown in Fig. P6.5, determine the relative angle of twist in degrees between (*a*) pulleys AD and (*b*) pulleys AC. Use $G = 80$ GPa and $L_1 = 0.5$ m, $L_2 = 1.5$ m, and $L_3 = 1$ m.

6.19   Determine the torques T_A, T_B, and T_C of the aluminum shaft ($G = 28$ GPa) in equilibrium, as shown in Fig. P6.19. The maximum shear stress in segment AB is 80 MPa and the rotation is 0.02 rad clockwise as viewed at A with respect to C.

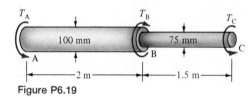

Figure P6.19

6.20   Redo Prob. 6.19, given that the relative angle of twist between A and C is zero.

6.21   A 1-in.-diameter solid steel shaft ($G = 12 \times 10^6$ psi) is subjected to end torques T that produce an angle of twist per unit length of 0.02 rad/ft. Determine (*a*) the maximum tensile stress and (*b*) the magnitude of the applied torque.

6.22   A disk is attached to a $1\frac{1}{2}$-in.-diameter, 20-in.-long steel shaft ($G = 12 \times 10^6$ psi), as shown in Fig. P6.22. In order to achieve a desired natural frequency of torsional vibrations, the stiffness of the system is specified so that the disk will rotate 2° under a torque of 10 kip·in. How deep (*x*) must a 1-in.-diameter hole be drilled to meet this requirement?

6.23   A 5-mm-diameter steel rod ($G = 80$ GPa) is rotated at a constant speed in a hole which imposes a frictional torque of 10 N·m/m of contact length (Fig. P6.23). Determine (*a*) the maximum contact length L if the shearing stress in the shaft is not to exceed 120 MPa and (*b*) the relative angle of twist between A and B.

6.24   A tapered round aluminum bar ($G = 28$ GPa) of length L is subjected to end torques, as shown in Fig. P6.24. Determine the angle of twist ϕ for $d_a = 25$ mm, $d_b = 75$ mm, $L = 2$ m, and $T = 80$ N·m.

6.25   If the top of a 200-mm-diameter steel well drill pipe ($G = 80$ GPa) rotates through 90° at a depth of 400 m, calculate the maximum shearing stress in the twisted pipe.

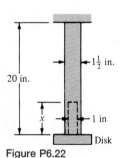

Figure P6.22

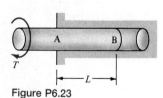

Figure P6.23

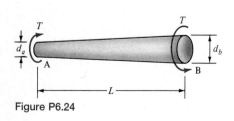

Figure P6.24

6.26 What is the magnitude of the largest allowable torque that a solid steel shaft ($G = 11.5 \times 10^6$ psi) 6 ft long and 2 in. in diameter can transmit if the maximum shear stress and the total angle of twist are limited to 14 ksi and 0.05 rad, respectively.

6.27 The circular shaft of diameter d shown in Fig. P6.27 is acted upon by a distributed torque of intensity $t(x)$ which varies linearly from zero at the free end to a maximum value of t_0 at the fixed end. (a) Determine the total angle of twist of the free end of the shaft in terms of t_0, d, L, and G. (b) Calculate the value of ϕ_A in degrees for $t_0 = 500$ N·m/m, $G = 28$ GPa, $L = 2$ m, and $d = 50$ mm.

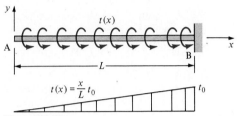

Figure P6.27

6.28 Redo Prob. 6.27, with the intensity per unit length of distributed torque held constant: $t(x) = t_0 = 200$ N·m/m.

6.29 As illustrated in Fig. P6.29, two steel shafts ($G = 80$ GPa) are connected by gears and subjected to a torque $T = 400$ N·m. Given $d_1 = 40$ mm and $d_2 = 30$ mm, determine (a) the angle of rotation in degrees at D and (b) the maximum shearing stress in each shaft.

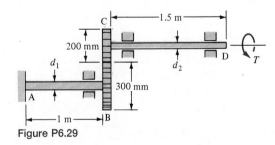

Figure P6.29

6.30 Calculate the torque T required to produce an angle of twist 2° at D of the gear-and-shaft system shown in Fig. P6.29. Let $G = 40$ GPa, $d_1 = 60$ mm, and $d_2 = 50$ mm.

Sec. 6.6

6.31 Determine the maximum torque T that may be applied to the composite shaft in Fig. 6.9 if the allowable shear stresses are 80 MPa in steel core and 60 MPa in the brass tube and if the angle of twist at the free end is not to exceed 3°. Use $c = 50$ mm, $b = 30$ mm, $L = 2$ m, $G_b = 40$ GPa, and $G_s = 80$ GPa.

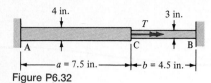

Figure P6.32

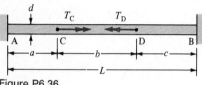

Figure P6.36

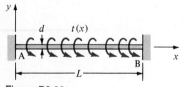

Figure P6.39

6.32 A solid steel shaft is built in at A and B, as shown in Fig. P6.32. Determine the permissible torque T if $\tau_{all} = 12$ ksi.

6.33 The round shaft AB in Fig. P6.32 is fixed to rigid walls at both ends and subjected at C to a 100-kip·in. torque. Segment AC is made of aluminum ($G_a = 4 \times 10^6$ psi) and segment CB is made of brass ($G_b = 6 \times 10^6$ psi). Calculate the maximum shear stress in each part.

6.34 Redo Prob. 6.33, given that the shaft AB is made of bronze for which $G = 6 \times 10^6$ psi.

6.35 Consider the composite shaft in Fig. 6.9 with $G_s = 80$ GPa and $G_b = 40$ GPa. If the brass tube carries 1.2 times as much torque as does the steel core, determine the ratio of the external to the internal diameter of the tube.

6.36 A solid round shaft with fixed ends is subjected to torques as shown in Fig. P6.36. (a) Determine the reactions in terms of T_C, T_D, a, b, c, and L. (b) Assuming that $T_C > T_D$, sketch the variation of torque with length along the bar. (c) Calculate the maximum shear stress in the shaft if $d = 75$ mm, $a = 300$ mm, $b = 200$ mm, $c = 100$ mm, and $T_C = 2T_D = 6$ kN·m.

6.37 Redo Prob. 6.36, given that the torque T_D acts in the opposite direction to that shown in Fig. P6.36.

6.38 The motor seen in Fig. P6.38 produces $T_A = 1200$ lb·in. of torque and is required to deliver $T_D = 900$ lb·in. of torque to a gear at D; the relative angle of twist of the shaft AD is 0.8°. Determine the frictional torques T_B and T_C at bearings B and C. Let $a = 1$ ft, $d = 1\frac{1}{4}$ in., and $G = 11.4 \times 10^6$ psi.

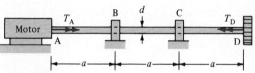

Figure P6.38

6.39 A solid circular shaft (Fig. P6.39) with fixed ends is subjected to distributed torque of intensity $t(x) = t_0$ N·m/m. Using $d = 30$ mm, $L = 4$ m, and $t_0 = 60$ kN·m/m, determine (a) the reactions at the walls and (b) the maximum shear stress in the shaft.

6.40 Redo Prob. 6.39, given that the intensity of the torque varies linearly, $t(x) = (x/L)t_0$ N·m/m, along the length of the shaft.

Secs. 6.7 and 6.8

6.41 A stepped shaft of diameters $D = 80$ mm and $d = 60$ mm is subjected to torque $T = 1.2$ kN·m (see Fig. 6.10). Calculate the maximum shearing stress in the shaft for (a) $r = 2$ mm and (b) $r = 6$ mm.

6.42 Two transmission shafts—one a hollow tube with an outer diameter of 100 mm and an inner diameter of 60 mm, the other solid with a diameter of 100 mm— are each to transmit 200 kW. If both operate at 6 Hz, compute the highest shearing stresses in each.

6.43 The circular shaft shown in Fig. P6.43 is subjected to torques at C and D. Given $G_b = 40$ GPa, $G_s = 80$ GPa, and $K = 1.5$ at the step in the shaft, determine (a) the angle of twist at D and (b) the maximum shearing stress in the shaft.

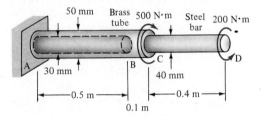

Figure P6.43

6.44 Consider the stepped shaft loaded as shown in Fig. P6.7, with $G = 4 \times 10^6$ psi, $L_1 = 2L_2 = 7.2$ ft, $d_1 = 2d_2 = 3$ in., $T_B = -3T_C$, and $K = 1.2$. Assuming that the applied torque produces a 1° rotation of the free end, determine the maximum shearing stress at section B of the shaft.

6.45 Given the maximum allowable shearing stress in a stepped shaft (Fig. 6.10) of 8 ksi, calculate the maximum power in horsepower (hp) that may be transmitted at 240 rpm. Use $D = 4$ in., $d = 2$ in., and $r = \frac{3}{16}$ in.

6.46 Design a solid aluminum shaft to transmit 300 hp at a speed of 1500 rpm, if the maximum shearing stress in the shaft is limited to 12 ksi. Assume a factor of safety $f_s = 1.5$.

6.47 Redo Prob. 6.46, with the additional factor that the angle of twist in a 6-ft length of the aluminum shaft ($G = 4 \times 10^6$ psi) is not to exceed 3°.

6.48 The compound shaft shown in Fig. P6.48 is attached to rigid supports at A and C. Segment AB is of aluminum ($G = 28$ GPa, $\tau_{all} = 40$ MPa), and the segment BC is of steel ($G = 80$ GPa, $\tau_{all} = 100$ MPa). Calculate the diameters d_a and d_s at which each material will simultaneously be stressed to its limit.

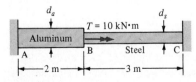

Figure P6.48

6.49 Determine the required diameter d_1 for the segment AB of the shaft shown in Fig. P6.7 if the permissible shearing stress is 50 MPa and the total angle of twist between A and C is not to exceed 0.02 rad. Let $G = 40$ GPa, $T_B = 2.5$ kN·m, $T_C = 1.5$ kN·m, $L_1 = 3$ m, $L_2 = 1$ m, and $d_2 = 30$ mm.

6.50 Redo Prob. 6.49 for the case in which the torque applied at B is $T_B = 3$ kN·m.

6.51 A stepped steel shaft is to transmit 200 hp from an engine to a generator at a speed of 150 rpm. If the allowable shearing stress is 14 ksi, determine the required fillet radius at the juncture of the 4-in.-diameter part with the 5-in. segment of the shaft.

6.52 The motor shown in Fig. P6.52 develops 130 hp at a speed of 600 rpm. Gears B and C deliver 95 and 40 hp, respectively, to operating machine tools. If the allowable shearing stress in the steel shaft ($G = 11.5 \times 10^6$ psi) is 10 ksi, determine (a) the required diameter of each segment and (b) the corresponding angle of twist in degrees between the ends A and C.

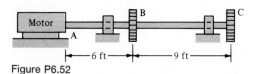

Figure P6.52

6.53 Determine the maximum power in kilowatts that may be transmitted across the shaft in Fig. P6.38 at 600 rpm, if the shearing stress is not to exceed 56 MPa. What is the corresponding total angle of twist between the ends A and D? Use $a = 3$ m, $d = 50$ mm, $G = 80$ GPa, and $T_B = T_C = 0$.

6.54 Calculate the diameter d_1 of shaft AB in Fig. P6.29 if the maximum shearing stress in each shaft is not to exceed 90 MPa. Use a safety factor $f_s = 1.5$ and $d_2 = 60$ mm.

Sec. 6.9

6.55 The uniform thin-walled rectangular shaft of Fig. 6.14 ($t_1 = t_2 = t_3 = t_4 = t$) is subjected to a torque $T = 1.4$ kN·m. Determine the smallest required wall thickness t if the shearing stress and the angle of twist per unit length are not to exceed 40 MPa and 0.03 rad/m, respectively. Use $G = 28$ GPa and $a = 2b = 100$ mm.

6.56 Rework Prob. 6.55, assuming that the cross section of the shaft is a square ($a = b = 80$ mm) box of uniform thickness t.

6.57 through 6.62. The aluminum shafts ($G = 28$ GPa) of cross sections shown in Figs. P6.57 through P6.62 are each subjected to a torque of 2 kN·m. Neglecting the effect of stress concentration, calculate, for each shape, (a) the maximum shearing stress and (b) the angle of twist in a 2-m length.

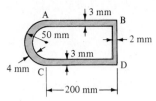

Figure P6.57

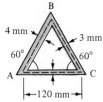

Figure P6.58

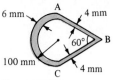

Figure P6.59

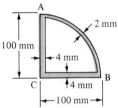

Figure P6.60

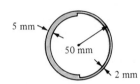

Figure P6.61

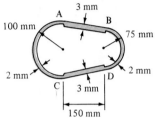

Figure P6.62

6.63 The cross section of an airplane fuselage made of an aluminum alloy ($G = 28$ GPa) is approximated as shown in Fig. P6.62. If the angle of twist per unit length is limited to $0.4°$ per meter of length of the fuselage, calculate the maximum allowable torque and the corresponding maximum shearing stress.

6.64 Redo Prob. 6.63 for the cross section shown in Fig. P6.57.

6.65 Redo Prob. 6.63 for the cross section shown in Fig. P6.59.

CHAPTER

7

STRESSES IN BEAMS

7.1 INTRODUCTION

The first part of this chapter contains material—procedures for determining the internal forces and moments in a beam—usually included in both statics and mechanics of materials texts. In later sections, consideration will be given to the stresses in a beam caused by the bending moment and shearing force. In analyzing such stresses, we assume that the beams are initially straight, not very thin, and stable under the applied loads. In addition, we assume that in most cases the beams are made of homogeneous and isotropic materials. But we shall also include a treatment of beams constructed of two materials and a discussion of stress concentrations. We shall conclude the chapter by considering beam design.

Here we are dealing with beams containing a longitudinal plane of symmetry, with applied loads acting in this plane. Bending deflections will therefore occur in the same plane, which is known as the *plane of bending*. This action produces only internal shear force V and bending moment M. Stresses produced by asymmetric and combined loadings will be discussed in the next chapter, and Chaps. 9 and 10 will present approaches for analysis of beam deflection under various conditions of support and applied loading.

We shall follow here a procedure similar to that employed in the investigation of torsion in Chap. 6. The three aspects of solid mechanics problems—statics, deformations, and geometry—are applied to derive the bending stress formula. To a certain degree, the results obtained are quite similar to those applicable to twisted bars. Just as in the cases of axially loaded rods and torsional members, disruptions in distributions of stress and strain may occur near supports, points of load application, and regions near geometric irregularities. It is assumed that these discontinuities lose significance at a distance approximately equal to the beam depth.

7.2   CLASSIFICATION OF BEAMS

A *beam* is a bar subjected to loads applied laterally or transversely to its axis. This flexural element is commonly used in structures and machines. Examples include the main members supporting floors of buildings, automotive axles and leaf springs, airplane wings, and brackets. Some beams carry transverse forces in combination with torsional and axial loading, but only the transverse loads and their effects on straight slender beams are treated in this chapter. Thus these beams are assumed to be *planar members*.

Beams are usually classified according to the way in which they are supported. Several commonly used beams are depicted in Fig. 7.1. The distance L between supports is termed the *span*. The beams shown in Fig. 7.1*a* through *c* are statically determinate; their reactions can be determined from the equilibrium conditions. However, the reactions of statically indeterminate beams (Figs. 7.1*d* through *f*) cannot be ascertained from the equilibrium equations alone, and the deflections associated with bending must be taken into consideration.

Beams are also classified according to their cross-sectional shapes. For example, an I beam and a T beam have cross sections geometrically formed like the letters I and T. Beams of steel, aluminum, and wood are manufactured in standard sizes; their dimensions and properties are listed in engineering handbooks. Tables B.2 through B.6 (App. B) present several common cases of steel sections. These include wide-flange shapes (W beams),

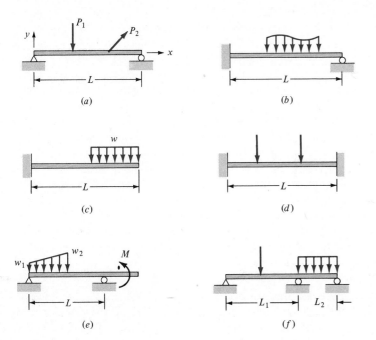

Figure 7.1   Types of beams: (*a*) simply supported beam; (*b*) cantilever beam; (*c*) overhanging beam; (*d*) fixed, simply supported beam; (*e*) fixed-end beam; and (*f*) continuous beam.

I shapes (also called S beams), C shapes (also referred to as channels), and L shapes, or angle sections. Note that the *web* is a thin vertical part of a beam. Thin horizontal parts of the beam are termed *flanges*. Interestingly, the cross section of a beam is described as *doubly* symmetric, *singly* symmetric, or asymmetric in accordance with whether it has two, one, or no axis of symmetry (see Fig. A.2).

Loads carried by beams may be of various kinds, as illustrated by the conventional diagrams in Fig. 7.1. Concentrated loads are forces such as P_1 and P_2. A couple load is represented by a moment such as M. The uniformly distributed, or uniform, load has constant intensity w per unit distance. The linearly varying load has an intensity changing from w_1 to w_2. Beam reactions to distributed loads can be calculated by placing the *equivalent* resultant load at the centroid of the distributed load and then applying the equations of statics. The applied forces and moments are *positive* if their vectors are in the direction of a positive coordinate axis. For example, in Fig. 7.1a and c, loadings P_1 and w_1 are negative and P_2 and M positive.

Sometimes two or more beams are connected by *hinges* or *pin joints* to form a continuous structure that is referred to as a *combined beam*, an example of which is shown in Fig. 7.2a. A hinge is capable of transmitting *only* a force with horizontal and vertical components; there will be no moment transmitted. Thus in many situations, the introduction of a hinge or hinges in a continuous beam makes the system statically determinate. To calculate the reactions, structures are separated into their parts at the hinges, as illustrated below.

EXAMPLE 7.1

Determine the support reactions for a combined beam loaded as shown in Fig. 7.2a.

Figure 7.2   Example 7.1: combined beam.

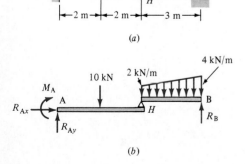

Solution   A free-body or load diagram for the combined beams is shown in Fig. 7.2b. Observe that the forces at the supports involve four unknowns and cannot be determined from this diagram. These forces can be calculated, however, by considering separately the free-body diagram of each beam (Fig. 7.2c and d): There are six unknowns and six equations of equilibrium available.

To begin with, the trapezoidal load is divided by the dashed line into two triangular loads (Fig. 7.2c). These, in turn, are replaced by two equivalent concentrated forces $(2 \times 3)/2 = 3$ kN and $(4 \times 3)/2 = 6$ kN, acting through the centroids of the distributed forces as marked by the vectors in the figure. Applying the equations of equilibrium, we have

$$\rightarrow \Sigma F_x = 0: \qquad\qquad\qquad\qquad H_x = 0$$

$$\circlearrowright \Sigma M_H = 0: \quad R_B(3) - 6(2) - 3(1) = 0 \qquad R_B = 5 \text{ kN} \uparrow$$

$$\uparrow \Sigma F_y = 0: \quad H_y - 3 - 6 + R_B = 0 \qquad H_y = 4 \text{ kN} \uparrow$$

The reactions H_x and H_y on HB act in the opposite directions on AH. Thus, using the free body of Fig. 7.2d, we obtain

$$\rightarrow \Sigma F_x = 0: \qquad\qquad\qquad\qquad R_{Ax} = 0$$

$$\uparrow \Sigma F_y = 0: \quad R_{Ay} - 10 - 4 = 0 \qquad R_{Ay} = 14 \text{ kN} \uparrow$$

$$\circlearrowright \Sigma M_A = 0: -M_A - 10(2) - 4(4) = 0 \qquad M_A = -36 \text{ kN·m} = 36 \text{ kN·m} \circlearrowright$$

It is noted that the positive signs of H_y, R_B, and R_{Ay} indicate that their directions have been correctly assumed in the diagrams. The inverse is the case of M_A, and the moment at A is counterclockwise.

7.3   SHEAR AND MOMENT IN BEAMS

In beams loaded by transverse loads in their plane, only two components of stress resultants occur: the shear force and bending moment (Sec. 1.7). These loading effects are sometimes referred to as *shear* and *moment* in beams. To determine the magnitude and sense of shearing force and bending at any section of a beam, the method of sections is applied, as illustrated in the following paragraphs.

Consider the uniformly loaded cantilever shown in Fig. 7.3a. Let us cut the beam at a cross section O located a distance x from the free end. To maintain the equilibrium of the two free bodies thus obtained, we must apply at the cross section the forces and moments directed in opposite senses but of the same magnitude (Fig. 7.3b). Note that positive V and M are placed on the exposed faces (refer to Fig. 1.2b).

We next replace the distributed load on the left-hand portion of the beam by the statically equivalent concentrated force wx located at $x/2$ (Fig. 7.3c). Application of the equilibrium conditions $\Sigma F_y = 0$ and $\Sigma M_O = 0$ to this free body leads to

$$V = wx \qquad M = -\tfrac{1}{2}wx^2$$

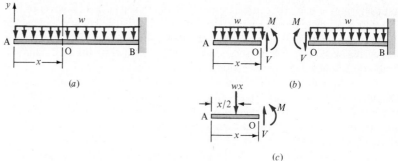

Figure 7.3 Shear force V and bending moment M.

This means that the shear force V is equal to the resultant of the transverse forces acting on the portion of the beam to either side of a section. Likewise, the bending moment M is equal to the algebraic sum of the moments of the external forces on the portion of the beam to either side of the section. This information permits the shear and moment to be determined without drawing free-body diagrams for each interval between the sections in a beam at which loads change, or *change-of-load points*.

Thus the equations for the stress resultants can be derived through the use of a free-body diagram and the conditions of equilibrium. Example 7.2 further illustrates the procedure. The variations of V and M along the beam owing to the change in loading can be shown by means of algebraic equations or by shear and bending-moment diagrams.

As pointed out in Sec. 1.7, the sign conventions adopted for internal forces and moments are associated with the deformations of a member. To illustrate this, consider the positive and negative shear forces V and bending moments M acting on segments of a beam cut out between two cross sections (Fig. 7.4a). We observe that a positive shear force tends to *raise* the right-hand face relative to the left-hand face of the segment, and a positive bending moment tends to bend the segment *concave upwards,* that is, so that it "retains water" (Fig. 7.4b). Similarly, a positive moment compresses the upper part of the segment and elongates the lower part.

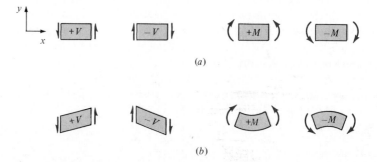

Figure 7.4 Sign conventions for beams: (a) definitions of positive and negative shear and moment and (b) deformations of beam segments caused by shear and moment.

EXAMPLE 7.2

Write shear and moment equations for an arbitrary section O in the interval CB of the beam loaded as shown in Fig. 7.5a. Calculate the values of V and M for O located 1 m from the right support.

Solution   A free-body diagram for the beam is shown in Fig. 7.5b. The reactions are calculated from the equilibrium conditions: $R_A = 22$ kN and $R_B = 28$ kN. The change-of-load points along the beam axis are designated A, B, C, and D. We make a cut at section O through the beam anywhere between C and B (Fig. 7.5c). Note that the unknown stress resultants V and M are indicated as positive values. Using the free body of Fig. 7.5c, we have, for the region $1 < x < 5$,

$$\uparrow \Sigma F_y = 0: \quad 22 - 20 - 5(x - 1) + V = 0$$

$$\circlearrowright \Sigma M_O = 0: \quad -22x + 20(x - 1) + 5(x - 1)\frac{x - 1}{2} + M = 0$$

from which

$$V = 5x - 7 \qquad M = -2.5x^2 + 7x + 17.5 \tag{a}$$

The equations for V and M in the intervals AC and BC can be similarly obtained. For $x = 4$, Eqs. (a) yield $V = 13$ kN and $M = 5.5$ kN·m.

Alternatively, V and M can be computed from a free-body diagram of the right-hand portion of the beam (Fig. 7.5d). Now the equations of statics $\Sigma F_y = 0$ and $\Sigma M_O = 0$ readily yield, for $x = 4$,

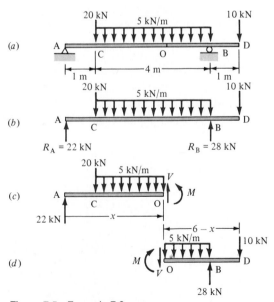

Figure 7.5   Example 7.2.

$$V = 28 - 5 - 10 = 13 \text{ kN}$$

$$M = 28(1) - 10(2) - 5(2 - 1)\frac{2-1}{2} = 5.5 \text{ kN·m}$$

as before.

7.4 LOAD, SHEAR, AND MOMENT RELATIONSHIPS

When a beam is subjected to a variety of loads, the approach of cutting the beam and determining shear and moment at a section by statics, as we have shown in the previous section, may prove quite cumbersome. A convenient alternative procedure employs the *load, shear,* and *bending-moment relations* to be derived in this section. The importance of these expressions will be evident as they are used in plotting the shear and moment diagrams, which we shall do in Sec. 7.5.

Consider the free-body diagram of an element of length *dx,* cut from a loaded beam (Fig. 7.6*a*). Note that the distributed load *w* per unit length, the shears, and the bending moments are depicted as positive (Fig. 7.6*b*). The changes in *V* and *M* from position *x* to *x* + *dx* are designated by *dV* and *dM,* respectively. Also indicated in Fig. 7.6*b* (dashed line) is the resultant *w dx* of the distributed load. Although *w* is not uniform, this is a permissible substitution for a very small length *dx.*

From consideration of equilibrium of the vertical forces acting on the element of Fig. 7.6*b,* we have

$$\uparrow \Sigma F_y = 0: \quad -V + (V + dV) + w\, dx = 0$$

The foregoing leads to

$$\frac{dV}{dx} = -w \tag{7.1}$$

which states that at any section in the beam, the *slope of the shear curve* is equal to $-w$. Integration of Eq. (7.1) between points A and B on the beam axis yields

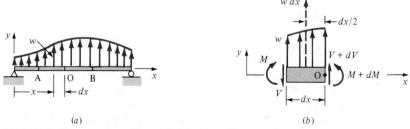

(a)   (b)

Figure 7.6 Beam and an element isolated from it.

$$V_B - V_A = -\int_A^B w\,dx \qquad (7.2a)$$

or

$$V_B - V_A = -(\text{area of load diagram between A and B}) \qquad (7.2b)$$

Note that Eq. (7.1) is not valid at the point of application of a concentrated load. Similarly, Eqs. (7.2) cannot be used when concentrated loads are applied between A and B, as the intensity of load w is undefined for a concentrated load.

For equilibrium, the sum of moments about O in Fig. 7.6b must also be zero:

$$\circlearrowleft \Sigma M_O = 0: \quad (M + dM) - M + V\,dx - w\,dx\,\frac{dx}{2} = 0$$

If second-order differentials are considered as negligible compared with differentials, the above reduces to

$$\frac{dM}{dx} = -V \qquad (7.3)$$

This relationship indicates that the *slope of the moment curve* is equal to $-V$. Thus the shear force is *inseparably* linked with a change in the bending moment along the length of the beam. Clearly, the *maximum* value of the moment occurs at the point where V (and hence dM/dx) is zero. Integrating Eq. (7.3) between A and B, we obtain

$$M_B - M_A = -\int_A^B V\,dx \qquad (7.4a)$$

or

$$M_B - M_A = -(\text{area of shear diagram between A and B}) \qquad (7.4b)$$

The differential equations of equilibrium, Eqs. (7.1) and (7.3), show that the shear and moment curves, respectively, will always be *one* and *two degrees* higher than the load curve. Note that Eq. (7.3) is not valid at the point of application of a concentrated load. Expressions (7.4) can be used even when concentrated loads are acting between A and B, but the relation is not valid if a couple is applied at a point between A and B.

7.5 SHEAR AND MOMENT DIAGRAMS

As observed in two preceding sections, the shear force and bending moment generally vary along the length of a beam. When designing a beam, it is useful to have available a graphical visualization of these variations. A *shear diagram* is a graph in which the shearing force is plotted against the horizontal distance (x) along a beam. Similarly, a graph depicting the bending moment plotted against the x axis is the *bending-moment diagram*.

The signs for shear V and moment M follow from general convention (Fig. 7.4). The definition of positive (or negative) moment shown in the figure is predominant in technical literature. However, in numerous texts on structural analysis, the direction of shear force is taken opposite to that adopted here. Thus *positive shears as defined in this text are plotted downward,* so that the outlines of the shear diagrams for the two sign conventions become identical. We note that for design purposes the sign of shear is usually unimportant.

It is convenient to place the shear and moment diagrams directly below the free-body, or load, diagram of the beam. The maximum and other significant values are generally marked on the diagrams. Here we take up the direct and summation methods of constructing shear and moment diagrams; both methods lend themselves to computer calculation.

The *direct approach* consists of calculating the reactions, writing algebraic expressions for the shear V and the moment M as described in Sec. 7.3, and constructing curves from these equations. The positive V and M are indicated on the cut sections. This permits all computed values of V and M to be plotted directly.

The *summation approach* involves drawing the shear diagram from the load curve and the moment diagram from the shear diagram, using the relationships obtained in the preceding section. This semigraphical method is a more efficient procedure than the direct procedure. It permits the rapid sketching of qualitative and quantitative shear and moment diagrams. The summation procedure for the construction of shear and moment diagrams is as follows:

1. Determine the reactions from the free-body diagram, or load diagram, of the entire beam.
2. Determine the value of shear at the change-of-load points, successively summing from the left end of the beam the vertical external forces or using Eq. (7.2).
3. Draw the shear diagram obtaining the shape from Eq. (7.1). Locate the points of zero shear.
4. Determine the values of moment at the change-of-load points and at the points of zero shear, either continuously summing the external moments from the left end of the beam or using Eq. (7.4), whichever is more appropriate.
5. Draw the moment diagram. The shape of the diagram is obtained from Eq. (7.3).

The two methods are illustrated in the solution of the following sample problems.

EXAMPLE 7.3

Draw the shear and moment diagrams for a simply supported beam with a uniformly distributed load of intensity w (Fig. 7.7a). Use (a) the direct approach and (b) the summation approach.

Solution   The equilibrium conditions for the entire beam yield $R_A = R_B = wL/2$ (Fig. 7.7b).

(a) Cut the beam at an arbitrary location x and draw the free-body diagram, with V and M positively directed relative to the exposed face (Fig. 7.7c). As the applied load is continuous, this cut typifies any section along the length of the beam. The equilibrium conditions, $\Sigma F_y = 0$ and $\Sigma M_O = 0$, result in

$$V = wx - \tfrac{1}{2}wL \qquad M = -\tfrac{1}{2}wx^2 + \tfrac{1}{2}wLx$$

The above indicates that the shear diagram consists of an inclined straight line and that the moment curve is a parabola. The plots of the foregoing expressions are shown in Fig. 7.7d and e. Note that the shear diagram crosses the horizontal line, for $V = 0$, at $x_1 = L/2$. The moment diagram is symmetrical about the middle of the beam.

(b) **Shear Diagram (Fig. 7.7d).**   The reaction at A must be plotted upward as the shear force just to the right of A is $V_A = -\tfrac{1}{2}wL$. The shear increases at the rate $dV/dx = w$. We thus have

$$V_B - V_A = wL \qquad V_B = \tfrac{1}{2}wL$$

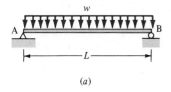

(a)

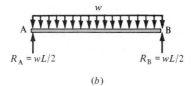

$R_A = wL/2$ \qquad $R_B = wL/2$

(b)

$wL/2$

(c)

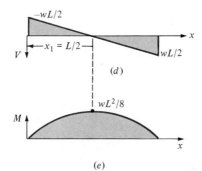

(d)

(e)

Figure 7.7 Example 7.3.

At B the upward reaction ($\frac{1}{2}wL$) closes the diagram and serves as a check of the analysis. The point of zero shear is located from the geometry of the diagram:

$$\text{Slope} = w = \frac{\frac{1}{2}wL}{x_1}$$

and $x_1 = \frac{1}{2}L$, as before.

Moment Diagram (Fig. 7.7e). The moment $M_A = 0$ at the left support. The shear, and hence the slope, of the moment diagram decreases uniformly from $\frac{1}{2}wL$ at A to zero at midlength. The maximum moment, the negative of the sum of the negative portion of the shear diagram, is therefore

$$M_{max} = \frac{1}{2}\frac{wL}{2}\frac{L}{2} = \frac{1}{8}wL^2 \qquad (a)$$

This represents the increase in the moment between A and midlength. Similarly, the moment decreases by $wL^2/8$ as we proceed from $x = \frac{1}{2}L$ to $x = L$, and at B the sum of the moments is zero.

EXAMPLE 7.4

A simply supported beam is subjected to a concentrated force P (Fig. 7.8a). Construct the shear and moment diagrams using the direct approach.

Solution The reactions at the supports are

$$R_A = \frac{Pb}{L} \uparrow \qquad R_B = \frac{Pa}{L} \uparrow$$

as calculated from the force diagram of the entire beam (Fig. 7.8b). Cut the beam at a location to the left of the applied load P, with both the shear force V and

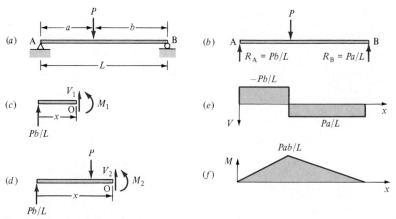

Figure 7.8 Example 7.4.

bending moment M directed as positive (Fig. 7.8c). The equilibrium conditions $\Sigma F_y = 0$ and $\Sigma M_O = 0$ for the region $0 < x < a$ then yield

$$V_1 = -\frac{Pb}{L} \qquad M_1 = \frac{Pb}{L}x$$

The above are plotted in the left portion of Fig. 7.8e and f. We next draw a free-body diagram of the right side of the load P (Fig. 7.8d). The equations of equilibrium, $\Sigma F_y = 0$ and $\Sigma M_O = 0$, for the region $a < x < L$ now result in

$$V_2 = \frac{Pb}{L} + P = \frac{Pa}{L}$$

$$M_2 = \frac{Pb}{L}x - P(x - a) = Pa\left(1 - \frac{x}{L}\right)$$

These expressions are plotted in the right-hand part of Fig. 7.8e and f.

We observe from the foregoing graphs that the shear is constant between the concentrated forces and that the bending moment varies linearly between the forces. However, at the point where P is applied, there is a *discontinuity* in the shear diagram and a corresponding change in the slope of moment diagram. Hence, it is necessary to consider sections on both sides of load P.

EXAMPLE 7.5

Redo Example 7.4 for the simple beam loaded symmetrically by two forces (Fig. 7.9a).

Solution   The reactions shown in the load diagram (Fig. 7.9b) are calculated from the equilibrium conditions. The change-of-load points are denoted A, B, C, and D. The free-body diagrams for the first two cuts are shown in Fig. 7.9c and d, where the unknown V and M are indicated as positively directed. The equations

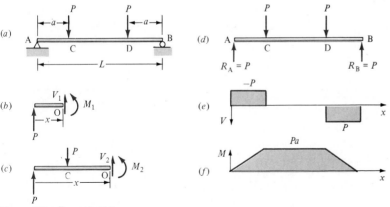

Figure 7.9   Example 7.5.

of statics, $\Sigma F_y = 0$ and $\Sigma M_O = 0$, then yield for the left-hand part of the beam $(0 < x < a)$

$$V_1 = -P \qquad M_1 = Px$$

and for the region $[a < x < (L - a)]$

$$V_2 = 0 \qquad M_2 = Pa$$

Owing to the symmetry about the midspan, the foregoing expressions suffice to construct the shear and moment diagrams of Fig. 7.9e and f.

It is noted that region CD has *no shear* force and is subjected only to a constant moment of Pa. Such a state of bending moment is termed *pure bending*. The regions AC and BD are in *nonuniform bending* because the bending moment M_1 is not constant and there are shear forces present.

Example 7.6

For the cantilever beam loaded as shown in Fig. 7.10a, draw shear and moment diagrams using the direct procedure.

Solution   The reactions at B are obtained from the free-body diagram of the entire beam (Fig. 7.10b):

$$R_B = 16 \text{ kips} \uparrow \qquad M_B = 100 \text{ kip·ft} \quad \circlearrowright$$

First the internal forces for the left portion of the beam are determined (Fig. 7.10c). From $\Sigma F_y = 0$ and $\Sigma M_O = 0$, for the region $0 < x < 4$, we have

Figure 7.10   Example 7.6.

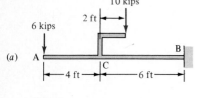

(a)

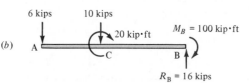

(b)

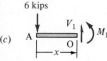

(c)

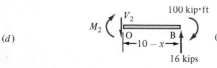

(d)

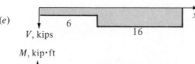

(e)

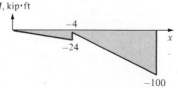

(f)

$$V_1 = 6 \qquad M_1 = -6x \qquad\qquad (b)$$

Next, consider as a free body the right part of the beam (Fig. 7.10d). Then, $\Sigma F_y = 0$ and $\Sigma M_O = 0$ lead to

$$V_2 = 16 \qquad M_2 = 16(10 - x) - 100 \qquad\qquad (c)$$

The above is valid in the region $4 < x < 10$.

Employing Eqs. (b) and (c), we can readily plot the shear and moment diagrams (Fig. 7.10e and f). Note that the couple of moment 20 kip·ft, applied at point C, introduces an abrupt change in the moment diagram but no change in the shear diagram.

EXAMPLE 7.7

Construct shear and moment diagrams for the beam loaded as shown in Fig. 7.11a. Employ (a) the direct approach and (b) the summation approach.

Solution The reactions are calculated from the conditions of equilibrium of forces acting on the entire beam (Fig. 7.11b):

$$R_A = 25 \text{ kN} \uparrow \qquad R_B = 35 \text{ kN} \uparrow$$

(a) The free-body diagrams of hypothetical segments cut at change-of-load points are depicted in Fig. 7.11c through e, with positive V and M indicated at the

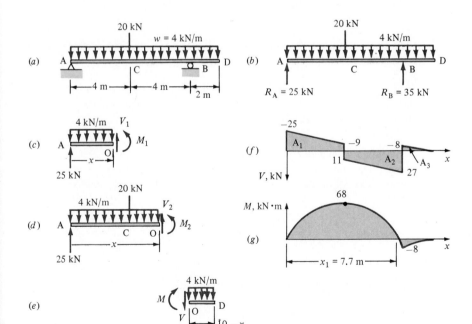

Figure 7.11 Example 7.7.

sections. Note that the segment with the *fewer* forces is chosen in Fig. 7.11*e*. Thus for the region $0 < x < 4$ (Fig. 7.11*c*),

$$\Sigma F_y = 0: \quad V_1 = 4x - 25$$
$$\Sigma M_O = 0: \quad M_1 = -2x^2 + 25x \tag{d}$$

For the region $4 < x < 8$ (Fig. 7.11*d*),

$$\Sigma F_y = 0: \quad V_2 = 4x - 5$$
$$\Sigma M_O = 0: \quad M_2 = -2x^2 + 5x + 80 \tag{e}$$

For the region $8 < x < 10$ (Fig. 7.11*e*),

$$\Sigma F_y = 0: \quad V_3 = 4x - 40$$
$$\Sigma M_O = 0: \quad M_3 = -2x^2 + 40x - 200 \tag{f}$$

The preceding algebraic expressions are plotted in Fig. 7.11*f* and *g*. We observe that the moment crosses the horizontal axis at $M = -2x^2 + 5x + 80 = 0$, that is, at $x_1 = 7.7$.

(*b*) **Shear Diagram (Fig. 7.11*f*).**   Close to the end A of the beam, $V_A = -R_A$ and hence 25 kN is plotted upward. The slope of the shear diagram is w between the concentrated forces. From Eq. (7.2), we obtain

$$V_C - V_A = (4)(4) = 16 \qquad V_C = -9 \text{ kN}$$

the value of the shear just to the left of C. At C the 20-kN force raises the value of the shear to $V_C = 11$ kN. Near the right of B,

$$V_B - V_C = (4)(4) = 16 \qquad V_B = 27 \text{ kN}$$

The shear reduces to $V_B = -8$ kN at B because of the 35-kN upward force acting there. Then,

$$V_D - V_B = (4)(2) = 8 \qquad V_D = 0$$

serves as a check. We observe from the shear diagram that the maximum force is 27 kN and occurs just to the left of support B.

Moment Diagram (Fig. 7.11*g*).   The bending moment is zero at each end of the beam. The areas of various parts of the shear diagram are computed as $A_1 = -68$ kN·m, $A_2 = 76$ kN·m, and $A_3 = -8$ kN·m. Equation (7.4) thus yields

$$M_C - M_A = 68 \qquad M_C = 68 \text{ kN·m}$$
$$M_B - M_C = -76 \qquad M_B = -8 \text{ kN·m}$$
$$M_D - M_B = 8 \qquad M_D = 0$$

Note that the slope is positive, and decreasing from A to C because of the negative and decreasing shear. Likewise, the slope is negative and increasing from C to B and positive and decreasing from B to D. The maximum moment is $M_{max} = 68$ kN·m.

EXAMPLE 7.8

Draw by the summation procedure the shear and moment diagrams for the beam shown in Fig. 7.12a.

Solution Application of the equilibrium conditions to the free-body diagram of the entire beam results in (Fig. 7.12b)

$$R_A = 12.8 \text{ kN} \uparrow \qquad R_B = 19.2 \text{ kN} \uparrow$$

Shear Diagram (Fig. 7.12c). Just to the right of C, V_C, the shear is zero. Then Eq. (7.2) gives

$$V_A - V_C = \tfrac{1}{2}(8)(1.5) = 6, \quad V_A = 6 \text{ kN}$$

the downward force near to the right of A. From C to A, the load increases linearly, and hence the shear curve is parabolic, having a positive and increasing slope. In the regions AD, DB, and BE, the slope of the shear curve is zero (that is, the shear is constant). At A the 12.8-kN upward force decreases the shear to -6.8 kN. The shear remains constant up to D where it increases by a 20-kN downward force to 13.2 kN. In a like manner, the value of the shear is lowered to -6 kN at B. No change in shear occurs until the point E, where the downward 6-kN force closes the diagram. The maximum shear $V_{max} = 13.2$ kN occurs in region BD.

Moment Diagram (Fig. 7.12d). The moment at end C is $M_C = 0$; Eq. (7.4) leads to

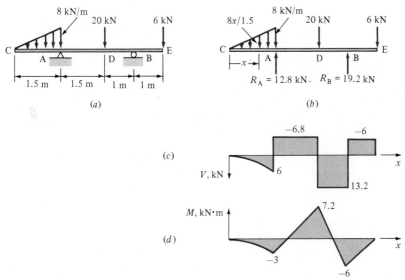

(a)

(b)

(c)

(d)

Figure 7.12 Example 7.8.

$$M_A - M_C = -\int_0^{1.5} \left(\frac{1}{2} \frac{8x}{1.5} x \right) dx \qquad M_A = -3 \text{ kN·m}$$

$$M_D - M_A = 6.8(1.5) \qquad\qquad M_D = 7.2 \text{ kN·m}$$

$$M_B - M_D = -13.2(1) \qquad\qquad M_B = -6 \text{ kN·m}$$

$$M_E - M_B = 6(1) \qquad\qquad M_E = 0$$

As M_E is known to be zero, a check on the calculations is provided. We find that, from C to A, the diagram takes the shape of a cubic curve concave downard with zero slope at C; in the regions AD, DB, and BE the diagram forms straight lines. The maximum moment, $M_{max} = 7.2$ kN·m, occurs at D.

7.6 BEAM BEHAVIOR IN PURE BENDING

A beam transmitting a constant bending moment is said to be in pure bending (recall Example 7.5). We now examine this simple case to establish the beam-deformation pattern. The assumptions relative to the behavior of slender beams subjected to bending and shear are introduced in the next section. It is to be noted here that, since the width of a beam of ordinary dimensions is small compared with its depth and since the loads act in a single (xy) plane, the bending of a beam is treated as a *two-dimensional* problem.

Consider the geometry of deformation of *pure bending of a beam* of uniform cross section with a vertical (y) axis of symmetry (Fig. 7.13a and b). Initially, the longitudinal axis of the beam (x axis) is a straight line. Note that the origin is arbitrarily placed at the left end of the beam. A number of segments, located some distance from the ends, between planes such as a-a and b-b and perpendicular to the x axis, are shown in the figure. Under the action of the end couples M, applied in the plane of symmetry, the beam axis deflects into a curve in the xy plane. This is called the *deflection curve,* or the *elastic curve,* of the beam.

As a result of the bending deformation depicted in Fig. 7.13c, the longitudinal top fibers shorten and are thus in compression, while the bottom fibers lengthen and are therefore in tension. Somewhere between the top and the bottom of the beam, there is a layer of fibers that does not change in length, although it is also curved. This layer or surface, indicated by the dotted line ss in the figure, is the *neutral surface* of the beam. Its intersection with a right section through the beam is termed the *neutral axis* of the cross section. (It is abbreviated N.A. in Fig. 7.13b.) Note that both "neutral surface" and "neutral axis" refer to a location of *zero strain or stress* in the member subjected to bending.

The location of the neutral axis in a beam will be determined in Sec. 7.9. This requires the consideration of the first two principles of analysis of deformable solids stated in Sec. 1.3—statics and deformations.

In pure bending, all elements will suffer the same deformation, and

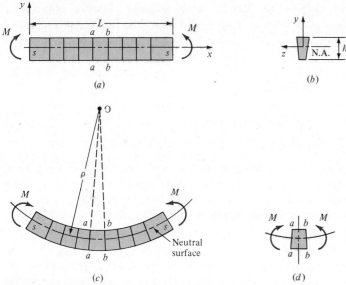

Figure 7.13 Beam in pure bending.

therefore the beam axis will be bent into a *circular* curve of radius ρ (Fig. 7.13*c*). The center of curvature O of the radius can be determined by extending to a point of intersection lines from sections such as *a-a* and *b-b*. Clearly, for nonuniform bending, the radius and center of curvature may differ from point to point along the length of the beam.

Since any diameter of a circle can serve as an axis of symmetry, it is concluded that all sections such as *a-a* and *b-b* remain straight in pure bending (Fig. 7.13*d*). This fact can be demonstrated experimentally. We may therefore state that *plane sections*, initially normal to the beam axis, *remain plane* and normal to it subsequent to bending. Geometric compatibility, which requires that the ends of the adjacent beam segments fit together, is satisfied by this pattern of deformation.

7.7 ASSUMPTIONS OF BEAM THEORY

The behavior of symmetric beams in pure bending, as described in the preceding section, is also assumed to be characteristic of other slender beams, including those having asymmetric cross section, arbitrary loading, and inelastic or nonhomogeneous material. This assumption, known as *Navier's hypothesis,* is the basis of the *technical theory of bending.* Therefore the derivation governing normal strain and stress will proceed as though the beam is deformed by pure bending. In the case of a nonuniform bending, the distribution of normal stresses in a given cross section is *not* significantly altered by the deformations caused by shearing stresses.

Consider a transversely loaded beam with cross section symmetrical about the y axis, and with the x axis representing the beam axis (Fig. 7.14a and b). Imagine a plane a-a passing through the beam perpendicular to x. The components of the displacement at point A occurring in the x and y directions will be denoted u and v, respectively. After bending, the beam axis at point A′ has deflection v and slope $dv/dx = \theta$ (Fig. 7.14c).

The fundamental assumptions of the technical theory for slender beams are based upon the geometry of deformation. Referring to Fig. 7.14, we can state them as follows:

1. The deflection of the beam axis is small compared with the span of the beam. The slope of the deflection curve is therefore very small and the square of the slope is negligible in comparison with unity. If the beam is slightly curved initially, the curvature is in the plane of the bending, and the radius of curvature is large in relation to its depth ($\rho \geq 10h$).

2. Plane sections initially normal to the beam axis remain plane and normal to that axis after bending (for example, a-a). This means that the shearing strains γ_{xy} are negligible. The deflection of the beam is thus associated *principally* with the axial or bending strains ε_x. The transverse normal strains ε_y and the remaining strains (ε_z, γ_{xz}, γ_{yz}) may also be ignored (as we shall see in Sec. 7.8).

3. The effect of the shearing stresses τ_{xy} on the distribution of the axial or bending stress σ_x is neglected. The stresses normal to the neutral surface, σ_y, are small compared with σ_x and may also be omitted. This supposition becomes unreliable in the vicinity of highly concentrated transverse loads.

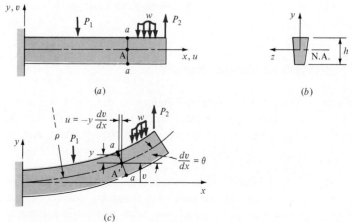

Figure 7.14 Beam subjected to transverse loading.

The foregoing assumptions apply whether the material behaves elastically or plastically and are analogous to those associated with the classical theory of plates. In the vast majority of engineering applications, adequate justification may be found for the simplifications cited with respect to the nature of deformation and stress. These simplifications permit the equations for the deflection curve and stress to be derived in a concise and straightforward manner. For purposes of clarity, the small deflections with which we are concerned will be shown greatly exaggerated on some diagrams.

In deep, short beams (where $L/h < 10$), shearing stresses are important. Such beams are treated by means of the theory of elasticity because assumptions 2 and 3 are no longer appropriate.

We note that in practice the span/depth ratio is approximately 10 or more for metal beams of compact section, 15 or more for beams with relatively thin webs, and 24 or more for rectangular timber beams.* Also, the slope of the deflection curve of the beam is almost always less than 5° or 0.087 rad and hence $(0.087)^2 = 0.0076 << 1$. Thus the formulas developed in this text generally yield results of good accuracy.

7.8 NORMAL STRAIN-CURVATURE RELATION

In order to gain insight into the beam-bending problem and to determine the strain distribution, we now examine further the geometry of deformation. As an initially straight beam deforms into some curved shape, it is necessary to describe the concept of curvature. The fundamental assumptions of the foregoing section, stated mathematically here, provide the basis for the derivations.

We begin by again considering the symmetrical beam (Fig. 7.14). The plane strain–displacement relations [Eqs. (3.4a)] become, as a consequence of the assumption 2,

$$\varepsilon_x = \frac{\partial u}{\partial x} \tag{7.5a}$$

$$\varepsilon_y = \frac{\partial v}{\partial y} = 0 \tag{7.5b}$$

$$\gamma_{xy} = \frac{\partial u}{\partial y} + \frac{\partial v}{\partial x} = 0 \tag{7.5c}$$

where ε_x, ε_y, and γ_{xy} denote the axial, transverse, and shearing strains, respectively. The above expressions specify how the idealized beam is to deform when it is loaded. Integrating $\varepsilon_y = \partial v/\partial y = 0$, we have

$$v = f(x) \tag{a}$$

*See W. C. Young, *Roark's Formulas for Stress and Strain,* 6th ed., McGraw-Hill, New York, 1989, chap. 7.

where $f(x)$, an arbitrary function, occurs because of the integration of a partial derivative. Equation (a) indicates that the lateral deflection does *not* vary over the beam depth. Then integration of Eq. (7.5c) with respect to y leads to

$$u = -y \frac{dv}{dx} + u_0(x)$$

Clearly, $u_0(x)$ represents the value of u on the neutral surface, which is free of deformation. It is concluded that $u_0 = 0$ and therefore

$$u = -y \frac{dv}{dx} \tag{7.6}$$

Equation (7.6) for u, represented in Fig. 7.14c at a section a-a passing through arbitrary point A′, shows that a cross section of the beam remains plane during bending. Substituting Eq. (7.6) into Eq. (7.5a), we have

$$\varepsilon_x = -y \frac{d^2v}{dx^2} \tag{7.7a}$$

This relationship provides the axial normal or *bending strain* at a distance y from the neutral axis.

From analytic geometry, the definition of the *curvature* κ (kappa) of the deflection curve at a distance x from the y axis is

$$\kappa = \frac{1}{\rho} = \frac{d^2v/dx^2}{[1 + (dv/dx)^2]^{3/2}} \tag{b}$$

where ρ is the radius of curvature (Fig. 7.14c). Because of assumption 1, the square of the slope may be considered negligible relative to unity. Therefore

$$\kappa = \frac{1}{\rho} = \frac{d^2v}{dx^2} \tag{7.8}$$

It is clear that κ represents the rate at which the slope varies over the beam axis; that is, $d(dv/dx)/dx$. If a positive slope dv/dx of the deflection curve becomes more positive as x increases, the curvature κ is positive, as shown in Fig. 7.14c. Thus the curvature of the beam axis is *positive* when it is bent *concave upward* and negative when it is bent concave downward. We observe from Fig. 7.13 that a positive bending moment produces positive curvature.

The strain-curvature relationship, on the basis of Eqs. (7.7a) and (7.8), may now be expressed in the form

$$\varepsilon_x = -y\kappa = -\frac{y}{\rho} \tag{7.7b}$$

The above states that the *normal strains* ε_x in the beam *vary linearly with the distance y from the neutral surface.* The situation described is analogous to that found in the torsion problem (Sec. 6.2), where the shearing strains vary linearly from the axis of a circular bar. However, in a beam the normal strains vary linearly from the neutral surface. It should be noted that as there are no material properties involved in deriving Eqs. (7.7), these relations apply to *inelastic* as well as to *elastic* situations.

The sign convention for the strain ε_x is the same as that used for normal strains in Chap. 3. Accordingly, if a fiber is above (below) the neutral surface, the distance y is positive (negative); then, for a positive curvature, ε_x will represent a negative (positive) strain, and there will be a shortening (elongation), as seen in Fig. 7.14c.

Transverse Strains. Interestingly, because of the effects of Poisson's ratio ν, the normal strains in the y and z directions are

$$\varepsilon_y = \varepsilon_z = -\nu\varepsilon_x = \frac{\nu y}{\rho} \qquad (c)$$

These relationships indicate that the compressed region ($y > 0$) of a beam subjected to positive bending moment will expand in both the y and z directions; the tensile zone ($y < 0$) contracts. Such transverse deformations of a segment of a bent, rectangular beam is shown in Fig. 7.15a. The original and deformed shapes of the cross section are depicted by dashed and solid lines in Fig. 7.15b, respectively. We observe that Poisson's effect deforms the neutral axis into a curve of large radius ρ_1, and the neutral surface becomes curved in two opposite directions. As a result, the top surface of the segment becomes saddle-shaped. Comparing $\varepsilon_z = \nu y/\rho$ and $\varepsilon_x = -y/\rho$, we conclude that the transverse radius of curvature is

$$\rho_1 = -\frac{\rho}{\nu} \qquad (d)$$

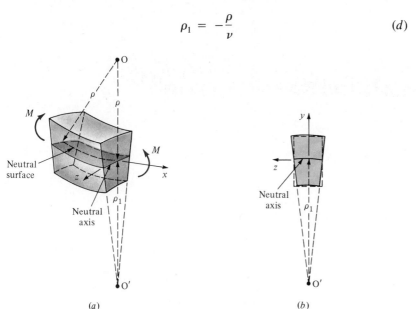

(a)   (b)

Figure 7.15   Segment of rectangular beam in pure bending.

The reciprocal of the above, $\kappa_1 = 1/\rho_1$, represents the curvature of the cross section and is termed the *anticlastic curvature*.

As stated in Sec. 7.7, however, in technical beam theory the transverse strains are neglected; the neutral surface is assumed to be curved in the xy plane *only* and remains flat in the z direction. The presence of the transverse strains and the associated changes in the shape of the cross section of the beams are not of much significance in most practical problems.

EXAMPLE 7.9

A bending moment M is applied at the free end of a steel cantilever of length $L = 8$ ft (Fig. 7.16a). Calculate the radius of curvature, the curvature, and the maximum deflection, knowing that the loading produces a strain at the bottom surface of the beam equal to the yield strain $\varepsilon_{yp} = 1200 \ \mu$. Let $h_1 = 3$ in. and $h = 5$ in.

Solution The beam is in pure bending and therefore the elastic curve is a circular arc (Fig. 7.16b). Upon substitution of the numerical values, Eqs. (7.7b) and (7.8) give

$$\rho = -\frac{y}{\varepsilon_{yp}} = -\frac{-3}{0.0012} = 2500 \text{ in.} = 208 \text{ ft}$$

$$\kappa = \frac{1}{\rho} = 0.0048 \text{ ft}^{-1}$$

Referring to Fig. 7.16b, we have

$$\sin \theta = \frac{L}{2\rho} \qquad v_{\text{max}} = \rho(1 - \cos \theta)$$

Substituting the data, we obtain $\theta = 0.0192$ rad, and the maximum deflection occurring at the midspan is

$$v_{\text{max}} = 2500(1 - 0.999815) = 0.461 \text{ in.}$$

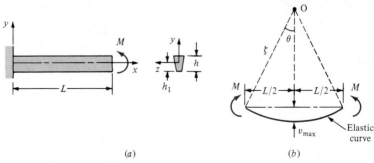

(a)   (b)

Figure 7.16 Example 7.9.

Interestingly,

$$\frac{h}{v_{max}} = \frac{5}{0.461} = 10.8 \qquad \frac{L}{v_{max}} = \frac{96}{0.461} = 208$$

For the slender beam under consideration, it is clear that, even for relatively large strains, the deflections are very small in relation to beam depth and span.

7.9 NORMAL STRESS: THE FLEXURE FORMULA

The stress acting normal to a beam cross section is referred to as the *bending*, or *flexure, stress*. When a beam is bent by transverse loads, such as shown in Fig. 7.14, there will usually be both a bending moment M and a shear force V acting on each cross section. The former and the latter are the resultants of the normal stresses and the shearing stresses, respectively.

On the basis of assumption 3 of Sec. 7.7, we shall ignore the shearing stresses for the time being. Calculation of the approximate shearing stresses in a beam will be considered in Sec. 7.12. It is significant to note that the normal strain and the normal stress relationships, Eqs. (7.7) and (7.12), yield the exact solutions for the special case of a slender beam subjected to pure bending. For transversely loaded slender beams, the results of these equations do not differ markedly from those given by more elaborate theory of elasticity, provided that solutions close to the ends are not required. The formulas become less accurate as the beam width increases relative to beam depth.

The distribution of the normal stress σ_x can be obtained by combining the strain ε_x defined by Eq. (7.7) with the stress-strain relation for the material and the conditions of equilibrium. We are here concerned only with *linearly elastic* material behavior. Thus, according to the Hooke's law,

$$\sigma_x = E\varepsilon_x = -E\kappa y \qquad (7.9)$$

The above shows that for an elastic beam, the flexure stress varies linearly with the distance y from the neutral axis (Fig. 7.17a). Note that the stresses are compressive above the neutral axis and tensile below the z axis when a positive bending moment M acts at the cross section of a beam as shown in the figure.

The conditions of equilibrium require that the resultant normal force produced by the stresses σ_x be zero, and that the moments of the stresses

Figure 7.17 Distribution of bending stress σ_x in a beam.

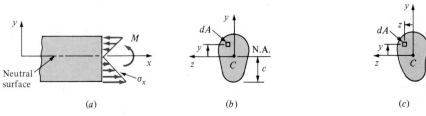

(a)   (b)   (c)

about the z axis be equal to the bending couple M acting on the section. Considering an element of area dA of the cross section at a distance y from the neutral axis (Fig. 7.17b), these conditions yield

$$\int \sigma_x \, dA = 0 \tag{a}$$

$$-\int (\sigma_x \, dA)y = M \tag{b}$$

where A is the cross-sectional area. The negative sign in the second expression means that a *positive* bending moment M is one which results in *compressive* (negative) stress at points of *positive* y. Introducing Eq. (7.9) into the above, we have

$$E\kappa \int y \, dA = 0 \tag{c}$$

$$E\kappa \int y^2 \, dA = M \tag{d}$$

As κ and E are constants, Eq. (c) indicates that the first moment of the cross-sectional area with respect to the z axis is zero,* $\int y \, dA = 0$. This requires that the *neutral axis z pass through the centroid C of the cross section.* Note that the symmetry of the cross section about the y axis means that the y and hence z axes are principal centroidal axes.

The moment of inertia I of the cross section about the centroidal axis is defined as

$$I = \int y^2 \, dA \tag{7.10}$$

Thus Eq. (d) can be expressed in the form

$$M = \kappa EI = \frac{EI}{\rho} \tag{7.11}$$

where EI is known as the *flexural rigidity* of the beam. The foregoing *moment-curvature* relationship will be used to determine beam deflections in Chaps. 9 and 10. Equation (7.11) indicates that when the bending moment is positive, the curvature is positive, as has already been observed in Sec. 7.8.

A formula for the normal stress can now be written by combining Eqs. (7.9) and (7.11):

$$\sigma_x = -\frac{My}{I} \tag{7.12}$$

This is usually called the elastic *flexure formula*. This important equation shows the stresses to vary linearly with the distance y from the neutral axis.

*See App. A for a discussion of moments of areas.

We observe that the stresses are proportional to the bending moment M at the section and inversely proportional to the moment of inertia I of the cross section. It is common practice to drop the sign in Eq. (7.12), as the sense of the stress can be found by inspection from the sense of the bending moment. The moments of inertia have dimensions of length to the fourth power. In the flexure formula, if M is measured in newton-meters, y in meters, and I in meters^4, the resulting normal stress σ_x will be expressed in pascals. If M is measured in pound-inches, y in inches, and I in inches^4, the units of stress become pounds per square inch.

The maximum stress σ_{max} occurs at the *outermost* fibers of the beam. Since, at a given section, M and I are constant, Eq. (7.12) yields (Fig. 7.16b)

$$\sigma_{max} = \frac{Mc}{I} = \frac{M}{S} \qquad (7.13)$$

where $|y_{max}| = c$, and $S = I/c$ is called the *elastic section modulus* of the beam. Section moduli have dimensions of length to the third power. We observe that for a given sectional area A (which is a measure of the relative weight of a uniform beam), S becomes *larger* as the beam shape is designed to concentrate as much of the area as possible *far* from the neutral axis. This consideration results in shapes such as I beams and wide-flange beams with large values of S/A to provide for high efficiency.

Useful properties of common areas are listed in the table inside the back cover of this book. For a beam of *rectangular* cross section with width b and height h, we have

$$I_z = \frac{bh^3}{12} \qquad S = \frac{bh^2}{6} \qquad (7.14)$$

For a *circular* cross section of radius r and diameter d, we have

$$I = \frac{\pi r^4}{4} = \frac{\pi d^4}{64} \qquad S = \frac{\pi r^3}{2} = \frac{\pi d^3}{32} \qquad (7.15)$$

Tables B.2 through B.6 (App. B) present area properties for some manufactured shapes. For cross-sectional forms not listed, we must determine the location of the neutral axis and the moment of inertia by using the techniques described in App. A.

Because of its simplicity, Eq. (7.13) is widely employed in practice. Note that this expression is analogous to the normal stress ($\sigma = P/A$) of axially loaded bars. The stress in the most remote fibers, computed from Eq. (7.13) for an experimentally obtained *ultimate* bending moment M_u (to be discussed in Sec. 12.6), is termed the *modulus of rupture in bending*:

$$\sigma_r = \frac{M_u}{S} \qquad (7.16)$$

This is a fictitious stress, as the proportional limit of the material is exceeded. However, it is often used as a measure of the bending strength of materials.

EXAMPLE 7.10

A simple cast-iron ($E = 175$ GPa) beam of rectangular cross section carries a load of 5 kN/m (Fig. 7.18a). Determine (a) the maximum tensile and compressive stresses at the midspan; (b) the normal stress and strain at point A; and (c) the radius of curvature ρ of the beam at B.

Solution   The neutral axis z passes through the centroid C and

$$I = \frac{bh^3}{12} = \frac{0.08(0.12)^3}{12} = 11.52 \times 10^{-6} \text{ m}^4$$

The section modulus of the cross-sectional area is thus

$$S = \frac{I}{c} = \frac{11.54 \times 10^{-6}}{0.06} = 192 \times 10^{-6} \text{ m}^3$$

(a) At the midspan, the bending moment is $M = 5(4)^2/8 = 10$ kN·m. Because M is positive, the maximum tensile and compressive stresses occur at the bottom and top fibers, respectively:

$$\sigma_{max} = \frac{M}{S} = \frac{10 \times 10^3}{192 \times 10^{-6}} = \pm 52.1 \text{ MPa}$$

These stresses act on infinitesimal elements at D and E, as shown in Fig. 7.18b.

(b) At a section through point A, the bending moment is $M = 10(1) - 5(1)^2/2 = 7.5$ kN·m, and we have

$$\sigma_A = -\frac{My_A}{I} = -\frac{7.5 \times 10^3(-0.02)}{11.52 \times 10^{-6}} = 13 \text{ MPa}$$

The normal strain at point A is thus

$$\varepsilon_A = \frac{\sigma_A}{E} = \frac{13 \times 10^6}{175 \times 10^9} = 74.3 \text{ } \mu$$

Figure 7.18 Example 7.10.

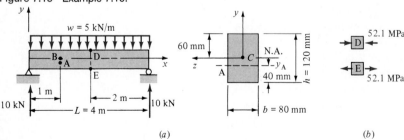

(a)   (b)

(c) Equation (7.7b) readily leads to

$$\rho = -\frac{y_A}{\varepsilon_A} = -\frac{-0.02}{74.3 \times 10^{-6}} = 269 \text{ m}$$

Alternatively, Eq. (7.11) also gives the above result. Clearly, the curvature increases along the axis of the beam toward the midspan at which $\rho_{min} = (175 \times 10^9)(11.52 \times 10^{-6})/10 \times 10^3 = 202$ m. Observe that ρ_{min} is many times larger than the span, and the deflection curve of the beam is therefore quite flat.

EXAMPLE 7.11

An overhanging beam of T-shaped cross section is loaded as shown in Fig. 7.19a. Determine the maximum tensile and compressive bending stresses.

Solution   The area is divided into two rectangular areas A_1 and A_2 for which the centroids are known (Fig. 7.19b). As the y axis is an axis of symmetry $\bar{z} = 0$ and Eq. (A.3b) results in

$$\bar{y} = \frac{A_1\bar{y}_1 + A_2\bar{y}_2}{A_1 + A_2} = \frac{20(60)70 + 60(20)30}{20(60) + 60(20)} = 50 \text{ mm}$$

Now we an use the transfer formula (A.9a) to calculate the centroidal moment of inertia $I_z = I$ of the entire cross section:

$$I = \sum \left(\frac{bh^3}{12} + Ad^2\right)$$

$$= \tfrac{1}{12}(60)(20)^3 + 20(60)(20)^2 + \tfrac{1}{12}(20)(60)^3 + 20(60)(20)^2$$

$$= 136 \times 10^4 \text{ mm}^4$$

The shear diagram (Fig. 7.19c) is drawn applying the procedure described in Sec. 7.5. Sections of zero shear or greatest moments are at $x = 1.25$ m and

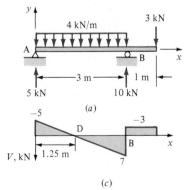

Figure 7.19   Example 7.11.

$x = 3$ m. Calculating the areas of the shear diagram, we readily obtain the bending moments at these sections: $M_D = 3.125$ kN·m and $M_B = -3$ kN·m.

The highest stress at a section occurs at the distance farther from the neutral surface. Owing to the positive moment at $x = 1.25$ m, the top fibers of the beam are in compression and the bottom fibers are in tension. Thus, from $\sigma = -My/I$, the flexural stresses are

$$\sigma_c = -\frac{3.125 \times 10^3(0.03)}{136 \times 10^{-8}} = -68.9 \text{ MPa}$$

$$\sigma_t = -\frac{3.125 \times 10^3(-0.05)}{136 \times 10^{-8}} = 114.9 \text{ MPa}$$

Similarly, the highest stresses at $x = 3$ m are found to be

$$\sigma_t = -\frac{-3 \times 10^3(0.03)}{136 \times 10^{-8}} = 66.2 \text{ MPa}$$

$$\sigma_c = -\frac{-3 \times 10^3(-0.05)}{136 \times 10^{-8}} = -110.3 \text{ MPa}$$

A comparison of the foregoing results shows that the maximum tensile stress is 114.9 MPa and occurs at the bottom of the beam at section D. The maximum compressive stress equals -110.3 MPa and is found at the top of the beam at section B at which the bending moment is *not* the largest. We thus conclude that to determine the critical bending stress in a beam of singly symmetric cross section (Fig. 7.19c), the *stresses at each section of zero shear* must be investigated. This is especially important for materials with different strengths in tension and compression.

Asymmetrical Sections. We have so far analyzed the bending of beams possessing at least one plane of symmetry. The applied loads were assumed to act in the plane of symmetry. If the cross section of a beam is *asymmetric* and the moment M is applied about either principal axis (for example, $M = M_z$ and $M_y = 0$), the derivation given above can be repeated identically. Figure 7.17a and c shows the situation described, where the y and z axes are assumed to be *principal axes*. As before, equilibrium condition $\Sigma F_x = 0$ leads to the requirement that the neutral axis (z) be a centroidal axis, and *equilibrium condition* $\Sigma M_z = 0$ leads to the flexure formula $\sigma_x = -My/I$. In addition, for an asymmetrical section, the stresses σ_x may build up an internal moment about the y axis. The equation of statics $\Sigma M_y = 0$, together with Eq. (7.9), thus results in

$$\int z(\sigma_x \, dA) = -E\kappa \int yz \, dA = -E\kappa I_{yz} = 0 \qquad (e)$$

where I_{yz} is the *product of inertia*. The above is satisfied only if $I_{yz} = 0$. For a symmetric or an asymmetric cross section, $I_{yz} = 0$ when the axes selected are the centroidal principal axes of the area, as is the case under consideration.

Therefore, the flexure formula applies to a beam with *any* cross-sectional shape, provided that I is a *principal* moment of inertia and M is a moment around a principal axis. When the bending moments do not act in the principal planes, the procedure to be described in Sec. 8.6 must be employed.

*7.10 STRESS CONCENTRATIONS

In situations involving beams of gradually varying cross section, the flexure formula which was derived for a uniform or *prismatic* beam still provides satisfactory solutions. On the other hand, if the cross section of the beam changes abruptly, Eq. (7.13) yields stress values which are usually too low. In the grooved flat bar seen in Fig. 7.20, for instance, the actual value of stress at the reduced section is greater than that predicted by the flexural theory.

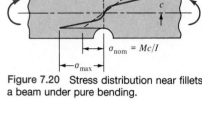

Figure 7.20   Stress distribution near fillets in a beam under pure bending.

The situation described above is similar to that treated earlier for stress in axial (Sec. 5.7) and torsional (Sec. 6.7) members. The highest bending stress is thus expressed in the form

$$\sigma_{max} = K\sigma_{nom} = K\,\frac{Mc}{I} \qquad (7.17)$$

where K is the *stress concentration factor*. Note that the nominal or average stress σ_{nom} is based upon the *reduced section*.

Figure 7.21 shows the values of the factor K for two cases of interest.

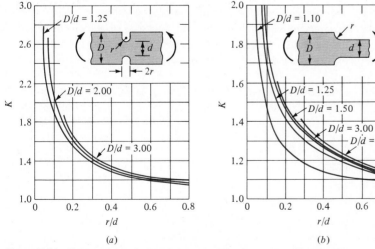

Figure 7.21   Stress-concentration factors for flat bars under bending. (*Source: M. M. Frocht, "Photoelastic Studies in Stress Concentration," Mechanical Engineering, August 1936, pp. 485–489.*)

A study of the data in this figure demonstrates the importance of using fillets and grooves of radius r that are as large as practical.

Consider, for example, a flat bar of thickness $b = 10$ mm having two grooves and subjected to bending moments $M = 360$ N·m. Assume that $D = 80$ mm, $d = 60$ mm, and $r = 6$ mm. Corresponding to the values given,

$$\frac{r}{d} = 0.1 \qquad \frac{D}{d} = 1.33$$

Interpolating in Fig. 7.21a, we have $K = 1.9$. The average stress is

$$\sigma_{\text{nom}} = \frac{6M}{bd^2} = \frac{6(360)}{0.01(0.06)^2} = 60 \text{ MPa}$$

and the highest stress σ_{max} is 1.9(60), or 114 MPa.

*7.11 BEAMS OF TWO MATERIALS

It is common practice to construct a beam of two or more materials having different moduli of elasticity to obtain a more efficient design. This is a so-called *composite beam*. Reinforced concrete beams and multilayered beams made by bonding together several sheets of material represent examples of such members. The assumptions of the technical theory of bending for a homogeneous beam (Sec. 7.7) are also valid for a beam of more than one material.

We shall employ the common *transformed-section method* to analyze a beam of symmetrical cross section built of two different materials (Fig. 7.22). The technique consists of transforming the cross section of either material into an *equivalent* cross section of the other material on which the resisting forces are the same as on the original section. The flexure formula is then applied to the transformed section. More complex cases can be treated in a like manner.

Geometric *compatibility of deformations* is satisfied by assumption (2) of Sec. 7.7: The cross sections of the beam remain plane during bending.

Figure 7.22 Beam of two materials: (a) composite section; (b) variation of strain; (c) stress distribution; and (d) transformed section.

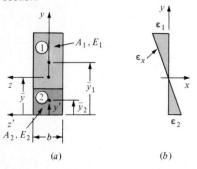
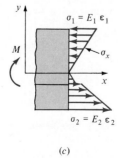
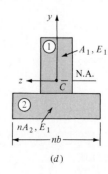

Thus the normal strains ε_x vary linearly from the top to the bottom of the beam (Fig. 7.22b). The location of the neutral axis is yet to be ascertained.

Both materials of the beam are assumed to obey Hooke's law, and their moduli of elasticity are denoted by E_1 and E_2. Then the *stress-strain relation* [Eq. (7.9)] leads to

$$\sigma_{x1} = E_1 \varepsilon_x = -E_1 \kappa y \qquad (7.18a)$$

$$\sigma_{x2} = E_2 \varepsilon_x = -E_2 \kappa y \qquad (7.18b)$$

The foregoing is depicted in Fig. 7.22c, where we assume that $E_2 > E_1$. Let us now define

$$n = \frac{E_2}{E_1} \qquad (7.19)$$

in which n is termed the *modular ratio*. While $n > 1$ in Eq. (7.19), this choice is arbitrary and the technique applies equally for $n < 1$.

Referring to the cross section of the composite beam (Fig. 7.22a), we find that the *equilibrium equations* (a and b) of Sec. 7.9 become

$$\int_{A_1} \sigma_{x1}\, dA + \int_{A_2} \sigma_{x2}\, dA = 0 \qquad (a)$$

$$-\int_{A_1} \sigma_{x1} y\, dA - \int_{A_2} \sigma_{x2} y\, dA = M \qquad (b)$$

where A_1 and A_2 designate the cross-sectional areas of materials 1 and 2, respectively. Inserting σ_{x1}, σ_{x2}, and n as given by Eqs. (7.18) and (7.19) into Eq. (a) gives

$$\int_{A_1} y\, dA + n \int_{A_2} y\, dA = 0 \qquad (c)$$

Using the bottom of the section as a reference (Fig. 7.22a), we obtain the following from Eq. (c), with $y = y' - \bar{y}$:

$$\int_{A_1} (y' - \bar{y})\, dA + n \int_{A_2} (y' - \bar{y})\, dA = 0$$

Inasmuch as

$$\int_{A_i} y'\, dA = \bar{y}_i \int_{A_i} \qquad \int_{A_i} \bar{y}\, dA = \bar{y} A_i$$

we have

$$A_1 \bar{y}_1 - A_1 \bar{y} + n A_2 \bar{y}_2 - n A_2 \bar{y} = 0$$

which yields an equivalent form of Eq. (c):

$$\bar{y} = \frac{A_1\bar{y}_1 + nA_2\bar{y}_2}{A_1 + nA_2} \tag{7.20}$$

This expression can be used to locate readily the *neutral axis* of a beam of two materials, provided that *only* the width of area A_2 is multiplied by n and that the dimensions of A_1 are unchanged. Thus the transformed section is composed of only material 1, as shown in Fig. 7.22d. Note that Eqs. (7.20) and (A.3) are identical.

In a like manner, condition (b), together with Eqs. (7.18) and (7.19), results in

$$M = \kappa E_1 \left(\int_{A_1} y^2 \, dA + n \int_{A_2} y^2 \, dA \right)$$

or

$$M = \kappa E_1 (I_1 + nI_2) \tag{7.21}$$

Here I_1 and I_2 represent the moments of inertia about the neutral axis of the cross-sectional areas 1 and 2, respectively. It is convenient to define

$$I_t = I_1 + nI_2 \tag{7.22}$$

as the moment of inertia of the *entire transformed area* about the neutral axis. The beam curvature, from Eq. (7.21), is thus

$$\kappa = \frac{1}{\rho} = \frac{M}{E_1 I_t} \tag{7.23}$$

The *flexure formulas* for a composite beam are now written by introducing the foregoing into Eqs. (7.18):

$$\sigma_{x1} = -\frac{My}{I_t} \qquad \sigma_{x2} = -\frac{nMy}{I_t} \tag{7.24}$$

where σ_{x1} and σ_{x2} are the stresses in materials 1 and 2, respectively. We observe that when $E_1 = E_2 = E$, Eqs. (7.24) reduce to the flexure formula for a homogeneous material, as expected.

EXAMPLE 7.12

A wood beam ($E_w = 10$ GPa) 50 mm wide 110 mm deep has a steel plate ($E_s = 200$ GPa) with a net section 25 mm by 10 mm securely fastened to its bottom face, as depicted in Fig. 7.23a. If the beam is subjected to a bending moment of 3 kN·m around a horizontal axis, calculate the maximum stresses in both materials using a transformed section of timber.

Solution The modular ratio $n = E_s/E_w = 20$. The centroid and the moment of inertia about the neutral axis of the transformed section (Fig. 7.23b) are

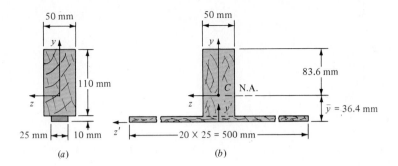

$$\bar{y} = \frac{50(110)(65) + 10(500)(5)}{50(110) + 10(500)} = 36.4 \text{ mm}$$

$$I_t = \tfrac{1}{12}(50)(110)^3 + 50(110)(28.6)^2 + \tfrac{1}{12}(500)(10)^3 + 500(10)(31.4)^2$$

$$= 15 \times 10^6 \text{ mm}^4$$

The maximum stresses in the wood and steel portions are therefore

$$(\sigma_w)_{max} = \frac{Mc}{I_t} = \frac{3 \times 10^3(0.0836)}{15 \times 10^{-6}} = 16.7 \text{ MPa}$$

$$(\sigma_s)_{max} = \frac{nMc}{I_t} = \frac{20(3 \times 10^3)(0.0364)}{15 \times 10^{-6}} = 145.6 \text{ MPa}$$

At the juncture of the two parts,

$$(\sigma_w)_{min} = \frac{My}{I_t} = \frac{3 \times 10^3(0.0264)}{15 \times 10^{-6}} = 5.3 \text{ MPa}$$

$$(\sigma_s)_{min} = n(\sigma_w)_{min} = 20(5.3) = 106 \text{ MPa}$$

Stresses at any other location may be obtained in a similar manner.

EXAMPLE 7.13

A composite beam is built of two pieces of brass ($E_b = 15 \times 10^6$ psi) bonded on each side to a steel plate ($E_s = 30 \times 10^6$ psi), as shown in Fig. 7.24a. The allowable stresses for brass and steel are 12 and 20 ksi, respectively. Determine the resisting moment for the beam.

Solution It is assumed that the steel will be transformed into brass. The modular ratio is $n = E_s/E_b = 2$ and the new width of the transformed steel section is $(\tfrac{3}{4})2 = 1\tfrac{1}{2}$ in. The equivalent areas are shown in Fig. 7.24b. The centroidal moment of inertia is

$$I_t = \frac{bh^3}{12} = \tfrac{1}{12}(2)(2.5)^3 = 2.604 \text{ in.}^4$$

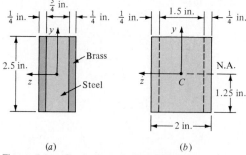

(a)   (b)

Figure 7.24   Example 7.13: (a) composite section and (b) equivalent brass section.

For the transformed steel section, the allowable stress is now $20/n = 10$ ksi, which is less than the 12 ksi for brass. Thus the stress in the steel is the *controlling stress*. From the flexure formula, it follows that

$$M_{max} = \frac{\sigma_{max} I_t}{c} = \frac{10 \times 10^3 (2.604)}{1.25} = 20.83 \text{ kip·in.}$$

EXAMPLE 7.14

Consider a concrete beam of width $b = 200$ mm and of effective depth $d = 350$ mm, reinforced with two steel bars providing a total cross-sectional area $A_s = 800$ mm^2 (Fig. 7.25a). Note that it is usual to use a 50-mm allowance to protect the steel from corrosion and fire. Given $n = E_s/E_c = 8$, determine the maximum stresses in the materials produced by a positive bending moment of 50 kN·m.

Solution   Only that portion of the cross section located a distance kd above the neutral axis is used in the transformed section (Fig. 7.25b); the *concrete is assumed to withstand no tension*. The transformed area of the steel nA_s is located by a single dimension from the neutral axis to its centroid. The compressive stress in the concrete is assumed to vary linearly from the neutral axis, while the steel is assumed to be uniformly stressed.

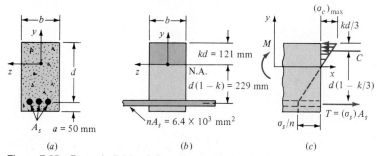

(a)   (b)   (c)

Figure 7.25   Example 7.14: reinforced concrete beam.

Inasmuch as the first moment of the transformed section with respect to the neutral axis must be zero, $\Sigma Q_z = 0$,

$$b(kd)\left(\frac{kd}{2}\right) - nA_s(d - kd) = 0$$

or

$$(kd)^2 + (kd)\frac{2n}{b}A_s - \frac{2n}{b}dA_s = 0 \qquad (7.25)$$

The neutral axis is located by solving this quadratic expression for kd. When we insert the data given, Eq. (7.25) becomes

$$(kd)^2 + 64(kd) - (22.4 \times 10^3) = 0$$

from which

$$kd = 121 \text{ mm} \qquad \text{and hence} \qquad 350 - kd = 229 \text{ mm} \qquad (d)$$

The moment of inertia of the transformed cross section about the neutral axis equals

$$I_t = \tfrac{1}{12}(0.2)(0.121)^3 + 0.2(0.121)(0.0605)^2 + 0 + 6.4 \times 10^{-3}(0.229)^2$$

$$= 453.7 \times 10^{-6} \text{ m}^4$$

Therefore

$$(\sigma_c)_{max} = \frac{Mc}{I_t} = \frac{50 \times 10^3(0.121)}{453.7 \times 10^{-6}} = 13.3 \text{ MPa}$$

$$\sigma_s = \frac{nMc}{I_t} = \frac{8(50 \times 10^3)(0.229)}{453.7 \times 10^{-6}} = 201.9 \text{ MPa}$$
(e)

The stresses act as depicted in Fig. 7.25c.

For an alternate solution, we determine $(\sigma_c)_{max}$ and σ_s from a free-body diagram of part of the beam.* Thus, referring to Fig. 7.25c, we find that the first statics condition, $\Sigma F_x = 0$, gives $C = T$. Here

$$C = \tfrac{1}{2}(\sigma_c)_{max}(bkd) \qquad \text{and} \qquad T = \sigma_s(A_s) \qquad (f)$$

represent the compressive and tensile stress resultants, respectively. Applying the second equilibrium condition, $\Sigma M_z = 0$, we have

$$M = Cd(1 - k/3) = Td(1 - k/3) \qquad (g)$$

Expressions (f) and (g) lead to

$$(\sigma_c)_{max} = \frac{2M}{bd^2k(1 - k/3)} \qquad \sigma_s = \frac{M}{A_sd(1 - k/3)} \qquad (7.26)$$

When we introduce the data and Eq. (d), the foregoing equations yield stress values identical with those of Eqs. (e).

---

*For details, see L. C. McCormac, *Design of Reinforced Concrete*, Harper & Row, New York, 1978.

7.12   THE SHEAR FORMULA: SHEAR FLOW

In beams subjected to nonuniform bending, transverse shear forces V, in addition to bending moments, act at the cross section. The resultant of the vertical stress distribution equals the shearing force at a cross-sectional plane. The vertical stress distribution is termed the *shearing stress* or *direct shearing stress*.

Some Preliminaries.   Determination of the shearing stress through the use of Hooke's law is not possible because, according to Eq. (7.5c), it is not related to the shearing strain. However, a relationship for the shearing stress can be obtained from the *equilibrium requirement* of force balance along the axis of the beam. Satisfaction of the geometric compatibility of deformation is *not* included in the analysis. The formula determined nevertheless yields results of acceptable accuracy for the beams of *ordinary* proportion.

The *vertical* shearing stress τ_{xy} at any point on the cross section must be equal to the *horizontal* shear stress τ_{yx} at the same point: $\tau_{xy} = \tau_{yx}$ (refer to Sec. 2.3). It is thus concluded that horizontal shear stresses must also exist in any beam subjected to a transverse loading. This can be demonstrated by considering a simply supported beam made of separate planks, as shown in Fig. 7.26a. When a transverse load P is applied to this composite beam, the planks slide relative to one another, and the resulting deformation will appear as seen in Fig. 7.26b. The fact that a solid beam does not exhibit this relative movement of longitudinal elements demonstrates the existence of shear stresses on horizontal (longitudinal) planes as well as on vertical (transverse) planes.

Consider an element of dx length, of symmetrical cross section, and isolated by two adjoining sections taken perpendicular to the axis of the beam (Fig. 7.27a). Such an element is shown as a free body in Fig. 7.27b. All internal forces shown acting on this element have a positive sense. Recall from Secs. 1.7 and 2.3 that the *direction* of the shearing stress at a section is the same as that of the shear force. No variation of shear force V occurs within dx, but a change in bending moment M does take place from one section to the next. Note that on the right side the moment is designated $M + dM$. The distribution of normal stresses σ_x on the exposed faces of the element—stresses produced by the bending moments—is also indicated in the figure.

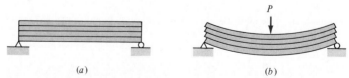

(a)   (b)

Figure 7.26   Bending of four separate planks.

Determination of the Shearing Stresses τ_{xy}. From the element of Fig. 7.27b, we isolate segment $mnm'n'$ (Fig. 7.27c). The fibers of this free body nearest to the neutral surface are located at an arbitrary distance y_1. Its bottom face is acted upon by the horizontal shearing stresses τ_{yx} present at this level of the beam. While the end faces of the segment are subjected to normal stresses as well as to the vertical shear stresses, its top face is the upper face of the beam and is free of stress. The stress distribution is assumed to be uniform across the width of the beam (Fig. 7.27d and e). Note that all shearing stresses are shown as *positive* values in the figure.

The horizontal shearing stress τ_{yx} distributed over area $b\,dx$ can be determined on the basis of equilibrium of forces acting on the beam segment, where b is the width of the beam at a distance y_1 from the neutral axis. The normal force distributed over the left face mm' on shaded area A^* of the cross section, from Eq. (7.12), is

$$\int_{-b/2}^{b/2}\int_{y_1}^{h_1}\sigma_x\,dy\,dz = -\int_{A^*}\frac{My}{I}\,dA$$

Similarly, an expression for the normal force on the right face nn' may be written as

$$-\int_{A^*}\frac{(M+dM)y}{I}\,dA$$

Referring to Fig. 7.27c and d, we find that the requirement of statics $\Sigma F_x = 0$ then yields

$$\int_{A^*}\frac{My}{I}\,dA - \int_{A^*}\frac{(M+dM)y}{I}\,dA - \tau_{yx}b\,dx = 0$$

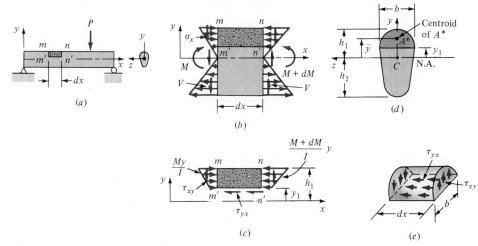

(a)

(b)

(c)

(d)

(e)

Figure 7.27 Shear stresses in a beam.

which reduces to

$$\tau_{yx} = -\frac{1}{Ib} \int_{A*} \frac{dM}{dx} y \, dA \qquad (a)$$

Note that the vertical shearing stresses acting on the side faces of the beam element develop oppositely directed vertical shear forces V. Thus the condition $\Sigma F_y = 0$ is also satisfied.

If $dM/dx = -V$ and $\tau_{yx} = \tau_{xy}$ are substituted into Eq. (a), the *horizontal* and *vertical shearing stresses* at any point of the beam section are expressed in the form

$$\tau_{xy} = \frac{V}{Ib} \int_{A*} y \, dA = \frac{VQ}{Ib} \qquad (7.27)$$

This is termed the *shear formula*. Here V is the total shear force at a section, I represents the moment of inertia of the *entire* cross section about the neutral axis, and b denotes the width of the beam at the level at which τ_{xy} is investigated. The integral represented by Q is the *first moment* of the *shaded* cross-sectional area $A*$ with respect to the neutral axis:

$$Q = \int_{A*} y \, dA = A*\bar{y} \qquad (7.28)$$

By definition, $\bar{y}$ is the distance from the neutral axis z to the centroid of $A*$ (see Sec. A.2). In the case of sections of regular geometry, $A*\bar{y}$ provides a convenient means of calculating Q.

Determination of the Shear Per Unit Length q.   The shear force acting across the width of the beam per unit length along the beam axis is determined by multiplying τ_{xy} of Eq. (7.27) by b (Fig. 7.27d). This quantity (in newtons per meter or pounds per inch) is designated q and is called the *shear flow*:

$$q = \frac{VQ}{I} \qquad (7.29)$$

Equation (7.29) is useful in the study of the connections in sections of a *built-up beam,* formed by joining two or more materials.

Shear in Thin-Walled Members.   Equations (7.27) and (7.29) may be used to obtain the horizontal shearing stress τ_{zx} and the shear flow between a portion such as *cdef* (Fig. 7.28a) and the remainder of a thin-walled beam, although the contact surface *c-c* is *vertical*. The procedure is the same as before and includes the computation of the Q about the neutral axis of area $A*$ shown shaded in the figure. Imagine that segment *cdef*, because of push or pull acting on it, attempts to slip relative to the entire beam. Now the shearing stress τ_{zx} is exerted in a vertical plane and directed perpendicularly to the plane of the paper. Thus the shear stress τ_{xz} acts horizontally: $\tau_{xz} =$

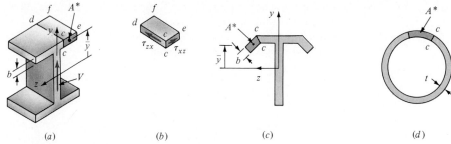

Figure 7.28   Shear stresses in thin-walled beams.

τ_{zx} (Fig. 7.28*b*). Determination of the sense of this stress will be described in Sec. 7.14. It is clear that τ_{xz} represents the *average* value of the horizontal shearing stress at any point of a transverse section of the flange.

Equations (7.27) and (7.29) may also be employed to determine the shearing stress and shear flow in other thin-walled members where the cutting planes *c-c* are *inclined* (Fig. 7.28*c* and *d*). Referring to Fig. 7.28*d*, we note that the slip between A^* and the solid area is resisted by *both* walls of the tube; thus $b = 2t$.

EXAMPLE 7.15

Four wooden planks are joined with nails to form the T beam shown in Fig. 7.29*a*. (Dimensions are in millimeters.) For an allowable shear force in each nail $F_n = 1.6$ kN and a vertical shear force in the beam $V = 4$ kN, determine (*a*) the necessary spacing s_1 of the nails for the upper portion; (*b*) the necessary spacing s_2 of the nails for the lower part; and (*c*) shearing stresses in joints *a-a* and *b-b* if the planks are glued together instead of being joined by nails. The moment of inertia of the entire area is $I = 9.1 \times 10^6$ mm⁴.

Solution   The horizontal shear forces transmitted by the three flanges and the web can be found from the shear-flow formula.

(*a*) In this case, Q_1 is the first moment of area of the upper flange about the neutral axis. Referring to Fig. 7.29*a*, we find

$$Q_1 = A_1^* \bar{y}_1 = 100(20)(42.9) = 85.8 \times 10^3 \text{ mm}^3 \qquad (b)$$

The shear flow at the section along line *a-a* is thus

$$q_1 = \frac{VQ_1}{I} = \frac{4000(85.8 \times 10^3)}{9.1 \times 10^6} = 37.7 \text{ N/mm}$$

which is resisted by the section of the nails. As the load capacity of the nails per unit length (F_n/s_1) is equal to the shear flow q_1, we write

$$s_1 = \frac{F_n}{q_1} = \frac{1600}{37.7} = 42.4 \text{ mm}$$

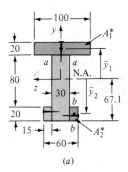

(*a*)

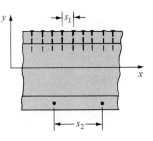

(*b*)

Figure 7.29   Example 7.15: a built-up beam.

This value—or as a practical matter, 40 mm—represents the maximum allowable spacing of nails at the upper flange (Fig. 7.29b).

(b) The section is now cut along line b-b and the contact surface is vertical. For the dimensions shown in Fig. 7.29a, we obtain

$$Q_2 = A_2^* \bar{y}_2 = 20(15)(57.1) = 17.1 \times 10^3 \text{ mm}^3 \qquad (c)$$

Then

$$q_2 = \frac{VQ_2}{I} = \frac{4000(17.1 \times 10^3)}{9.1 \times 10^6} = 7.52 \text{ N/mm}$$

$$s_2 = \frac{F_n}{q_2} = \frac{1600}{7.52} = 212.8 \text{ mm}$$

For convenience in construction, the nails would probably be spaced at 210-mm intervals. This applies to both lower flanges.

(c) The widths of the glued joints are $b_1 = 0.03$ m and $b_2 = 0.02$ m at a-a and b-b, respectively. Substituting Eqs. (b) and (c) into the shear formula, Eq. (7.27), we then have

$$(\tau_{xy})_a = \frac{VQ_1}{Ib_1} = \frac{4000(85.8 \times 10^{-6})}{9.1 \times 10^{-6}(0.03)} = 1.26 \text{ MPa}$$

$$(\tau_{xy})_b = \frac{VQ_2}{Ib_2} = \frac{4000(17.1 \times 10^{-6})}{9.1 \times 10^{-6}(0.02)} = 376 \text{ kPa}$$

The required shear strength of the glue is thus 1.26 MPa.

The shearing stresses at any distance from the z axis are obtained similarly, as illustrated in the following two sections for some commonly used cross-sectional areas.

7.13 SHEAR-STRESS DISTRIBUTION IN RECTANGULAR BEAMS

The shearing stresses at any point in a transversely loaded beam may be determined from the shear formula $\tau = VQ/Ib$. For a determination of the variation of stress over the beam depth, it is necessary to examine the variation of Q/b because V and I are *constants* for a given cross section. In our analysis, we shall assume the shear stress to be uniform across the width of the beam.

Consider a beam of rectangular section of width b and depth h transmitting a vertical shear force V (Fig. 7.30a). The shearing stress at the distance y_1 from the neutral axis is

$$\tau_{xy} = \frac{V}{Ib} \int_{-b/2}^{b/2} \int_{y_1}^{h/2} y \, dy \, dz = \frac{V}{2I}\left(\frac{h^2}{4} - y_1^2\right) \qquad (7.30)$$

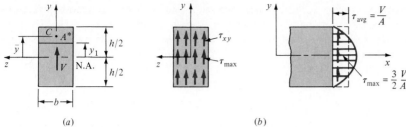

Figure 7.30   Shear stresses in a beam of rectangular cross section.

showing that the distribution of the shear stress in a transverse section of a rectangular beam is parabolic (Fig. 7.30b). The shear stress is zero at the top and bottom of the section ($y_1 = \pm h/2$) and has its maximum value at the neutral axis ($y_1 = 0$):

$$\tau_{max} = \frac{Vh^2}{8I} = \frac{Vh^2}{8bh^3/12} = \frac{3}{2}\frac{V}{A} \qquad (7.31)$$

where $A = bh$ is the cross-sectional area of the beam.

Alternatively, the result given by Eq. (7.31) may be obtained readily by computing the first moment Q of one-half the cross-sectional area about the neutral axis of the beam:

$$\tau_{max} = \frac{V}{Ib}A^*\bar{y} = \frac{V(bh/2)(h/4)}{Ib} = \frac{3}{2}\frac{V}{A}$$

Note that the *maximum* shearing stress is 50 percent larger than the *average* shear stress $\tau_{avg} = V/A$, as indicated by the dashed lines in Fig. 7.30b.

It should be noted that the maximum shear stress does not always occur at the neutral axis, z. For example, for a cross section having nonparallel sides, such as a *triangular section*, the maximum value of Q/b, and hence τ_{max}, occurs at midheight ($h/2$), while the z axis is located at $h/3$ from the base (see Prob. 7.105).

For narrow beams, Eq. (7.30) yields solutions which are in good agreement with the "exact" stress distributions found by the methods of the theory of elasticity. Equation (7.31) is particularly useful, as beams of rectangular cross-sectional form are often employed in practice.

EXAMPLE 7.16

A box beam is loaded as shown in Fig. 7.31a. The moment of inertia I about the neutral axis is 10.5×10^{-6} m^4 (Fig. 7.31b). Calculate (a) the shearing and bending stresses acting at point E and (b) the maximum shear and bending stresses.

194 MECHANICS OF MATERIALS

Solution After we have determined the reactions by application of statics, the shear and bending-moment diagrams are drawn as shown in Fig. 7.31c. The area of the cross section above the neutral axis is resolved into three rectangles (Fig. 7.31d).

(a) At the transverse section through point E, we have $V_E = -3$ kN and $M_E = 4.5$ kN·m. Referring to Fig. 7.31d, we have

$$Q_1 = A^*_1 \bar{y}_1 = 0.08(0.04)(0.0367) = 117.4 \times 10^{-6} \text{ m}^3$$

Then,

$$\tau_E = \frac{V_E Q_1}{Ib} = \frac{-3000(117.4 \times 10^{-6})}{10.5 \times 10^{-6}(2 \times 0.02)} = -839 \text{ kPa}$$

$$\sigma_E = -\frac{M_E c_E}{I} = -\frac{4500(0.0567 - 0.04)}{10.5 \times 10^{-6}} = -7.16 \text{ MPa}$$

The state of stress at point E is represented in Fig. 7.31e.

(b) The maximum shear force and the bending moment in the beam are $V_{CB} = 7$ kN and $M_B = -8$ kN·m (Fig. 7.31c). Referring to Fig. 7.31d, we obtain

$$Q_{max} = A^*_1 \bar{y}_1 + 2A^*_2 \bar{y}_2 = 117.4 \times 10^{-6} + 2(0.0167)(0.02)(0.0084)$$
$$= 123 \times 10^{-6} \text{ m}^3$$

Hence

Figure 7.31 Example 7.16.

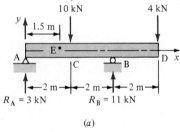

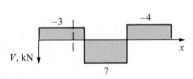

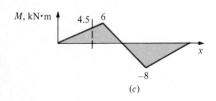

(c)

(a)

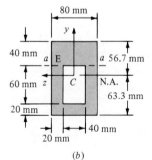

(b)

(d)

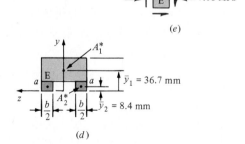

(e)

$$\tau_{max} = \frac{V_{BC}Q_{max}}{Ib} = \frac{7000(123 \times 10^{-6})}{10.5 \times 10^{-6}(2 \times 0.02)} = 2.05 \text{ MPa}$$

$$\sigma_{max} = \frac{M_B c}{I} = \frac{8000(-0.0633)}{10.5 \times 10^{-6}} = -48.2 \text{ MPa}$$

where the minus sign indicates compression.

Clearly τ_{max} occurs at the neutral axis in the interval BC while σ_{max} applies to the bottom fibers of the beam at the support B. We note that in the computation of Q, it is immaterial as to whether we take the area above or below a transverse line such as a-a (Fig. 7.31b). For instance, Q_{max} using the area below the neutral axis is $0.08(0.0633)(0.0633/2) - 0.04(0.0433)(0.0433/2) = 123 \times 10^{-6} \text{ m}^3$, as before.

*7.14 SHEAR-STRESS DISTRIBUTION IN FLANGED BEAMS

The distribution of shearing stresses in beams with flanges (W beams and I beams) may be determined by applying Eq. (7.27). The same procedure previously applied to rectangular beams is employed here. As before, it is assumed that the shearing stresses are uniform across the width of the beam.

Consider a flanged beam transmitting a vertical shear force V (Fig. 7.32a). The shear stress τ_{xy} is treated as a constant over the web thickness. Then, for $0 \le y_1 \le c_1$,

$$\tau_{xy} = \frac{V}{Ib}A^*\bar{y} = \frac{V}{It}\left[b(c - c_1)\left(c_1 + \frac{c - c_1}{2}\right) + t(c_1 - y_1)\left(y_1 + \frac{c_1 - y_1}{2}\right)\right]$$

Upon simplifying, the above yields the following expression for the *shearing stress in the web* of the beam:

$$\tau_{xy} = \frac{V}{2It}[b(c^2 - c_1^2) + t(c_1^2 - y_1^2)] \qquad (7.32)$$

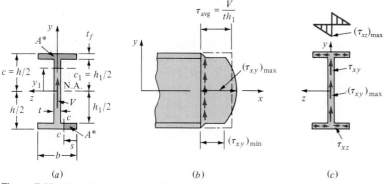

Figure 7.32 Shearing stresses in flanged beams.

The moment of inertia of the entire cross-sectional area about the neutral axis is

$$I = \frac{b(2c)^3}{12} - \frac{(b - t)(2c)^3}{12} = \tfrac{2}{3}(bc^3 - bc_1^3 + tc_1^3) \qquad (7.33)$$

Equation (7.32) shows that τ_{xy} varies *parabolically* throughout the height of the web (Fig. 7.32b). The extreme values of shear stress found at the neutral axis ($y_1 = 0$) and juncture with the flange ($y_1 = \pm c_1$) are, from Eq. (7.32),

$$(\tau_{xy})_{max} = \frac{V}{2It}(bc^2 - bc_1^2 + tc_1^2)$$

$$(\tau_{xy})_{min} = \frac{Vb}{2It}(c^2 - c_1^2)$$

$$(7.34)$$

Note that, when $t << b$, the values of these stresses do not differ appreciably. Example 7.17 will compare the numerical values of the maximum and minimum stresses for an ordinary I-beam section.

An expression for the shearing stress in the beam flanges can be written similarly. Thus, for $c_1 \le y_1 \le c$, we have

$$\tau_{xy} = \frac{V}{Ib}\left[b(c - y_1)\left(y_1 + \frac{c - y_1}{2}\right)\right] = \frac{V}{2I}(c^2 - y_1^2) \qquad (7.35)$$

which represents a *parabolic* variation of stress in the flanges, indicated by the dashed lines in Fig. 7.32b.

Referring again to Fig. 7.32b, we observe that for a thin flange, the stress is *very small* as compared with the shear stress in the web. Thus the *average* or *approximate maximum shear stress* in the beam may be obtained by dividing V by the web area ($h_1 t$):

$$\tau_{avg} = \frac{V}{A_{web}} \qquad (7.36)$$

The above is shown by the dotted lines in the figure.

The stress distribution given by Eq. (7.35) is fictitious because the inner planes of flanges must be free of shearing stress, since they are load-free boundaries of the beam. This contradiction cannot be resolved by the technical theory; the theory of elasticity must be applied to obtain a correct solution. However, this does not result in a serious error, since, as already pointed out, the web carries almost all of the shear force. The distribution of shearing stresses τ_{xy} at the junction of the web and the flange is quite complex, and beams are provided with fillets at these points to reduce concentrations of stress.

The magnitude of the horizontal shearing stress τ_{xz} across any section c-c of the flange (Fig. 7.32a) may be obtained using Eq. (7.27), as was indicated in Sec. 7.12. The cut c-c is located a distance s from the free edge

e. The first moment of area $A^* = st_f$ about the z axis, $Q = st_f(c - t_f/2)$. The shearing stress at c-c is thus

$$\tau_{xz} = \frac{VQ}{It_f} = \frac{V}{I}\left(c - \frac{t_f}{2}\right)s \tag{7.37}$$

The distribution of shear stress on the flange is *linear* with s, as Eq. (7.37) indicates. Maximum shearing stress occurs at $s = b/2$:

$$(\tau_{xz})_{max} = \frac{V}{2I}\left(c - \frac{t_f}{2}\right)b \tag{7.38}$$

The variation of vertical and horizontal shearing stresses, or *shear flow*, over the entire cross-sectional area may be depicted as shown in Fig. 7.32c. The *sense* of the horizontal shear stress τ_{xz} may easily be obtained from the sense of the vertical shearing stress τ_{xy}. In the web, the latter is the *same* as the sense of the shear force V. Note that the vertical shear stresses τ_{xy} in the flanges are small compared with the horizontal shear stresses τ_{xz} and are not shown in the figure.

EXAMPLE 7.17

An I beam of dimensions $h = 254$ mm, $h_1 = 230$ mm, $b = 120$ mm, and $t = 8$ mm is subjected to a shear force of 60 kN (Fig. 7.32a). Calculate (a) the maximum and minimum shearing stresses in the web; (b) the maximum vertical shear stress in the flanges; (c) the maximum horizontal shearing stress in the flanges; and (d) the average vertical shearing stress in the beam.

Solution Substitution of the data given into Eq. (7.33) yields

$$I = \tfrac{2}{3}[120(127)^3 - 120(115)^3 + 8(115)^3] = 50.3 \times 10^6 \text{ mm}^4$$

(a) The stresses in the web are, from Eqs. (7.34),

$$(\tau_{xy})_{max} = \frac{60 \times 10^3}{2(50.3 \times 10^{-6})(0.008)}$$

$$\times [0.12(0.127)^2 - 0.12(0.115)^2 + 0.008(0.115)^2]$$

$$= 33.8 \text{ MPa}$$

$$(\tau_{xy})_{min} = \frac{60 \times 10^3}{2(50.3 \times 10^{-6})(0.008)}[0.12(0.127)^2 - 0.12(0.115)^2]$$

$$= 26 \text{ MPa}$$

(b) The maximum vertical shearing stress in the flanges occurs at $y_1 = \pm0.115$ m. From Eq. (7.35), we obtain

$$(\tau_{xy})_{max} = \frac{60 \times 10^3}{2(50.3 \times 10^{-6})}[(0.127)^2 - (0.115)^2] = 1.73 \text{ MPa}$$

(*c*) The use of Eq. (7.38) results in

$$(\tau_{xz})_{\text{max}} = \frac{60 \times 10^3}{2(50.3 \times 10^{-6})}(0.127 - 0.006)(0.12) = 8.69 \text{ MPa}$$

(*d*) Applying Eq. (7.36), we have

$$\tau_{\text{avg}} = \frac{60 \times 10^3}{(0.008)(0.23)} = 32.6 \text{ MPa}$$

Thus this stress differs by about 4 percent from the 33.8 MPa found by the accurate equation in part (*a*) above.

7.15 COMPARISON OF SHEAR AND BENDING STRESSES

The deformational behavior of beams was illustrated through the assumptions discussed in Sec. 7.7. It was considered that the shearing action was negligible and the transverse loads applied to a slender beam were supported by bending action. This may be justified by comparing the magnitudes of the beam flexural and shearing stresses.

Consider, for example, the bending of a simply supported rectangular beam subjected to a central load, as shown in Fig. 7.33. The shear and moment diagrams (refer to Fig. 7.8) indicate that the maximum bending moment occurring at midspan has a value of $PL/4$ and the shear force has a constant value of $P/2$ between each support and load P. Thus, using Eq. (7.31), we have

$$\tau_{\text{max}} = \frac{3}{2}\frac{V}{A} = \frac{3}{2}\frac{P/2}{bh} = \frac{3}{4}\frac{P}{bh} \qquad (a)$$

and, from Eq. (7.13), we obtain

$$\sigma_{\text{max}} = \frac{Mc}{I} = \frac{(PL/4)(h/2)}{bh^3/12} = \frac{3}{2}\frac{PL}{bh^2} \qquad (b)$$

The ratio of the maximum shearing stress to the maximum bending stress provides a measure of beam slenderness:

$$\frac{\tau_{\text{max}}}{\sigma_{\text{max}}} = \frac{1}{2}\left(\frac{h}{L}\right) \qquad (c)$$

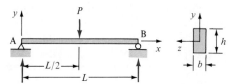

Figure 7.33 Simply supported beam with a concentrated central load.

If, for instance, $L = 10h$, the quotient is only $\frac{1}{20}$. For a slender beam, $h \ll L$, and the shearing stress τ_{xy} is therefore much smaller than the bending stress σ_x. The *applied load* is thus *resisted mainly by the bending stresses* in the beam.

The foregoing conclusion applies, *in most cases,* to beams of various cross-sectional shapes that are subjected to any type of loading. Of course, the factor of proportionality in Eq. (*c*) will be different for beams of different cross-sectional forms and for different kinds of loads in the same beam (see Probs. 7.111 and 7.112). However, it is possible for a beam to be loaded near a support such that the shearing stress is even greater than the bending stress.

*7.16   DESIGN OF PRISMATIC BEAMS

Our concern in this section will be the elastic design of prismatic beams for strength. We shall select the dimensions of a beam section such that it carries safely applied loads without exceeding the allowable stresses in both flexure and shear. Thus the design of the member will be controlled by the largest normal and shearing stresses developed at the critical sections where the maximum value of the bending moment and shear force occur. Shear and bending-moment diagrams are very helpful for locating these critical sections. Sometimes beam design is based upon the plastic moment capacity; this so-called *limit design* is considered in Chap. 12.

In heavily loaded short beams, design is usually governed by shear, while in slender beams the flexure stress, as discussed in the previous section, generally predominates. Shearing is more significant in wood than in steel beams, as wood has relatively low shear strength parallel to the grain.

Application of the rational procedure in design, outlined in Sec. 1.9, to a beam of ordinary proportions is often as follows:

1. *Evaluate the mode of possible failure*. It is assumed that failure results from yielding or from fracture, and flexural stress is considered to be most closely associated with structural damage.
2. *Determine the relationships between load and stress*. The significant value of the bending stress is $\sigma = M_{max}/S$.
3. *Determine the maximum usable value of stress*. The maximum usable value of σ to avoid failure, σ_{max}, is the yield strength σ_{yp} or the ultimate strength σ_u.
4. *Select the factor of safety*. A factor of safety f_s is applied to σ_{max} to obtain the allowable stress: $\sigma_{all} = \sigma_{max}/f_s$. The required *section modulus* of a beam is then

$$S = \frac{M_{max}}{\sigma_{all}} \tag{7.39}$$

Note that when the allowable stress is the same in tension and compression, a doubly symmetric section should be chosen. However, if σ_{all} differs in tension and compression, a singly symmetric section (for example, a T beam) should be selected such that the distances to the extreme fibers in tension and compression are in a ratio nearly identical to the respective σ_{all}'s.

Good engineering design should lead to a beam of minimum weight, thereby saving material. There are generally several different beam sizes with the required value of S. It is common practice to select the one with the *lightest weight* per unit length (or the smallest sectional area) from the tables of beam properties.

We next check the *shear-resistance requirement* of the beam tentatively selected. After substituting the suitable data for Q, I, b, and V_{max} into Eqs. (7.27), we obtain the maximum shearing stress in the beam from

$$\tau_{max} = \frac{V_{max}Q}{Ib} \tag{7.40}$$

If the value calculated for τ_{max} is smaller than the allowable shearing stress τ_{all}, the beam is acceptable. Otherwise, a stronger beam should be chosen and the process repeated.

EXAMPLE 7.18

Select a wide-flange steel beam to carry the loads shown in Fig. 7.34a. The allowable bending and shear stresses are 24 and 14 ksi, respectively.

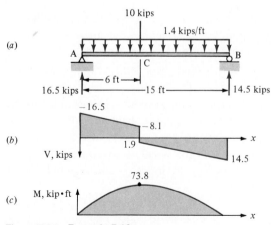

Figure 7.34 Example 7.18.

Solution Referring to the shear and bending-moment diagrams (Fig. 7.34b, and c), we determine that M_{max} = 73.8 kip·ft and V_{max} = 16.5 kips. Equation (7.39) thus yields

$$S = \frac{73.8 \times 12}{24} = 36.9 \text{ in.}^3$$

From Table B.2 we select the *lightest* member which has a section modulus larger than this value of S: a 12-in. W beam weighing 35 lb/ft (S = 45.6 in.³). As will often be the case, the weight of the beam (35 × 15 = 525 lb) is very small compared with the applied load (31 kips), so it is disregarded.

From Table B.2, the area of the web of a W 12 × 35 section is 12.5 × 0.3 = 3.75 in.², which is substituted into Eq. (7.36):

$$\tau_{avg} = \frac{16.5 \times 10^3}{3.75} = 4.4 \text{ ksi}$$

As this stress is well within the allowable limit of 14 ksi, the beam is satisfactory.

EXAMPLE 7.19

Determine the critical span length L of a simple beam of rectangular section ($b \times h$) subjected to a central load P (see Fig. 7.33), given that the shearing and flexure stresses reach their allowable values, τ_{all} and σ_{all}, simultaneously. Use a factor of safety of f_s.

Solution From Eqs. (a) and (b) of Sec. 7.15, we know that the allowable stresses are

$$\tau_{all} = \frac{\tau_{max}}{f_s} = \frac{3}{4}\frac{P}{bhf_s} \qquad \text{or} \qquad P = \frac{4}{3}bhf_s\tau_{all}$$

and

$$\sigma_{all} = \frac{\sigma_{max}}{f_s} = \frac{3}{2}\frac{PL}{bh^2 f_s} \qquad \text{or} \qquad P = \frac{2bh^2 f_s}{3L}\sigma_{all}$$

Therefore

$$\frac{4}{3}bhf_s\tau_{all} = \frac{2bh^2 f_s}{3L}\sigma_{all}$$

The foregoing yields the critical length

$$L = \frac{h}{2}\frac{\sigma_{all}}{\tau_{all}} \qquad\qquad (7.41)$$

If the length of the beam is larger than that obtained above, bending governs the design; for shorter values, shear controls.

---

*7.17 DESIGN OF BEAMS OF CONSTANT STRENGTH

If every element of a beam is stressed to a prescribed allowable value, presumed uniform throughout, then clearly the beam material will be used in the most efficient manner. For a given material, such a design is of minimum weight. What is sought is a depth variation $h(x)$, where width b is constant, such that $\sigma = \sigma_{all}$ everywhere in the beam. At any cross section, the *required section modulus* is then given by

$$S = \frac{M}{\sigma_{all}} \qquad (7.42)$$

in which M is the bending moment on an *arbitrary* section. Nonprismatic or tapered beams designed in this way are called *beams of constant strength*. It should be noted that shear controls the design at those beam locations where the moment is small.

Examples of beams of uniform strength include leaf springs and certain forged or cast machine components. For a structural member, fabrication and design constraints make it impractical to produce a beam of constant stress. For this reason, welded cover plates are often used for parts of prismatic beams in which the moment is large—for example, in a bridge girder.

It should be noted that when the angle between the sides of a tapered beam is small, the flexure formula results in little error. On the other hand, the results obtained by the use of the shear formula, Eqs. (7.27), may not be sufficiently accurate for nonprismatic beams. A modified form of this formula is often employed for design purposes. The exact stress distribution in a rectangular wedge is found by applying the theory of elasticity.

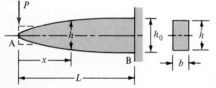

Figure 7.35 Example 7.20: beam of constant strength.

EXAMPLE 7.20

Design a cantilever beam of uniform strength and width to support a concentrated load P at its free end (Fig. 7.35).

Solution At a distance x from A, we have $M = Px$ and $S = bh^2/6$. Equation (7.42) is therefore

$$\frac{bh^2}{6} = \frac{Px}{\sigma_{all}} \qquad (a)$$

Similarly, at a fixed end ($x = L$ and $h = h_0$),

$$\frac{bh_0^2}{6} = \frac{PL}{\sigma_{all}} \qquad (b)$$

Division of Eq. (a) by Eq. (b) results in

$$h = h_0 \sqrt{\frac{x}{L}} \qquad (7.43)$$

Thus the depth of the beam varies parabolically from the free end. Obviously, the cross section of the beam near end A must be designed to resist the shear force.

PROBLEMS

Secs. 7.1 to 7.3

7.1 through 7.9 Write shear and moment expressions for an arbitrary section O in the interval CB of the beams of Figs. P7.1 through P7.9. (Place the origin at point A.) Calculate the values of V and M for a section O located at midspan. Also draw the internal-axial-force diagrams where applicable.

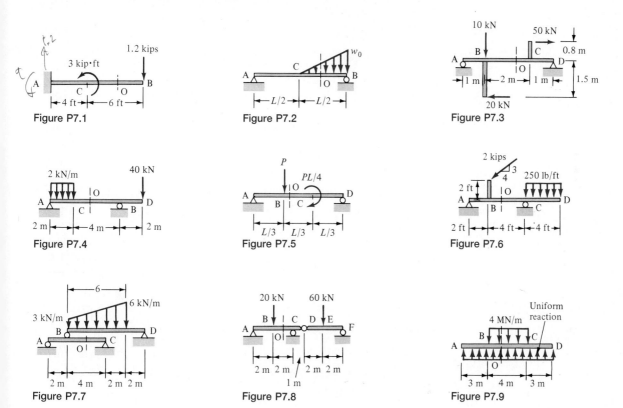

Figure P7.1

Figure P7.2

Figure P7.3

Figure P7.4

Figure P7.5

Figure P7.6

Figure P7.7

Figure P7.8

Figure P7.9

Secs. 7.4 and 7.5

Note: **In the following problems determine all critical ordinates of shear and moment diagrams.**

7.10 through 7.12 Draw the shear and moment diagrams of the beams shown in Figs. P7.1 through P7.3. Use the summation approach.

7.13 through 7.15 Draw the shear and moment diagrams of the beams shown in Figs. P7.4 through P7.6. Use the summation approach.

7.16 through 7.18 Draw the shear and moment diagrams of the beams shown in Figs. P7.7 through P7.9. Use the summation approach.

7.19 through 7.27 Construct the shear and moment diagrams of the beams of Figs. P7.19 through P7.27. Employ the summation approach.

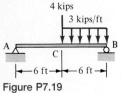

Figure P7.19

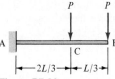

Figure P7.20

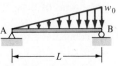

Figure P7.21

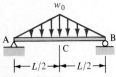

Figure P7.22

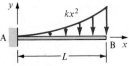

Figure P7.23

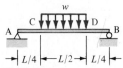

Figure P7.24

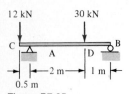

Figure P7.25

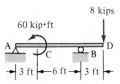

Figure P7.26

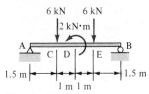

Figure P7.27

7.28 through 7.30 Redo Probs. 7.19 through 7.21, using the direct approach. Place the origin at point A.

7.31 through 7.33 Redo Probs. 7.22 through 7.24, using the direct approach. Place the origin at point A.

7.34 For the transmission shaft of Fig. P7.34, construct (a) the torque diagram; (b) the moment diagram for the xy plane; and (c) the moment diagram for the xz plane.

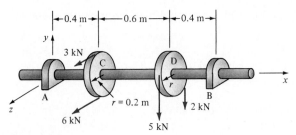

Figure P7.34

7.35 through 7.40 The shear diagrams for the beams supported at A and B are shown in Figs. P7.35 through P7.40. Determine the loading on the beams and draw the moment diagrams. Assume that no couples act on the beams.

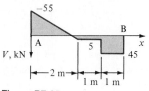

Figure P7.35

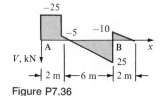

Figure P7.36

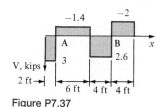

Figure P7.37

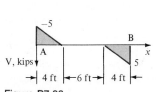

Figure P7.38

Figure P7.39

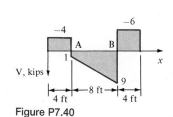

Figure P7.40

Secs. 7.6 to 7.9

7.41 A copper wire $\frac{1}{32}$ in. in diameter is coiled around a spool of 12-in. diameter. Given $E = 17 \times 10^6$ psi, compute the maximum stress and the bending moment in the wire.

7.42 An aluminum bar ($E = 70$ GPa), 10 mm wide, 30 mm deep, and 2 m long, is acted upon by a bending moment so that the midpoint deflection is 10 mm (see Fig. 7.13c). Determine the maximum strain and the bending moment in the bar.

7.43 A steel beam ($E = 200$ GPa and $\sigma_{all} = 160$ MPa) experiences pure bending (Fig. P7.43). Calculate the maximum moment M that may be applied and the corresponding radius of curvature ρ if the cross section is (a) a circle of diameter $d = 12$ mm and (b) an equilateral triangle of sides $b = 15$ mm.

Figure P7.43

7.44 If a circular beam and equilateral triangular beam are to resist the same maximum stress and bending moment (Fig. P7.43), what is the ratio of their cross-sectional areas?

Figure P7.45

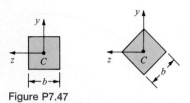

Figure P7.47

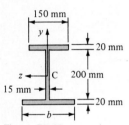

Figure P7.49

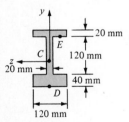

Figure P7.51

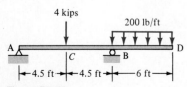

Figure P7.53

7.45 A steel rod of semicircular cross section is acted upon by a $M_z = -1.5$ kip·in. moment (Fig. P7.45). Given $E = 30 \times 10^6$ psi, calculate (a) the maximum tensile and compressive strains in the rod and (b) the radius of curvature of the rod.

7.46 As shown in Fig. P7.45, a brass bar ($E = 15 \times 10^6$ psi) of semicircular cross section is bent over a cylinder of 18-ft radius. What is the bending moment M_z in the bar?

7.47 Moment M acts about the horizontal (z) axis of a square bar in the two different positions shown in Fig. P7.47. Given the modulus of elasticity E of the material, determine the maximum stress and the radius of curvature of the bar.

7.48 An S 150 × 26 steel I beam (see Table B.3) is simply supported on a 10-m span and carries a uniform load of intensity w. Given $\sigma_{all} = 120$ MPa, (a) determine the maximum value of w; (b) determine the width of a solid rectangular section of the same 152-mm depth to replace the I shape; and (c) compare the areas of the two sections.

7.49 An unsymmetrical W beam is acted upon by a bending moment M_z, as shown in Fig. P7.49. For a ratio of bending stresses of $\frac{5}{8}$, at the top and bottom of the beam, determine the width b of the bottom flange.

7.50 A rectangular beam of width b and depth h is to be cut from a circular bar of diameter D, as shown in Fig. P7.50. Determine the ratio of h/b so that the beam will carry a maximum moment M in pure bending.

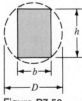

Figure P7.50

7.51 If a beam of the cross section shown in Fig. P7.51 is subjected to the bending moment $M_z = 20$ kN·m, what is the total force acting on the lower flange of the beam?

7.52 A beam of the cross section depicted in Fig. P7.51 is subjected to a bending moment $M_z = 10$ kN·m. Calculate the total force acting on the top flange of the beam.

7.53 A W 6 × 16 steel beam (see Table B.2) is loaded as shown in Fig. P7.53. On a section located 3.6 ft from A, calculate (a) the bending stress at a point 1 in. above the bottom of the beam and (b) the maximum bending stress on the section.

7.54 Redo Prob. 7.53 for an S 6 × 12.5 steel beam (see Table B.3).

7.55 A W 8 × 35 steel beam (see Table B.2) is loaded as shown in Fig. P7.53. On a section 4.2 ft to the right of C, calculate (a) the flexural stress at a point located $\frac{3}{4}$ in. below the top of the beam and (b) the maximum bending stress on the section.

7.56 A rectangular beam, 80 mm wide by 120 mm deep, is loaded as depicted in Fig. P7.56. Calculate the magnitude and location of the maximum bending stress in the beam.

7.57 Redo Prob. 7.56 for the case in which an additional downward force of 4 kN acts at C.

7.58 A beam of the cross section shown in Fig. P7.51 is loaded as shown in Fig. P7.56. On the center section, calculate the bending stress at (a) point E and (b) point D.

7.59 As depicted in Fig. P7.59, two vertical forces are applied to an S 380 × 64 steel beam (see Table B.3) resting on the ground. Determine the largest bending stress in the beam.

7.60 A W 410 × 60 wide-flange beam (see Table B.2) is reinforced by two 220 × 40-mm steel cover plates of the cross section shown in Fig. P7.60. The beam carries a uniformly distributed load w on a simply supported span of 10 m. Given an allowable bending stress of 120 MPa, calculate the permissible value for w.

7.61 Repeat Prob. 7.60, with w = 0, for a beam carrying a concentrated load P applied on a section 3 m from the left support.

7.62 An overhanging beam is loaded as shown in Fig. P7.62. Given the allowable stresses of σ_t = 7 ksi and σ_c = 18 ksi, determine the maximum permissible value of w.

7.63 For the beam loaded as shown in Fig. P7.62, using w = 600 kips/ft, determine, on a section 3 ft to the right of A, the bending stress at (a) point E and (b) point F.

7.64 A simply supported beam 15 ft long is constructed of two C 10 × 20 steel channels (see Table B.4) riveted back to back. Using an allowable bending stress of 22 ksi, determine the central concentrated load P that can be applied if (a) the webs are horizontal (Fig. P7.64a) and (b) the webs are vertical (Fig. P7.64b).

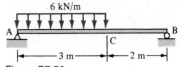

Figure P7.56

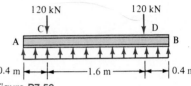

Figure P7.59

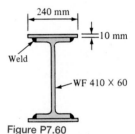

Figure P7.60

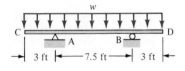

Figure P7.62

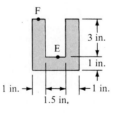

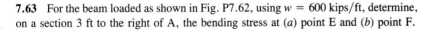

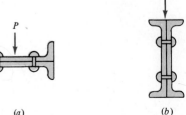

(a)

(b)

Figure P7.64

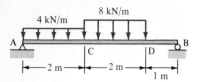

Figure P7.65

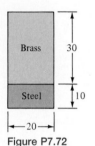

Figure P7.72

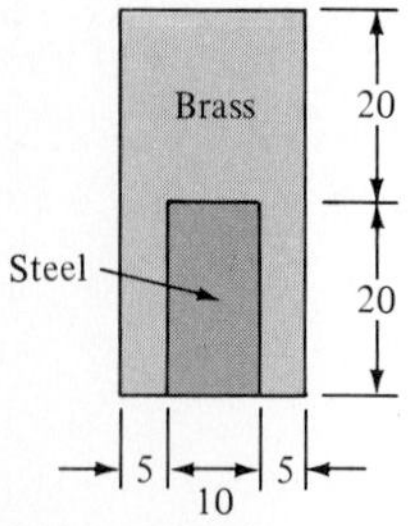

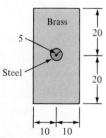

Figure P7.73

Figure P7.74

7.65 An S 200 × 34 steel beam (see Table B.3) is loaded as shown in Fig. P7.65. Determine (*a*) the bending stress on section C and at a point 10 mm below the top of the beam and (*b*) the maximum bending stress in the beam.

7.66 Rework Prob. 7.65 for the case in which end A is free and end B is fixed.

Secs. 7.10 and 7.11

7.67 Redo Prob. 7.56, given that the beam has two vertical U-shaped grooves opposite each other on its edges at C (use Fig. 7.21*a*). The grooves are 12 mm deep and have a radius of (*a*) $r = 5$ mm and (*b*) $r = 10$ mm.

7.68 If the beam loaded as shown in Fig. 7.18 has two 15-mm-deep vertical U-shaped grooves opposite each other on its edges (use Fig. 7.21*a*) at midlength and $\sigma_{\text{all}} = 150$ MPa, calculate the smallest permissible radius of the grooves.

7.69 For the cantilever spring shown in Fig. P7.69, $b = \frac{1}{2}$ in. and $P = 80$ lb. Calculate the largest bending stress for a fillet radius r of (*a*) $\frac{3}{16}$ in. and (*b*) $\frac{3}{4}$ in.

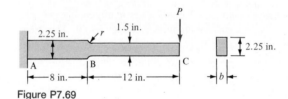

Figure P7.69

7.70 Determine the largest permissible load P that the cantilever spring shown in Fig. P7.69 can carry for $\sigma_{\text{all}} = 20$ ksi, radius $r = \frac{3}{8}$ in., and width $b = \frac{5}{8}$ in.

7.71 Based upon a factor of safety of $f_s = 1.5$, determine the minimum permissible width b of the beam shown in Fig. P7.69 for $P = 90$ kips, $\sigma_{\text{max}} = 18$ ksi, and $r = \frac{5}{16}$ in.

7.72 through 7.74 The composite beams of Figs. P7.72 through P7.74 are fabricated by bonding a steel bar ($E_s = 200$ GPa) and a brass bar ($E_b = 100$ GPa) and are subjected to the bending moment $M_z = 200$ N·m. (Dimensions are in millimeters.) Determine the maximum bending stress in the brass and steel.

7.75 through 7.77 For the composite beams with cross sections as shown in Figs. P7.72 through P7.74, determine the maximum permissible bending moments for $(\sigma_b)_{\text{all}} = 120$ MPa, $(\sigma_s)_{\text{all}} = 140$ MPa, $E_b = 100$ GPa, and $E_s = 200$ GPa.

7.78 A 6-in.-wide by 10-in.-deep wood cantilever beam ($E_w = 1.5 \times 10^6$ psi) that is 5 ft long is reinforced with 6-in.-wide and $\frac{1}{8}$-in.-deep steel plates ($E_s = 30 \times 10^6$ psi) on its top and bottom faces. If the beam is subjected to a uniform load of 150 lb/ft over its entire length, determine the maximum bending stresses in each material.

7.79 Redo Prob. 7.78, given that the beam is simply supported.

7.80 Rework Prob. 7.78, given that the wood cantilever beam is reinforced with 10-in.-deep and $\frac{1}{8}$-in.-wide steel plates on its side faces.

7.81 Determine the maximum permissible value of P for the loading shown in Fig. P7.81 for $(\sigma_c)_{\text{all}} = 8.4$ MPa, $(\sigma_s)_{\text{all}} = 140$ MPa, and $n = 8$.

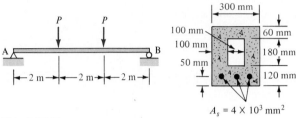

Figure P7.81

7.82 Consider the reinforced concrete beam of Fig. 7.25a, with $b = 400$ mm, $d = 600$ mm, $A_s = 2500$ mm^2, and $n = 8$. For $(\sigma_c)_{\text{all}} = 10$ MPa and $(\sigma_s)_{\text{all}} = 140$ MPa, determine the largest bending moment that may be applied to the beam.

7.83 An aluminum plate ($E_a = 70$ GPa) and a steel plate ($E_s = 210$ GPa) are bonded together to create the composite beam shown in Fig. P7.83. Design the cross section of the beam for $(\sigma_a)_{\text{all}} = 80$ MPa and $(\sigma_s)_{\text{all}} = 240$ MPa.

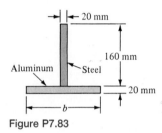

Figure P7.83

7.84 The beam of cross section shown in Fig. 7.25a is to support a moment of $M_z = 40$ kN·m. Given $(\sigma_s)_{\text{all}} = 150$ MPa, $(\sigma_c)_{\text{all}} = 7.5$ MPa, $d = 500$ mm, and $n = 10$, calculate the minimum permissible value of (a) width b and (b) area A_s.

Secs. 7.12 to 7.15

7.85 Given a beam loaded as shown in Fig. P7.62, with $w = 1.2$ kips/ft, determine (a) the shearing stress at a point 1.5 ft to the left of B and $\frac{3}{4}$ in. below the top of the beam and (b) the maximum shear stress in the beam.

7.86 Determine the ratio of the largest vertical shear force (V_y) to the maximum horizontal shear force (V_z) that can be carried *separately* by the I beam of the cross section shown in Fig. P7.86.

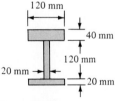

Figure P7.86

7.87 A beam of the cross section shown in Fig. P7.86 is loaded as shown in Fig. P7.87. Determine (a) the shearing stress at point E and 40 mm from the top of the beam and (b) the maximum shearing stress in the beam.

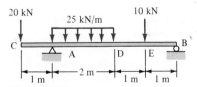

Figure P7.87

7.88 Given the beam and the loading shown in Fig. 7.19, calculate (*a*) the shear stress at a point 2 m from A and 25 mm from the top of the beam and (*b*) the greatest shearing stress.

7.89 through 7.91 A vertically symmetric beam of 10-mm uniform wall thickness, as shown in Figs. P7.89 through P7.91, is acted upon by a shear force of $V = 20$ kN. Determine (*a*) the shearing stress at the sections indicated and (*b*) the maximum shearing stress in the beam. (Dimensions are in millimeters.)

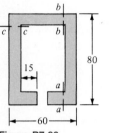

Figure P7.89

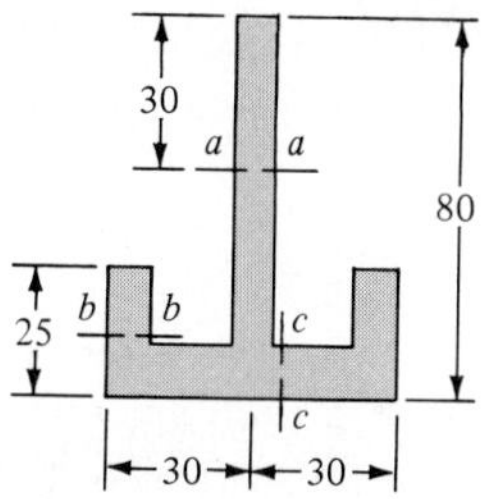

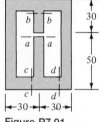

Figure P7.90

Figure P7.91

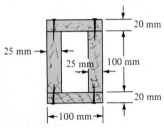

Figure P7.93

7.92 A beam of the cross section shown in Fig. P7.89 is loaded as seen in Fig. P7.65. Determine the shearing stress at point C, 10 mm from the bottom of the beam.

7.93 As shown in Fig. P7.93, a beam is constructed of four wood boards nailed together, two 20 × 100 mm and two 25 × 100 mm. Given a longitudinal spacing between each nail of $s = 100$ mm and an allowable vertical shear force in the beam of $V = 3$ kN, calculate (*a*) the shear force in each nail and (*b*) the shearing stress 40 mm from the top of the beam.

7.94 Redo Prob. 7.93, given that the cross section of the beam is turned 90°.

7.95 For the beam of Prob. 7.93, if $V = 2$ kN and the allowable shear force in each nail is 250 N, calculate the largest permissible longitudinal spacing between nails.

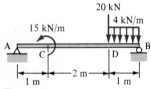

Figure P7.97

7.96 Repeat Prob. 7.95 for the beam cross section turned 90°.

7.97 A beam of the cross section shown in Fig. P7.93 is loaded as shown in Fig. P7.97. Calculate the maximum shearing stress in the beam.

7.98 A built-up cantilever beam is fabricated by gluing four planks as shown in Fig. P7.98 and is loaded as shown in Fig. P7.87. Determine the maximum shearing stress (*a*) in the beam; (*b*) at joint E; and (*c*) at joint F.

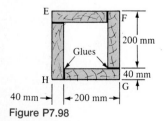

Figure P7.98

7.99 A laminated beam is composed of three glued boards (see Fig. P7.99). Given allowable shearing stresses of 120 psi in the glue and 500 psi in the wood, calculate (*a*) the highest permissible value for P and (*b*) the corresponding maximum flexure stress in the beam.

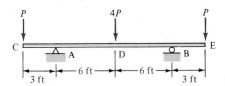

Figure P7.99

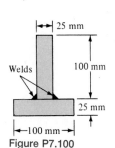

Figure P7.100

7.100 A beam is composed of two 20 × 100-mm steel plates welded together as shown in Fig. P7.100. Given an allowable shear force per unit length in each weld of 200 kN/m, determine (*a*) the largest permissible vertical shear force in the beam and (*b*) the corresponding maximum shearing stress in the beam.

7.101 Resolve Prob. 7.100 for a built-up beam of the cross section shown in Fig. P7.60.

7.102 Redo Prob. 7.100 for a symmetrical built-up beam of the cross section shown in Fig. P7.102.

7.103 A beam is loaded and supported as shown in Fig. P7.62. Given allowable tensile and compressive stresses of 6 and 10 ksi, respectively, and a shearing stress not to exceed 1.2 ksi, determine the maximum permissible load *w*.

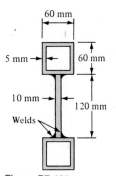

Figure P7.102

7.104 and 7.105 The maximum shearing stress in a beam acted upon by a vertical force V may be expressed as follows:

$$\tau_{max} = k \frac{V}{A}$$

Here A is the cross-sectional area of the beam (Figs. P7.104 and P7.105) and *k* represents a numerical factor, or *shear coefficient*. Determine (*a*) the location of the point at which τ_{max} occurs; (*b*) the value of *k*; and (*c*) the maximum vertical shear force V the section may carry for *c* = 2 in., *b* = *h* = 4 in., and τ_{max} = 9 ksi.

Figure P7.104

Figure P7.105

7.106   A box beam supports the loads shown in Fig. P7.106. Compute the maximum value of P for $\sigma_{all} = 5$ MPa and $\tau_{all} = 1.2$ MPa.

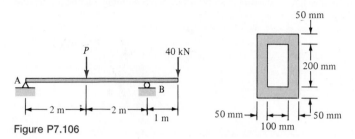

Figure P7.106

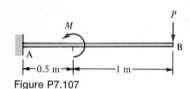

Figure P7.107

7.107   A cantilever beam of the cross section shown in Fig. P7.98 is loaded as shown in Fig. P7.107. If $P = 800$ N and $M = 200$ N·m, determine (a) the maximum bending stress in the beam and (b) the maximum shearing stress in the beam.

7.108   As shown in Fig. P7.108, a rectangular beam 6 in. wide and 12 in. deep supports a total distributed load of W pounds and a concentrated load of $2W$ pounds. If $\sigma_{all} = 1800$ psi and $\tau_{all} = 120$ psi, determine the maximum value of W.

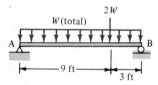

Figure P7.108

7.109   A beam of built-up cross section is loaded as shown in Fig. P7.109. Assume that *abcd* is a plank and that the allowable shear stress in each bolt is 60 MPa. Determine (a) the maximum bending stress at section B and (b) the bolt size to be used with 15-mm spacing along the beam.

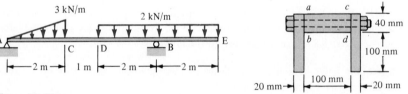

Figure P7.109

7.110   Consider the cantilever beam subjected to a variable load of $w = kx^3$ kN/m, as shown in Fig. P7.110. Determine (a) the maximum bending stresses and (b) the maximum shearing stress at the neutral axis.

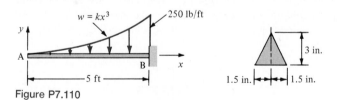

Figure P7.110

7.111   Redo the sample problem of Sec. 7.15 for the beam subjected to a uniformly distributed load of intensity w and $P = 0$.

7.112 Rework the sample problem of Sec. 7.15 for the beam cross section shown in Fig. P7.104.

Secs. 7.16 and 7.17

7.113 A wooden beam is loaded as shown in Fig. P7.113. For σ_{all} = 8.4 MPa, τ_{all} = 1.2 MPa, and h = $3b$ = 150 mm, determine the maximum permissible (a) length and (b) distributed load w.

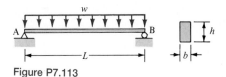

Figure P7.113

7.114 Redo Prob. 7.113 for the case in which end A of the beam is fixed and end B is free.

7.115 Design the cross section of the wooden beam shown in Fig. P7.113, given that w = 600 lb/ft, L = 12 ft, b = 6 in., σ_{all} = 1800 psi, and τ_{all} = 150 psi.

7.116 Determine the depth h needed in the rectangular beam loaded as seen in Fig. P7.56, given that σ_{all} = 10 MPa, τ_{all} = 600 kPa, and width b = 150 mm.

7.117 If σ_{all} = 140 MPa and τ_{all} = 80 MPa, select the wide-flange section (see Table B.2) that should be used to support the loading shown in Fig. P7.107. Use P = 16 kN and M = 2 kN·m.

7.118 Redo Prob. 7.117 for the case in which an S beam (see Table B.3) is to be selected.

7.119 For σ_{all} = 18 ksi, calculate the minimum width b of the steel beam carrying the loads shown in Fig. P7.119.

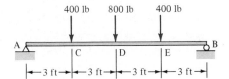

Figure P7.119

7.120 Given that σ_{all} = 9 MPa and τ_{all} = 1.1 MPa, design the cross section of a rectangular wooden beam loaded as shown in Fig. P7.120. Assume that the beam is to be twice as deep as it is wide (h = $2b$).

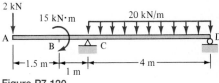

Figure P7.120

7.121 Redo Example 7.20 for the case in which the beam is acted upon by a uniformly distributed load of intensity w and P = 0.

7.122 Design a simply supported beam of constant strength and width, loaded as shown in Fig. P7.122.

7.123 Repeat Example 7.20 for the case in which the beam is subjected to a linearly varying loading of intensity zero at the free end and w_0 at the fixed end.

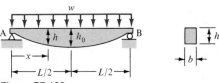

Figure P7.122

CHAPTER
8

COMBINED STRESSES

8.1 INTRODUCTION

The basic relationships governing the influence of the fundamental types of loads—axially centric, torsional, and flexural—have been developed in the preceding chapters. However, structural and machine members must often simultaneously resist more than one type of loading. For example, a shaft in torsion may at the same time, because of its own weight, be subjected to bending. In this chapter we shall consider the effects of such *combined loadings*.

Combined stresses can be determined by superposition of the stresses due to each load acting separately. This implies that the presence of one load does not affect the stresses and strains contributed by another. The *principle of superposition* is applicable in cases involving linearly elastic material behavior where deformations are small. Once the combined stresses have been calculated, the methods of stress analysis of Chap. 4 are generally employed to investigate the principal stresses and the maximum shearing stress.

In Secs. 8.2 through 8.5, consideration will be given to stresses arising from various combinations of fundamental loads applied to members with a vertical axis of symmetry. Next to be treated are the combined normal stresses caused by asymmetrical or skew bending and an axial force (Secs. 8.6 and 8.7). Then in Sec. 8.8 we will discuss the shear center of thin-walled members under asymmetric loading. The last two sections of this chapter deal with prediction of failure in a structure and the design of transmission shafts subjected to simultaneously applied forces.

8.2 AXIAL AND TORSIONAL LOADS

A shaft or other machine member is often subjected to the simultaneous action of axial and torsional loads. Examples of each members include a drill rod and the propeller shaft of a ship or airplane. Inasmuch as radial and circumferential normal stresses are absent, these loadings produce a state of plane strain. Clearly the axial stress is identical at every point, and the maximum shearing stress is found at the outer fibers. Thus the critical stresses occur on elements at the surface of the member.

Consider the solid circular cantilever rod shown in Fig. 8.1a. The rod is loaded at its free end by an axial force P and a twisting couple T. In Fig. 8.1b the axial and shearing stresses are shown to be acting on a small element A at the rod's surface. Note that individual stresses are first depicted separately and then superposed to represent the combined stress. For the case under discussion,

$$\sigma_x = \frac{P}{A} = \frac{4P}{\pi d^2} \qquad \tau = -\frac{Tc}{J} = -\frac{16T}{\pi d^3} \qquad (a)$$

where d is the rod diameter.

We can now proceed to obtain the critical stresses acting at a point A at inclined directions. The principal stresses and the maximum shearing stress, Eqs. (4.6) and (4.8), for members subjected to *combined normal and shear* stresses become, respectively,

$$\sigma_{1,2} = \frac{\sigma_x}{2} \pm \sqrt{\left(\frac{\sigma_x}{2}\right)^2 + \tau^2} \qquad (8.1)$$

$$\tau_{max} = \sqrt{\left(\frac{\sigma_x}{2}\right)^2 + \tau^2} \qquad (8.2)$$

and the principal directions are, applying Eq. (4.5),

$$\tan 2\theta_p = \frac{2\tau}{\sigma_x} \qquad (8.3)$$

Here the θ_p's are measured from the longitudinal axis of the rod.

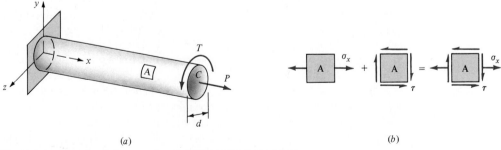

(a)   (b)

Figure 8.1   Combined stresses due to axial and torsional loads.

EXAMPLE 8.1

A solid circular shaft 2 in. in diameter is subjected to an axial tension of 25 kips and to a twisting moment of 10 kip·in. (Fig. 8.1a). Calculate the maximum tensile stress.

Solution Substituting the data given into Eqs. (a) yields (Fig. 8.1b):

$$\sigma_x = \frac{4(25 \times 10^3)}{\pi(2)^2} = 7.96 \text{ ksi}$$

$$\tau = -\frac{16(30 \times 10^3)}{\pi(2)^3} = -6.37 \text{ ksi}$$

From Eqs. (8.1), we have

$$\sigma_{1,2} = \frac{7.96}{2} \pm \sqrt{\left(\frac{7.96}{2}\right)^2 + (-6.37)^2}$$

or

$$\sigma_1 = 11.49 \text{ ksi} \qquad \sigma_2 = -3.53 \text{ ksi}$$

The foregoing stresses occur on planes defined by Eq. (8.3):

$$2\theta_p = \tan^{-1}\frac{2(-6.37)}{7.96} = -58° \qquad \text{or} \qquad \theta_p = -29° \text{ and } -119°$$

Insertion of the first of these values into Eq. (4.4a) gives $\sigma_{x'} = 11.49$ ksi. It is thus observed that the maximum stress, $\sigma_1 = \sigma_{x'}$, occurs on a plane oriented 29° (clockwise) with respect to the geometric axis of the shaft.

*8.3 DIRECT SHEAR AND TORSIONAL LOADS: HELICAL SPRINGS

An example typical of a *direct*, or *cross*, *shear load* combined with a *torsional load* is found in a closely coiled helical spring. A spring of this kind is formed by wrapping a wire or rod around a circular cylinder in such a manner that the rod forms a helix of uniformly and closely spaced coil turns. We are here concerned with a tension or compression spring composed of a slender rod of radius c wound into a helix of mean radius R, where $c/R \ll 1$. Any one coil of the spring will be assumed to lie in a plane approximately perpendicular to the spring axis.

In the case of a closely coiled spring, a section taken perpendicular to the axis of the spring's rod can be considered to be nearly vertical. Hence, load P applied to the spring is resisted by a transverse shear force $V = P$ and a torque $T = PR$ acting on the cross section of the coil (Fig. 8.2a). The direct shearing stress is customarily assumed to be uniform over the cross-sectional area $A = \pi d^2/4$ of the rod: $\tau_d = 4P/\pi d^2$ (Fig. 8.2b). The torsional

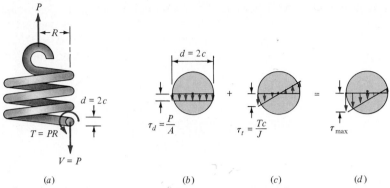

Figure 8.2 Combined stresses due to direct shear and torsion.

shearing stress, neglecting the initial curvature of the rod and using $J = \pi d^4/32$, is $\tau_t = 16PR/\pi d^3$ (Fig. 8.2c).

The superposition of the individual stresses leads to the resultant combined state of stress shown in Fig. 8.2d. It is observed in the figure that the critical shearing stress occurs on an element at the inside surface of the coil and is of magnitude

$$\tau_{max} = \frac{P}{A} + \frac{Tc}{J} = \frac{4P}{\pi d^2} + \frac{16PR}{\pi d^3} \qquad (8.4a)$$

This may be written

$$\tau_{max} = \frac{16PR}{\pi d^3}\left(1 + \frac{d}{4R}\right) \qquad (8.4b)$$

For a slender rod, the ratio c/R is small, and it is clear that the maximum shearing stress is caused primarily by torsion.

In a heavy spring (for example, a railroad-car spring), where $2c/R$ is large, the length differential of the longitudinal surface elements must be considered, and there is a stress concentration at a critical point of the rod. In these springs, the effect of cross shearing stress cannot be disregarded. The following more accurate relationship, known as the *Wahl formula*, takes into account the initial curvature of the spring wire and the exact distribution of the direct shear stress over the cross section:*

$$\tau_{max} = \frac{16PR}{\pi d^3}\left(\frac{4m - 1}{4m - 4} + \frac{0.615}{m}\right) \qquad (8.5)$$

Here $m = 2R/d$. When $c/R = \frac{1}{10}$, for example, the maximum stress computed from Eq. (8.5) is approximately 8 percent greater than that found using Eq. (8.4).

*See A. M. Wahl, *Mechanical Springs,* McGraw-Hill, New York, 1963.

In determining the deflection of a closely coiled spring, it is common practice to ignore the effect of cross shear. Thus the twist causes one end of the rod segment to rotate an angle $d\phi$ relative to the other, where $\phi = TL/JG$ (Sec. 6.5). This corresponds to a deflection $d\delta$ at the axis of the spring:

$$d\delta = R\, d\phi = R\frac{T\, dL}{GJ}$$

The *total deflection* of an n coil spring of length $L = 2\pi Rn$ is therefore

$$\delta = \frac{RTL}{GJ} = \frac{64nPR^3}{d^4 G} \tag{8.6}$$

The elastic behavior of a spring may be expressed conveniently by the *spring constant k*:

$$k = \frac{P}{\delta} \tag{8.7}$$

For the case under consideration, $k = d^4 G/64nR^3$.

8.4 AXIAL, TRANSVERSE, AND TORSIONAL LOADS

The discussion of Sec. 8.2 concerning the principal stresses resulting from centric loads and torsion is now extended to include bending. As we shall see, under this combined loading the critical points may not be readily located. Thus it may be necessary to examine the stress distribution in some detail.

Consider, for example, a solid circular cantilever bar subjected to a transverse force R, a torque T, and a centric force P at its free end (Fig. 8.3a). Every section experiences an axial force P, a torque T, a bending moment M, and a shear force $R = V$. The corresponding stresses may be obtained using the applicable basic stress relationships as follows:

$$\sigma'_x = \frac{P}{A} \qquad \tau_t = -\frac{Tc}{J} \qquad \sigma''_x = \frac{Mc}{I} \qquad \tau_d = -\frac{VQ}{Ib}$$

Figure 8.3 Combined stresses due to torsion, tension, bending, and direct shear.

In Fig. 8.3b and c, the stresses shown are those acting on an element

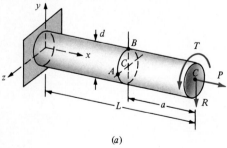

(a)

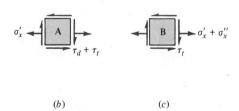

(b)   (c)

B at the top of the bar and an element A on the side of the bar at the neutral axis. Clearly the flexural stress is maximum at the greatest distance from the centroidal axis at a section where the bending moment is maximum. As M has its peak value (RL) at the fixed end, the normal stresses are largest for element B located at the fixed end of the bar. Note that the direct shearing stress is maximum at the centroidal z axis of the cross section, and the torsional stress is maximum at the outer fibers of the bar. The total shearing stress acting on element A is the algebraic sum of these two shears. Hence B (located at the support) and A are the critical points at which the most severe stresses occur. The method of Sec. 4.3 or 4.4 can now be used to obtain the principal stresses and the maximum shearing stress at a critical point.

EXAMPLE 8.2

An 80-mm-diameter cantilever aluminum bar is loaded as shown in Fig. 8.3a. Given allowable stresses of 90 MPa in tension and 50 MPa in shear on a section 320 mm from the free end, determine the largest value of R. Let $T = 0.2R$ N·m and $P = 20R$ N.

Solution The geometric properties of the section are

$$A = \pi c^2 = 16\pi(10^{-4}) \text{ m}^2$$
$$I = \tfrac{1}{4}\pi c^4 = \tfrac{1}{4}\pi(0.04)^4 = 64\pi(10^{-8}) \text{ m}^4$$
$$J = 2I = 128\pi(10^{-8}) \text{ m}^4$$

The normal stress at all points of the bar is

$$\sigma'_x = \frac{P}{A} = \frac{20R}{16\pi(10^{-4})} = \frac{12{,}500R}{\pi} \tag{a}$$

and the torsional stress at the outer fibers of the bar is

$$\tau_t = -\frac{Tc}{J} = -\frac{0.2R(0.04)}{128\pi(10^{-8})} = -\frac{6250R}{\pi} \tag{b}$$

The maximum tensile bending stress occurs at point B of the section considered. Therefore, for $a = 320$ mm, we have

$$\sigma''_x = \frac{Mc}{I} = \frac{0.32R(0.04)}{64\pi(10^{-8})} = \frac{20{,}000R}{\pi} \tag{c}$$

Inasmuch as $Q = A_s\bar{y} = (\pi c^2/2)(4c/3\pi) = 2c^3/3$ and $b = 2c$, the maximum direct shearing stress at point A is

$$\tau_d = -\frac{VQ}{Ib} = -\frac{4V}{3A} = -\frac{4R}{3(16\pi \times 10^{-4})} = -\frac{833R}{\pi} \tag{d}$$

The maximum principal stress and the maximum shearing stress at point A (Fig. 8.3b), using Eqs. (4.6), (4.8), (a), (b), and (d), are

$$(\sigma_1)_A = \frac{12{,}500R}{2\pi} + \sqrt{\left(\frac{12{,}500R}{2\pi}\right)^2 + \left(-\frac{7083R}{\pi}\right)^2}$$

$$= \frac{6250R}{\pi} + \frac{9446R}{\pi} = \frac{15{,}696R}{\pi}$$

$$(\tau_{max})_A = \frac{9446R}{\pi}$$

Similarly, at point B (Fig. 8.3c),

$$(\sigma_1)_B = \frac{32{,}500R}{2\pi} + \sqrt{\left(\frac{32{,}500R}{2\pi}\right)^2 + \left(-\frac{6250R}{\pi}\right)^2}$$

$$= \frac{16{,}250R}{\pi} + \frac{17{,}411R}{\pi} = \frac{33{,}661R}{\pi}$$

$$(\tau_{max})_B = \frac{17{,}411R}{\pi}$$

Note that the stresses at B are more severe than those at A. Substituting the given data into the foregoing, we have

$$90(10^6) = \frac{33{,}661R}{\pi} \qquad \text{or} \qquad R = 8.4 \text{ kN}$$

and

$$50(10^6) = \frac{17{,}411R}{\pi} \qquad \text{or} \qquad R = 9 \text{ kN}$$

The magnitude of the largest permissible load is thus $R = 8.4$ kN.

8.5 DIRECT SHEAR AND BENDING LOADS: PRINCIPAL STRESSES IN BEAMS

Methods for determining the normal and shearing stresses acting at any point in a beam were presented in Secs. 7.9 and 7.12, and procedures for locating the sections of maximum M and V were established in Sec. 7.5. However, a detailed study requires the calculation of the principal stresses and the maximum shearing stress at various locations of the beam due to *combined shear and bending loads.*

Recall that the bending stress is a maximum at the outer edges of a section and is zero at the neutral axis. On the other hand, the shearing stress is zero at the top or bottom edge and is usually a maximum at the neutral axis. Thus the maximum flexural stresses acting on the surface elements at a section are also the principal stresses, whereas the element at the neutral surface experiences only pure shear. At any point of the cross section, an element is subjected simultaneously to flexural and shearing stresses. To ascertain the largest stresses at such locations, either of the methods of analysis presented in Chap. 4 may be employed.

We shall demonstrate in Example 8.3 that, for a beam of rectangular cross section, the maximum normal stress may be determined from the flexure formula, Eq. (7.12). This also applies to many beams of nonrectangular solid sections. Note, however, that for wide-flange sections, the high normal and shearing stresses (σ_x and τ_{xy}) occurring at the flange-web junction may sometimes result in a maximum principal stress at the surface of the flange, as we shall see in Example 8.4.

EXAMPLE 8.3

Calculate the principal stresses at points A, B, C, D, and E for the simply supported rectangular beam loaded as shown in Fig. 8.4a.

Solution The bending moment and shear force at section AB are obtained from the equilibrium conditions for a segment of the beam (Fig. 8.4b). The required geometric properties of the section are

$$A = 0.025 \times 0.12 = 0.003 \text{ m}^2$$
$$I = \tfrac{1}{12}(0.025)(0.12)^3 = 3.6 \times 10^{-6} \text{ m}^4$$
$$Q_B = Q_D = (0.025)(0.03)(0.045) = 33.75 \times 10^{-6} \text{ m}^3$$

The cross-sectional stresses at each point can now be readily calculated:

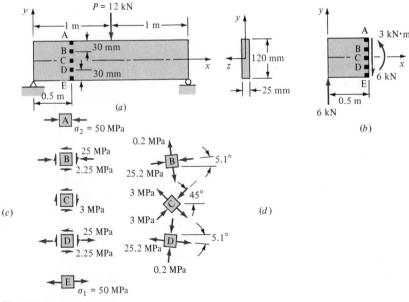

Figure 8.4 Example 8.3.

$$\sigma_{A,E} = \frac{Mc}{I} = \frac{3000(0.06)}{3.6 \times 10^{-6}} = \pm 50 \text{ MPa}$$

$$\sigma_{B,D} = \tfrac{1}{2}\sigma_A = \pm 25 \text{ MPa}$$

$$\sigma_C = 0$$

$$\tau_{B,D} = \frac{VQ_B}{Ib} = \frac{6000(33.75 \times 10^{-6})}{3.6 \times 10^{-6}(0.025)} = 2.25 \text{ MPa}$$

$$\tau_C = \tau_{\max} = \frac{3V}{2A} = \frac{1.5(6000)}{0.003} = 3 \text{ MPa}$$

These stresses are depicted in Fig. 8.4c as acting on elements having horizontal and vertical faces.

The principal stresses and their directions at points B and C are, from Eqs. (4.6) and (4.7),

$$(\sigma_1)_B = -\frac{25}{2} + \sqrt{\left(-\frac{25}{2}\right)^2 + (2.25)^2} = -12.5 + 12.7 = 0.2 \text{ MPa}$$

$$(\sigma_2)_B = -25.2 \text{ MPa}$$

$$(\sigma_1)_C = 3 \text{ MPa}$$

$$(\sigma_2)_C = -3 \text{ MPa}$$

and

$$(\theta_p)_B = \frac{1}{2}\tan^{-1}\frac{-2(2.25)}{-25} = 5.1°$$

$$(\theta_p)_C = \frac{1}{2}\tan^{-1}\frac{2(3)}{0} = 45°$$

Similarly, we obtain $(\sigma_1)_D = 25.7$ MPa, $(\sigma_2)_D = -0.2$ MPa, and $(\theta'_p)_D = -5.1°$. The results are shown in Fig. 8.4d. We observe from the figure that the magnitude of the algebraically larger principal stress (σ_1) diminishes continuously from a maximum value at the bottom of the beam to zero at the top, while its direction gradually rotates through 90°. Identical comments apply to the compressive principal stress (σ_2).

The stresses at many cross sections of the beam may be investigated in a like manner. Two systems of curves drawn with their slope at each point in the direction of the principal stress are known as *stress trajectories*. Such trajectories are illustrated in Fig. 8.5 for the beam of Fig. 8.4a. Here the directions of the tensile stresses are represented by the solid lines, while the compressive stresses are depicted by the dashed lines. (Note that Fig. 8.5 neglects to show the effects of any stress concentrations.) Obviously, the two types of curves always intersect at right angles. The trajectories become either horizontal or vertical at the top and bottom surfaces of the beam. Stress trajectories are particularly important in photoelastic studies.

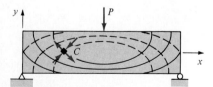

Figure 8.5 Principal stress trajectories for a rectangular beam.

EXAMPLE 8.4

A cantilever W 150 × 24 steel beam is loaded as shown in Fig. 8.6a. Calculate the maximum normal stress and the maximum shearing stress at the flange-web junction for two cases: (a) L = 300 mm and (b) L = 500 mm.

Solution From Table B.2 we find that the centroidal moment of inertia of the section is $I = 13.4 \times 10^{-6}$ mm⁴. The first moment of the sectional areas at the junction of flange-web and the neutral axis, respectively, are

$$Q_B = 102(10.3)(74.85) = 78.6 \times 10^3 \text{ mm}^3$$

$$Q_C = Q_B + 6.6(69.7)(34.85) = 94.6 \times 10^3 \text{ mm}^3$$

The shear V = 60 kN is constant over the entire beam.

(a) The bending moment M at section AE is $60 \times 10^3(0.3) = 18$ kN·m (Fig. 8.6a). The normal stresses at points A and B are thus

$$\sigma_A = \frac{Mc}{I} = \frac{18(10^3)(0.08)}{13.4 \times 10^{-6}} = 107.5 \text{ MPa}$$

$$\sigma_B = \frac{69.7}{80}(107.5) = 93.7 \text{ MPa}$$

as shown in Fig. 8.6b. The shearing stresses at points B and C (Fig. 8.6c) are

$$\tau_B = \frac{VQ_B}{Ib} = \frac{60(10^3)(78.6 \times 10^{-6})}{13.4(10^{-6})(0.0066)} = 53.3 \text{ MPa}$$

$$\tau_C = \frac{VQ_C}{Ib} = \frac{60(10^3)(94.6 \times 10^{-6})}{13.4(10^{-6})(0.0066)} = 64.2 \text{ MPa}$$

Applying Eqs. (4.6) and (4.8), we obtain the maximum principal stress and the maximum shearing stress at the junction of flange and web:

$$(\sigma_1)_B = \frac{93.7}{2} + \sqrt{\left(\frac{93.7}{2}\right)^2 + (53.3)^2} = 46.9 + 71 = 117.9 \text{ MPa}$$

$$(\tau_{max})_B = 71 \text{ MPa}$$

It is observed that the maximum normal stress does *not* occur at the extreme fibers.

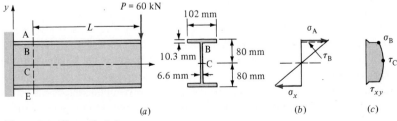

(a)   (b)   (c)

Figure 8.6 Example 8.4.

(b) The bending moment M at section AE is $60 \times 10^3(0.5)$, or 30 kN·m. The flexural stresses are thus

$$\sigma_A = \frac{Mc}{I} = \frac{30 \times 10^3(0.08)}{13.4 \times 10^{-6}} = 179.1 \text{ MPa}$$

$$\sigma_B = \frac{69.7}{80}(179.1) = 156 \text{ MPa}$$

while the shearing stresses remain constant. Next, at the juncture of the web with the flanges, we obtain

$$(\sigma_1)_B = \frac{156}{2} + \sqrt{\left(\frac{156}{2}\right)^2 + (53.3)^2} = 78 + 94.5 = 172.5 \text{ MPa}$$

$$(\tau_{max})_B = 94.5 \text{ MPa}$$

In this case, the maximum normal stress occurs at the extreme fibers of the beam.

In general, only for very short wide-flange beams and unusual situations need the principal stresses be analyzed. Otherwise, the maximum flexural stresses are an order of magnitude *larger* than the maximum shearing stresses that are probably found at the same point.

*8.6 ASYMMETRIC BENDING

The analysis of bending in the preceding two sections was limited to beams with at least one longitudinal plane of symmetry and subjected to loads acting on that plane. We now consider the asymmetric bending of linearly elastic beams that occurs when the cross sections are not symmetric or when the loads do not act in a plane of symmetry. Such a case is depicted in Fig. 8.7a. The bending plane of the applied bending moment M makes an angle α with the horizontal z axis. It is assumed that the y and z axes are the centroidal *principal axes* of the cross section and that the premises of the technical theory of bending (Sec. 7.7) apply.

To derive expressions for the flexural stresses, we begin by resolving the applied moment into components M_y and M_z along the y and z axes, respectively:

$$M_y = M \sin \alpha \qquad M_z = M \cos \alpha \tag{a}$$

Two alternative representations of the above are shown in Fig. 8.7b and c.

Note that when the bending moment of a beam is caused by applied *transverse forces*, it is often more convenient to use the following equivalent procedure. First these forces are resolved into two components, one in each centroidal principal plane. Next the moments (M_y and M_z) caused by these components about the respective axes are found.

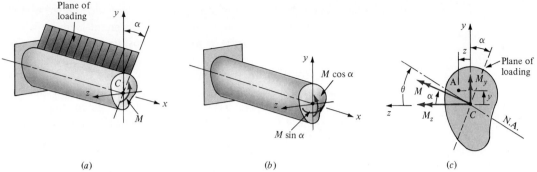

Figure 8.7 Asymmetric bending of a beam.

        (a)   (b)   (c)

As the bending moments M_y and M_z act about the principal axes, the corresponding bending stresses can be determined from the flexure formula, Eq. (7.12). The *combined flexural stress* at a point A having coordinates y and z in the cross section (Fig. 8.7c) is then obtained by applying the principle of superposition:

$$\sigma_x = -\frac{M_z y}{I_z} + \frac{M_y z}{I_y} \qquad (8.8)$$

Here I_y and I_z are the moments of inertia of the cross section about the principal centroidal axes y and z, respectively. The sign of the first term on the right side agrees with that of Eq. (7.12). On the other hand, the second term is taken positive to obtain the correspondence in sign among the normal stresses and the sense of the positive moment M_y. Note that in most problems, by considering physical action on the beam, we can directly assign the sign of each term in the above formula.

The equation of the *neutral axis* of the cross section is found by setting $\sigma_x = 0$ in Eq. (8.8). In so doing, we have

$$y = \frac{M_y I_z}{M_z I_y} z \qquad (8.9)$$

The angle ϕ that this straight line forms with the z axis (Fig. 8.7c) is given by the relation

$$\tan \phi = \frac{y}{z} = \frac{M_y I_z}{M_z I_y} \qquad (8.10a)$$

After we substitute for M_y and M_z from Eqs. (a), Eq. (8.10a) becomes

$$\tan \phi = \frac{I_z}{I_y} \tan \alpha \qquad (8.10b)$$

The largest bending stress occurs at a point located *farthest* from the neutral axis. Inasmuch as I_y and I_z are both positive, ϕ and α are of identical signs. These angles (measured from the z axis) are positive in the clockwise direction.

We observe from Eq. (8.10b) that $\phi > \alpha$ if $I_z > I_y$, and $\phi < \alpha$ if $I_z < I_y$. Thus the neutral axis is always located between the applied moment vector M and the principal axis associated with the minimum moment of inertia. Clearly, unless $I_z = I_y$, the neutral axis is not perpendicular to the plane of loading.

EXAMPLE 8.5

A rectangular, simply supported beam of length $L = 2$ m carries a load of $P = 10$ kN whose plane of action is inclined at $\alpha = 30°$ to the y axis, as shown in Fig. 8.8. Determine (a) the maximum bending stress in the beam and (b) the orientation of the neutral axis.

Solution   The centroidal principal moments of inertia of the cross section are

$$I_z = \tfrac{1}{12}(0.06)(0.09)^3 = 3.64 \times 10^{-6} \text{ m}^4$$

$$I_y = \tfrac{1}{12}(0.09)(0.06)^3 = 1.62 \times 10^{-6} \text{ m}^4$$

(a) The load components are $P \cos \alpha$ in the negative y direction and $P \sin \alpha$ in the positive z direction. Thus the maximum bending moments occur on the cross section at midspan and are expressed by

$$M_z = \tfrac{1}{4}PL \cos \alpha \qquad M_y = \tfrac{1}{4}PL \sin \alpha \qquad (b)$$

Substituting the data, we have (Fig. 8.9a)

$$M_z = \tfrac{1}{4}(10)(2) \cos 30° = 4.3 \text{ kN·m}$$

$$M_y = \tfrac{1}{4}(10)(2) \sin 30° = 2.5 \text{ kN·m}$$

Now application of Eq. (8.8) leads to

$$\sigma_A = -\frac{M_z y}{I_z} + \frac{M_y z}{I_y} = -\frac{4300(-0.045)}{3.64 \times 10^{-6}} + \frac{2500(0.03)}{1.62 \times 10^{-6}}$$

$$= 53.2 + 46.3 = 99.5 \text{ MPa}$$

$$\sigma_B = 53.2 - 46.3 = 6.9 \text{ MPa}$$

$$\sigma_D = -53.2 + 46.3 = -6.9 \text{ MPa}$$

$$\sigma_E = -53.2 - 46.3 = -99.5 \text{ MPa}$$

Figure 8.8   Example 8.5.

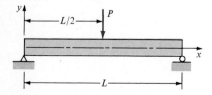

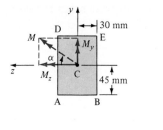
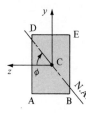
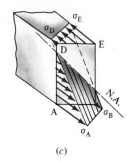

(a)   (b)   (c)

Figure 8.9   Example 8.5.

The maximum flexural stress is therefore 99.5 MPa.

(b) The angle ϕ that the neutral surface forms with the horizontal plane (Fig. 8.9b) is calculated from Eq. (8.10):

$$\tan \phi = \frac{I_z}{I_y} \tan \alpha = \frac{3.64}{1.62} \tan 30° = 1.297$$

$$\phi = 52.4°$$

The stress distribution across the cross section at the midspan is depicted in Fig. 8.9c.

EXAMPLE 8.6

A positive bending moment of 6 kN·m acts in a vertical plane and is applied to a steel beam having the Z-shaped cross section (Fig. 8.10a). Dimensions are in millimeters. The moments of inertia with respect to the y and z axes are $I_y = 6.93 \times 10^6$ mm^4, $I_z = 24.03 \times 10^6$ mm^4, and $I_{yz} = 9.45 \times 10^6$ mm^4. Determine (a) the bending stress at point A and (b) the orientation of the neutral axis.

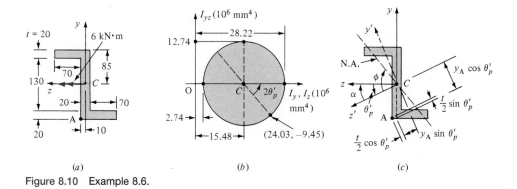

(a)   (b)   (c)

Figure 8.10   Example 8.6.

Solution The principal moments of inertia and their orientation can be determined by using Mohr's circle (Fig. 8.10b), from which we obtain

$$I_1 = 28.22 \times 10^6 \text{ mm}^4$$
$$I_2 = 2.74 \times 10^6 \text{ mm}^4$$
$$\theta_p' = 23.9°$$

The orientation of principal axes y' and z' is shown in Fig. 8.10c. It is observed that $I_1 = I_{z'}$ and $I_2 = I_{y'}$.

(a) The loading is resolved into components parallel to the y' and z' axes as follows:

$$M_{y'} = M_z \sin \theta_p' = 6000 \sin 23.9° = 2431 \text{ N·m}$$
$$M_{z'} = M_z \cos \theta_p' = 6000 \cos 23.9° = 5486 \text{ N·m}$$

Also, the coordinates y_A' and z_A' of point A are (Fig. 8.10c)

$$y_A' = -y_A \cos \theta_p' + (t/2) \sin \theta_p' \tag{c}$$
$$= -0.085 \cos 23.9° + 0.01 \sin 23.9° = -73.7 \text{ mm}$$

$$z_A' = y_A \sin \theta_p' + (t/2) \cos \theta_p' \tag{d}$$
$$= 0.085 \sin 23.9° + 0.01 \cos 23.9° = 43.6 \text{ mm}$$

The stress at point A, calculated from Eq. (8.8), is

$$\sigma_A = -\frac{M_{z'} y_A'}{I_{z'}} + \frac{M_{y'} z_A'}{I_{y'}} = -\frac{5486(-0.0737)}{28.22 \times 10^{-6}} + \frac{2431(0.0436)}{2.74 \times 10^{-6}}$$

$$= 14.32 + 38.68 = 53 \text{ MPa}$$

(b) Applying Eq. (8.10b), we obtain the angle ϕ that locates the neutral axis:

$$\tan \phi = \frac{I_{z'}}{I_{y'}} \tan \theta_p' = \frac{28.22}{2.74} \tan 23.9° = 4.564$$

from which $\phi = 77.6°$, as shown in Fig. 8.10c.

8.7 ECCENTRIC AXIAL LOADS

In this section we shall analyze the distribution of stresses when the line of action of an axial load does *not* pass through the centroid of the cross section. This *off-center*, or *eccentrically loaded*, case is shown in Fig. 8.11a for an axial tensile force P acting at distances y_0 and z_0 from the principal centroidal axes z and y, respectively.

The eccentric load P is statically equivalent to a force P applied at the centroid plus two bending moments,

$$M_y = Pz_0 \qquad M_z = Py_0 \tag{a}$$

about the centroidal axes (Fig. 8.11b). The *combined normal stress* at any point (y, z) of the cross section can now be obtained by adding an axial

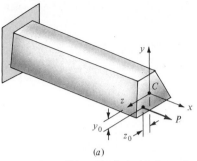
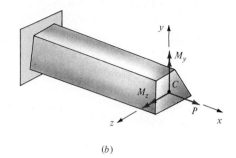

(a)   (b)

Figure 8.11 Eccentrically loaded member.

stress term to Eq. (8.9). Hence we have

$$\sigma_x = \frac{P}{A} - \frac{M_z y}{I_z} + \frac{M_y z}{I_y} \tag{8.11}$$

If the axial load is compressive, the value of P above is negative. Clearly, in this equation y and z can assume both positive and negative values, depending upon the location of σ_x. Equation (8.11) applies to members of any cross-sectional shape, provided that the y and z are the principal axes.

The *equation of the neutral axis* of the section can be determined by setting $\sigma_x = 0$ in Eq. (8.11):

$$\frac{P}{A} - \frac{M_z y}{I_z} + \frac{M_y z}{I_y} = 0 \tag{8.12}$$

This *straight line* is analogous to the neutral axis appropriate to pure bending. It is the line about which the plane section rotates. With $P \neq 0$, the neutral axis does *not* pass through the centroid of a section. For small eccentricities, y_0 and z_0 of the applied axial load P, the neutral axis lies outside the cross section. Equation (8.11) shows that the distribution of stresses across the section is linear. The largest stress occurs at the point most remote from the neutral axis.

Note that occasionally eccentric axial loading occurs in a *plane of symmetry* (such as xy). In this case, one of the bending moment components (M_y) is zero, and Eq. (8.11) for *combined normal stress* reduces to

$$\sigma_x = \frac{P}{A} - \frac{My}{I} \tag{8.13}$$

Then Eq. (8.12) of the *neutral axis* becomes

$$\frac{P}{A} - \frac{My}{I} = 0 \tag{8.14}$$

Here M and I are, respectively, the applied bending moment and the moment of inertia of the section about the principal centroidal axes.

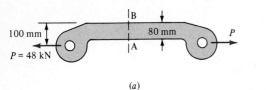

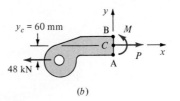

(a)

(b)

Figure 8.12 Example 8.7.

EXAMPLE 8.7

A 20-mm and 80-mm link of rectangular cross section is loaded by a force of $P = 48$ kN, as shown in Fig. 8.12a. Calculate (a) the maximum normal stress on section A-B and (b) the location of the neutral axis.

Solution A free body of the left portion of the link is shown in Fig. 8.12b. The internal forces in the cross section are equivalent to a centric force $P = 48$ kN and a bending moment $M = 2.88$ kN·m.
(a) The axial stress, uniformly distributed in cut A-B (Fig. 8.13a), is

$$\sigma_a = \frac{P}{A} = \frac{48 \times 10^3}{0.02 \times 0.08} = 30 \text{ MPa}$$

The stress distribution owing to bending moment M is linear, with its maximum value at the edges of section A-B (Fig. 8.13b):

$$\sigma_b = \frac{Mc}{I} = \frac{2880(0.04)}{0.02(0.08^3)/12} = 135 \text{ MPa}$$

Hence, using Eq. (8.13), or superposing the two distributions, we obtain the greatest tensile and compressive stresses, respectively, for the elements at the edges:

$$\sigma_A = \sigma_a + \sigma_b = 30 + 135 = 165 \text{ MPa}$$

$$\sigma_B = \sigma_a - \sigma_b = 30 - 135 = -105 \text{ MPa}$$

The resultant stress distribution for section A-B is depicted in Fig. 8.13c.
(b) The distance y_n between the centroid and the neutral axis (Fig. 8.13c) can be obtained from Eq. (8.14):

$$0 = \frac{P}{A} - \frac{My_n}{I} = 30(10^6) - \frac{2880y_n}{0.02(0.08^3)/12}$$

The foregoing yields $y_n = 8.9$ mm.

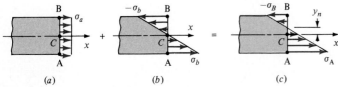

(a)

(b)

(c)

Figure 8.13 Example 8.7.

EXAMPLE 8.8

What is the largest force P that can be supported by the block shown in Fig. 8.14a if the allowable stresses on plane ABDE are $(\sigma_c)_{all} = 80$ MPa in compression and $(\sigma_t)_{all} = 20$ MPa in tension?

Solution The eccentric load is replaced by a centric force P and two moments, $M_y = 0.05P$ N·m and $M_z = 0.03P$ N·m (Fig. 8.14b). The area and the moments of inertia of the cross section are

$$A = 0.06 \times 0.1 = 0.006 \text{ m}^2$$
$$I_y = \tfrac{1}{12}(0.06)(0.1)^3 = 5 \times 10^{-6} \text{ m}^4$$
$$I_z = \tfrac{1}{12}(0.1)(0.06)^3 = 1.8 \times 10^{-6} \text{ m}^4$$

The stress due to the centric axial load P is

$$\sigma_a = \frac{P}{A} = \frac{-P}{0.006} = -166.67P$$

The maximum values of stress owing to bending moments M_y and M_z are, respectively,

$$\sigma_b' = \frac{M_y z_{max}}{I_y} = \frac{0.05P(0.05)}{5 \times 10^{-6}} = 500P$$

$$\sigma_b'' = \frac{M_z y_{max}}{I_z} = \frac{0.03P(0.03)}{1.8 \times 10^{-6}} = 500P$$

Hence Eq. (8.11) yields the combined normal stresses for the corner elements:

$$\sigma_x = \sigma_a \pm \sigma_b' \pm \sigma_b''$$

In the foregoing situation, the signs must be assigned by referring to *sense of the moments* depicted in Fig. 8.14b. Thus the largest values of the compressive and tensile stresses at corners A and E are

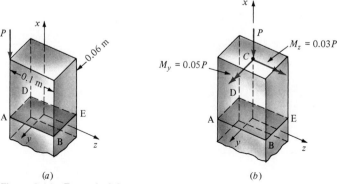

(a) (b)

Figure 8.14 Example 8.8.

$$\sigma_A = -166.67P - 500P - 500P = -1166.67P$$
$$\sigma_E = -166.67P + 500P + 500P = 833.33P$$

Substitution of the permissible stresses into the above results in

$$P = \frac{-80 \times 10^6}{-1166.67} = 68.6 \text{ kN} \qquad P = \frac{20 \times 10^6}{833.33} = 24 \text{ kN}$$

The magnitude of the maximum allowable eccentric load is the smaller of these values, $P = 24$ kN.

EXAMPLE 8.9

A cantilever having a channel section is loaded as shown in Fig. 8.15a. Calculate (a) the normal stress at A and B and (b) the principal stresses and the maximum shearing stress at D.

Solution We replace the 25-kN load by an equivalent 15-kN horizontal force and 450-N·m bending moment at the centroid C and a 20-kN vertical force at the free end. Next the stress resultants are found at section A-B (Fig. 8.15b). Referring to Secs. A.1 and A.3, one readily calculates

$$A = 0.0024 \text{ m}^2 \qquad \bar{y} = 50 \text{ mm} \qquad I = 136 \times 10^{-8} \text{ m}^4$$

Note that the properties of this channel section about the z axis are the same as those of the T section (see Fig. 7.19). Both sections possess an axis of symmetry.
(a) Applying Eq. (8.13), we find that the combined normal stresses for the corner elements are

$$\sigma_A = -\frac{P}{A} + \frac{My_A}{I} = -\frac{15 \times 10^3}{0.0024} + \frac{1950(0.03)}{136 \times 10^{-8}}$$
$$= -6.25 + 43.01 = 36.8 \text{ MPa}$$

and

$$\sigma_B = -\frac{P}{A} - \frac{My_B}{I} = -6.25 - \frac{1950(0.05)}{136 \times 10^{-8}} = -77.9 \text{ MPa}$$

(b) The normal and shearing stresses at point D (Fig. 8.15b) are, respectively,

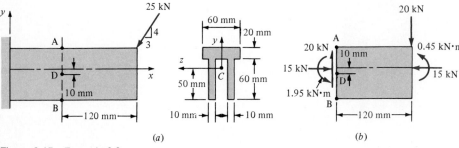

(a)

(b)

Figure 8.15   Example 8.9.

$$\sigma_D = -\frac{P}{A} - \frac{My_D}{I} = -6.25 - \frac{1950(0.01)}{136 \times 10^{-8}} = -20.6 \text{ MPa}$$

$$\tau_D = \frac{VQ}{Ib} = \frac{20 \times 10^3 [2(0.01)(0.04)(0.03)]}{136 \times 10^{-8}(0.02)} = 17.6 \text{ MPa}$$

The principal stresses are determined from Eq. (4.6):

$$\sigma_{1,2} = -10.3 \pm \sqrt{(-10.3)^2 + (17.6)^2} = -10.3 \pm 20.4$$

Therefore $\sigma_1 = 10.1$ MPa, $\sigma_2 = -30.7$ MPa, and $\tau_{max} = 20.4$ MPa.

*8.8 THE SHEAR CENTER

For any cross-sectional shape, a single point may be found in the plane of the cross section through which passes the resultant of the transverse shearing stresses. This point is called the *shear center*. A transverse load applied on a beam must act through the shear center if only bending is to take place. However, when the load is not applied through the shear center, both bending and twisting actions occur. In that case, torsional stresses must be investigated in addition to flexural and direct shearing stresses.

The shear center for a section with *one* axis of symmetry lies on this axis. For a section possessing *two* axes of symmetry, the shear center is located at the intersection of the two axes, that is, at the centroid. Note that for *solid* sections and *closed* hollow sections, the shear center is often near the centroid. Such sections have high torsional rigidity. Thus the effects of twisting can usually be neglected when the transverse load is applied at or near the centroid.

We shall confine our treatment to *thin-walled* open sections—for example, I beams, Z sections, channels, and angles. Here shear stresses are assumed to be distributed *uniformly* over the wall thickness and to act tangentially to the beam surface. To find the shear center S for a typical section, as shown in Fig. 8.16a, we begin by calculating the shearing stresses as described in Sec. 7.12:

$$\tau_{xy} = \frac{V_y Q_z}{I_z b} \qquad (a)$$

where $Q_z = \int_{A*} y \, dA$ and $A*$ is the area beyond the point at which τ_{xy} is obtained. The y and z axes are the principal centroidal axes of the cross section. The moment M_x of these stresses with respect to arbitrary point A is next obtained. The external moment due to the applied shear load V_y about A equals $V_y e_z$. The distance between A and the shear center is found by applying the principle of moments:

$$e_z = \frac{M_x}{V_y} \qquad (8.15)$$

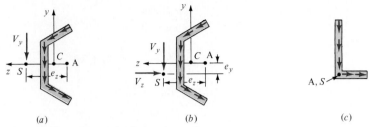

Figure 8.16 Shear center for angle sections.

If the applied force is z-directed, the line of action may be located in a manner similar to that described above. It is noted that the shearing stress formula is now of the form

$$\tau_{xz} = \frac{V_z Q_y}{I_y b} \qquad (b)$$

in which $Q_y = \int_{A*} z \, dA$. In the situation where *both* V_y and V_z exist, the shear center is at the intersection of the two lines of action (Fig. 8.16b). The shearing stresses due to these *combined loads* may be obtained by using the principle of superposition.

The determination of M_x is facilitated by judicious selection of point A, such as illustrated in Fig. 8.16c where we observe that the moment M_x of the shear stresses about A is zero; point A coincides with the shear center S. Clearly, for any section consisting of two intersecting rectangular elements, the shear center is located at the point of intersection.*

EXAMPLE 8.10

Determine the distribution of shearing stresses and the shear center S of a channel section of uniform thickness loaded as a cantilever by a vertical shear force V_y (Fig. 8.17a).

Solution The shearing stress in flange DE at distance s from E is found first (Fig. 8.17a). At E, the shear stress is zero. The first moment of area st about the z axis is $Q_z = stc$. Equation (a) therefore becomes

$$\tau_{xy} = V_y \frac{sc}{I_z} \qquad (c)$$

with a peak value at $s = b$:

$$\tau_D = V_y \frac{bc}{I_z} \qquad (8.16)$$

---

*Several references are available for readers seeking a more thorough treatment; see, for example, P. Kuhn, *Stresses in Aircraft and Shell Structures*, McGraw-Hill, New York, 1956.

Owing to the symmetry of the section, $\tau_A = \tau_D$. The stress varies parabolically over the web, as in the case of the I beam discussed in Sec. 7.14. The largest shearing stress

$$\tau_{max} = \frac{3V_y(4b + h)}{2th(6b + h)} \tag{8.17}$$

occurs at the neutral axis (see part c, Prob. 8.76). The distribution of shear stress over the entire channel section is sketched in Fig. 8.17b.

As the shearing stress varies along the flange length, the flange force, using Eq. (8.16), is

$$F_{AB} = F_{DE} = \tfrac{1}{2}\tau_D bt = V_y \frac{b^2 ct}{2I_z} \tag{d}$$

The vertical shear force transmitted by the flange is negligibly small (see Example 7.17), and we can therefore take $F_{AB} = V_y$ (Fig. 8.17c).

Next, applying Eq. (8.15) at A, we have $M_x = 2F_{DE}c = Pe$. Substitution of Eq. (d) into this expression yields

$$e = \frac{b^2 c^2 t}{I_z} = \frac{b^2 h^2 t}{4I_z} \tag{e}$$

Here the moment of inertia I_z of the channel section is

$$I_z = \tfrac{1}{12}th^3 + 2(\tfrac{1}{12}bt^3 + \tfrac{1}{4}bth^2) \tag{f}$$

Then, neglecting the terms containing t^3 as very small, we turn Eq. (f) into

$$I_z = \tfrac{1}{12}th^3 + \tfrac{1}{2}bth^2 \tag{8.18}$$

and the shear center is therefore located by

$$e = \frac{3b^2}{6b + h} = \frac{b}{2 + (h/3b)} \tag{8.19}$$

Note that e is independent of t and may vary from 0 to $b/2$, according to the value of ratio $h/3b$.

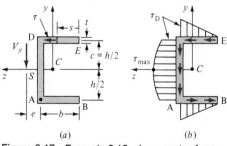

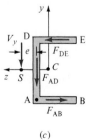

(a)   *(b)*   *(c)*

Figure 8.17 Example 8.10: shear center for a channel section.

*8.9 YIELD AND FRACTURE CRITERIA

The mechanical behavior of materials subject to *uniaxial loads* or *pure tension* is graphically presented on stress-strain diagrams. In these cases, the onset of failure by yielding or fracture can be predicted readily with acceptable accuracy. However, most structures are subject to a variety of *combined loads*. Several theories of failure have been developed for predicting failure of materials in these latter situations.

Detailed discussion of fracture criterion is beyond the scope of this text, and we consider here only a simple fracture theory and two yield theories. In our discussion of these theories, we denote the ultimate stress obtained in a tension test by σ_u and in a compression test by σ_u'. The yield-point stress determined from a tensile test is designated by σ_{yp}. In the development which follows, we consider an element subjected to triaxial principal stresses where $\sigma_1 > \sigma_2 > \sigma_3$. (Recall that subscripts 1, 2, and 3 refer to the principal directions.)

Maximum Principal Stress Theory. The maximum principal stress theory, or *Rankine theory*, states that fracture occurs when the principal stress at a point in the structure reaches the ultimate stress in simple tension or compression for the material. It follows that fracture impends when

$$|\sigma_1| = \sigma_u \qquad \text{or} \qquad |\sigma_3| = \sigma_u' \qquad (8.20)$$

For materials possessing the same ultimate strength in tension and compression ($\sigma_u = \sigma_u'$), in the case of *plane stress* ($\sigma_3 = 0$), Eq. (8.20) becomes

$$|\sigma_1| = \sigma_u \qquad \text{or} \qquad |\sigma_2| = \sigma_u \qquad (8.21)$$

The above expression is plotted in Fig. 8.18. Failure will occur for any combination of stresses on or outside the boundaries; no fracture occurs for a combination of stresses inside the square.

Experiments show that this theory can predict fracture failures reasonably well for brittle materials (particularly in quadrant 1 in Fig. 8.18), and the Rankine theory is generally accepted in design practice for such materials.

Maximum Shear Stress Theory. The maximum shear stress theory, or the *Tresca* or *Guest theory*, asserts that yielding begins when the maximum shearing stress equals the maximum shearing stress at the yield point in a simple tension test. The largest value of the shear stress is $\tau_{max} = \frac{1}{2}|\sigma_1 - \sigma_3|$. In uniaxial tension, $\sigma_2 = \sigma_3 = 0$ and $\tau_{max} = \sigma_{yp}/2$. Therefore, at the onset of yielding,

$$\tau_{max} = \sigma_{yp} \qquad \text{or} \qquad |\sigma_1 - \sigma_3| = \sigma_{yp} \qquad (8.22)$$

In the case of *plane stress* ($\sigma_3 = 0$), when σ_1 and σ_2 are of *opposite sign* (that is, one tensile, the other compressive), the yield condition is given by

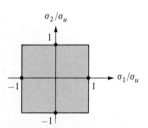

Figure 8.18 Fracture criterion based on maximum principal stress.

$$|\sigma_1 - \sigma_2| = \sigma_{yp} \tag{8.23a}$$

When σ_1 and σ_2 carry the *same sign,* the maximum shearing stress is half the numerically larger principal stress σ_1 or σ_2. Thus, the criterion corresponding to this situation is

$$|\sigma_1| = \sigma_{yp} \quad \text{or} \quad |\sigma_2| = \sigma_{yp} \tag{8.23b}$$

Equations (8.23) are depicted in Fig. 8.19. Note that Eq. (8.23a) applies to the second and fourth quadrants. In the first and third quadrants the criteria are expressed by Eqs. (8.23b). The boundaries of the hexagon mark the onset of yielding, with points outside the shaded region representing the yielded state.

The maximum shear stress theory is frequently applied in machine design because it is slightly conservative and is easy to apply. Good agreement with experiment has been realized for ductile materials.

Maximum Energy of Distortion or von Mises Theory. This theory states that yielding occurs when the root mean square of the difference between the principal stresses for a three-dimensional state of stress reaches the same value in a tensile test. Thus at the beginning of yielding,

$$\{\tfrac{1}{3}[(\sigma_1 - \sigma_2)^2 + (\sigma_2 - \sigma_3)^2 + (\sigma_3 - \sigma_1)^2]\}^{1/2}$$
$$= \{\tfrac{1}{3}[(\sigma_{yp} - 0)^2 + (0 - 0)^2 + (0 - \sigma_{yp})^2]\}^{1/2} = (\tfrac{2}{3})^{1/2}\sigma_{yp}$$

or

$$(\sigma_1 - \sigma_2)^2 + (\sigma_2 - \sigma_3)^2 + (\sigma_3 - \sigma_1)^2 = 2\sigma_{yp}^2 \tag{8.24}$$

This result is also obtained by equating the energy of distortion for a general state of stress to the limiting value determined from a tensile test. [These relations will be presented in Eqs. (11.6) and (11.7c).]

In the case of *plane stress* ($\sigma_3 = 0$), Eq. (8.24) reduces to

$$\sigma_1^2 - \sigma_1\sigma_2 + \sigma_2^2 = \sigma_{yp}^2 \tag{8.25}$$

The foregoing defines the ellipse shown in Fig. 8.20. Points within the surface represent the states of nonyielding.

The von Mises theory agrees best with the test data for ductile materials and is in common use in design.

Note that a convenient comparison of the above-described theories may be made by means of a superposition of Figs. 8.18 through 8.20.

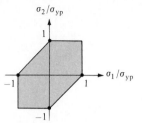

Figure 8.19 Yield criterion based on maximum shearing stress.

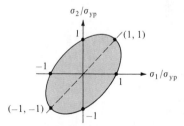

Figure 8.20 Yield criterion based on maximum distortion energy.

EXAMPLE 8.11

A closed-end cylinder, 2 ft in diameter and $\tfrac{1}{2}$-in. thick, is fabricated of steel of tensile strength $\sigma_{yp} = 36$ ksi. Calculate the allowable pressure the shell can carry based upon a factor of safety $f_s = 2$. Apply the two yielding theories of failure discussed above.

Solution   The circumferential, axial, and radial stresses are given by

$$\sigma_1 = \frac{pr}{t} = 24p \qquad \sigma_2 = \frac{pr}{2t} = 12p \qquad \sigma_3 = 0$$

Insertion of these expressions into Eqs. (8.22) and (8.25) will provide the critical pressures: For the *maximum shearing stress theory*:

$$24p - 0 = \tfrac{1}{2}(36 \times 10^3)$$
$$p = 750 \text{ psi}$$

For the *maximum energy of distortion theory*:

$$p(24^2 - 24 \times 12 + 12^2)^{1/2} = \tfrac{1}{2}(36 \times 10^3)$$
$$p = 866 \text{ psi}$$

The permissible value of the internal pressure is thus limited to 750 psi.

EXAMPLE 8.12

A circular shaft of diameter d and tensile yield strength σ_{yp} is subjected to combined axial tensile force P and torque T. Determine the moment M that can also be applied simultaneously to the shaft. Use a factor of safety f_s and employ the maximum energy of distortion theory of failure. Compute the value of M for the following data: $P = 100$ kN, $T = 5$ kN·m, $d = 60$ mm, $\sigma_{yp} = 300$ MPa, and $f_s = 1.2$.

Solution   For the situation described, the outer-fiber stresses in the shaft are

$$\sigma = \frac{My}{I} + \frac{P}{A} = \frac{32M}{\pi d^3} + \frac{4P}{\pi d^2}$$
$$\tau = \frac{Tr}{J} = \frac{16T}{\pi d^3} \tag{a}$$

On the basis of Eqs. (8.25) and (8.1) and the maximum energy of distortion theory, and using $\sigma_{all} = \sigma_{yp}/f_s$, we obtain

$$\sigma^2 + 3\tau^2 = (\sigma_{all})^2 \tag{8.26}$$

Introducing Eqs. (a) into Eq. (8.26), we derive the governing expression:

$$\left(\frac{32M}{\pi d^3} + \frac{4P}{\pi d^2} \right)^2 + 3 \left(\frac{16T}{\pi d^3} \right)^2 = (\sigma_{all})^2 \tag{8.27}$$

Upon simplification and rearrangement, the foregoing results in the quadratic equation in M:

$$M^2 + (0.25Pd)M + [0.75T^2 + 0.0156P^2 d^2 - 0.0096(\sigma_{all})^2 d^6] = 0$$

The valid solution is therefore

$$M = -\tfrac{1}{2}c_1 + \tfrac{1}{2}[c_1^2 - 4(c_2 + c_3 - c_4)]^{1/2} \tag{b}$$

where $c_1 = 0.25Pd$, $c_2 = 0.75T^2$, $c_3 = 0.0156P^2d^2$, and $c_4 = 0.0096(\sigma_{all})^2d^6$. (This formulation is suitable for computer solution.) When we substitute the numerical values, Eq. (b) leads to $M = 2.29$ kN·m.

*8.10   DESIGN OF TRANSMISSION SHAFTS

Shafts are used in a wide variety of machine applications. The design process for circular torsion members has been described in Sec. 6.8; we are now concerned with the members carrying loads of *combined bending and torsion*, which is the case for most transmission shafts. Though in practice design usually includes consideration of load fluctuation and stress concentration, as well as calculations for associated keys and couplings, these will be neglected in the ensuing *direct analytical procedure*, an approach of great significance in the design of power equipment.

Consider a solid circular shaft of diameter d, acted upon by bending moment M and torque T. To begin with, we determine the maximum bending and torsional shearing stresses occurring in the outer fibers at a *critical section*:

$$\sigma = \frac{32M}{\pi d^3} \qquad \tau = \frac{16T}{\pi d^3} \tag{a}$$

The principal stresses and the maximum shear stress, Eqs. (8.1) and (8.2), are then

$$\sigma_{1,2} = \frac{16}{\pi d^3}(M \pm \sqrt{M^2 + T^2}) \tag{8.28a}$$

and

$$\tau_{max} = \frac{16}{\pi d^3}\sqrt{M^2 + T^2} \tag{8.28b}$$

Here the moment, expressed in terms of its components in the two coordinate planes, is

$$M = \sqrt{M_y^2 + M_z^2} \tag{8.29}$$

Finally, Eqs. (8.28) are used with a selected failure criterion.

It follows that by assigning $\tau_{all} = \tau_{max}/f_s$, a design formula is obtained that is based upon the *maximum shearing stress theory* of failure:

$$d = \sqrt[3]{\frac{16}{\pi \tau_{all}}\sqrt{M^2 + T^2}} \tag{8.30}$$

where f_s represents the factor of safety. Similarly, setting $\sigma_{all} = \sigma_1/f_s$, in accordance with the *maximum principal stress theory* of failure, we have the following design formula:

$$d = \sqrt[3]{\frac{16}{\pi\sigma_{\text{all}}}\left(M + \sqrt{M^2 + T^2}\right)} \qquad (8.31)$$

Yet another expression, this one based upon the von Mises theory, may be obtained in a like manner.

It should be noted that the ASME code for design of transmission shafts makes proper allowance for shock and fatigue effects by inserting constants into Eqs. (8.30) and (8.31).* The code also recommends that the allowable stress in shear be taken at 56 MPa for ordinary steel and that when there is a keyway at the critical section, this stress be reduced by 25 percent. For the complete requirements, reference should be made to the current edition of the code.

EXAMPLE 8.13

A solid shaft is fitted with 300-mm-diameter pulleys, supported by frictionless bearings at A and B, and loaded as shown in Fig. 8.21a. If $\tau_{\text{all}} = 60$ MPa, calculate the required diameter of the shaft according to the maximum shear stress theory.

Solution A complete free-body diagram of the shaft is shown in Fig. 8.21b. The determination of the largest value of $(M_z^2 + M_y^2 + T^2)^{1/2}$ is facilitated by use of moment and the torque diagrams (Fig. 8.21c). At point C, we find

*For more detailed discussion, consult M. F. Spotts, *Design of Machine Elements*, 5th ed., Prentice-Hall, Englewood Cliffs, N.J., 1978.

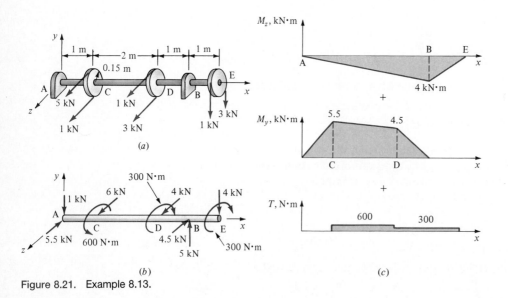

Figure 8.21. Example 8.13.

$$(M_z^2 + M_y^2 + T^2)^{1/2} = (1^2 + 5.5^2 + 0.6^2)^{1/2} = 5.62 \text{ kN·m}$$

while at D and B, we find 5.44 kN·m and 4.01 kN·m, respectively. Hence the critical section is at C.

Applying Eq. (8.30), we thus obtain

$$d = \sqrt[3]{\frac{16(5.62 \times 10^3)}{\pi(60 \times 10^6)}} = 78.1 \text{ mm}$$

A commercial-size shaft of 80-mm diameter should be used.

EXAMPLE 8.14

A circular cast-iron shaft ($\sigma_u = 140$ MPa and $\sigma_u' = 560$ MPa), rotating at 600 rpm and transmitting 40 kW is subjected to a moment of 400 N·m. Using a safety factor of 2, determine the required diameter of the shaft on the basis of the maximum principal stress theory.

Solution Noting that $f = 600$ rpm $= 10$ Hz, we calculate the shaft torque:

$$T = \frac{159 \text{ kW}}{f} = \frac{159(40)}{10} = 636 \text{ N·m}$$

The allowable stresses in tension and compression are 70 and 280 MPa, respectively. Introducing the data given into Eq. (8.31), we obtain

$$d = \sqrt[3]{\frac{16}{\pi(70 \times 10^6)}} (400 + \sqrt{400^2 + 636^2}) = 44 \times 10^{-3} \text{ m} = 44 \text{ mm}$$

and

$$d = \sqrt[3]{\frac{16}{\pi(280 \times 10^6)}} (400 + \sqrt{400^2 + 636^2}) = 28 \times 10^{-3} \text{ m} = 28 \text{ mm}$$

Hence the allowable diameter is 44 mm.

PROBLEMS

Secs. 8.1 to 8.5

8.1 A cylinder with 4-in. and 6-in. internal and external diameters is subjected to an axial compressive load of 20π kips and a torque of 15π kip·in. Calculate the maximum principal stress and the maximum shearing stress. Show the results on a properly oriented element.

8.2 A hollow shaft with 50-mm outer diameter and 30-mm inner diameter is acted upon by an axial tensile load of 40 kN, a torque of 500 N·m, and a bending moment of 200 N·m. Use Mohr's circle to determine the principal stresses and the planes on which they act.

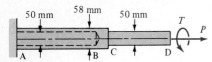

Figure P8.4

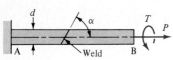

Figure P8.9

8.3   Redo Prob. 8.2 for a 40-mm-diameter shaft made of solid steel and subjected to a zero axial load.

8.4   Determine the largest value of the axial load P that can be carried by the stepped steel shaft shown in Fig. P8.4 for $\sigma_{all} = 100$ MPa, $\tau_{all} = 60$ MPa, and $T = 0.01P$ N·m.

8.5   Repeat Prob. 8.4 for the case in which the 50-mm-diameter hole in the shaft is eliminated.

8.6   Determine the maximum shearing stress and its orientation in the stepped shaft shown in Fig. P8.4. Use $P = 30\pi$ kN and $T = 0.01P$ kN·m.

8.7   A closed-ended cylinder 10 in. in internal radius and $\frac{1}{2}$ in. thick is subjected to an internal pressure of 500 psi and an axial tension of 100 kips. Determine for the shell wall (*a*) the maximum tensile stress and (*b*) the maximum shearing stresses and the orientation of their planes.

8.8   A thin-walled cylinder of 250-mm inner radius and 10-mm wall thickness is acted upon by an internal pressure of 3 MPa and a torque of 40 kN·m. Calculate the largest shearing stress and its orientation in the wall of this closed-ended vessel.

8.9   Calculate the normal and shearing stresses on the spiral weld of the steel shaft loaded as shown in Fig. P8.9. The data are $P = 200$ kN, $T = 2$ kN·m, $d = 50$ mm, and $\alpha = 60°$.

8.10   A cantilever steel shaft is loaded as shown in Fig. P8.9. Given $\sigma_{all} = 20$ ksi and $\tau_{all} = 12$ ksi on the spiral weld, $d = 4$ in., $\alpha = 50°$, and $P = 40\pi$ kips, calculate the permissible value of T.

8.11   A thin-walled cylindrical pressure vessel of 0.4-m radius and 8-mm wall thickness has a welded spiral seam at an angle of 60° with the axial direction. The shell is subjected to an internal pressure of p pascals and an axial tensile load of 10π kilonewtons applied through rigid end plates. Calculate the permissible value of p if the normal and shear stresses acting simultaneously in the plane of welding are not to exceed 25 and -10 MPa, respectively.

8.12   A helical spring of 4-in. mean radius is fabricated of 1-in.-diameter steel bar ($\sigma_{all} = 40$ ksi). Determine the largest value of P using (*a*) Eq. (8.4) and (*b*) Eq. (8.5).

8.13   Redo Prob. 8.12 for $d = \frac{1}{4}$ in. and $D = 2\frac{1}{2}$ in.

8.14   A helical brass spring ($G = 35$ GPa) is subjected to a load of $P = 800$ N. Given the number of coils $n = 15$, $d = 10$ mm, and $R = 40$ mm (see Fig. 8.2*a*), calculate (*a*) the maximum shearing stress in the rod using Eqs. (8.4) and (8.5); (*b*) the deflection of the spring; and (*c*) the spring constant.

8.15   Rework Prob. 8.14 for a helical spring made of copper ($G = 41$ GPa) and with an R of 30 mm.

8.16 A steel helical spring fits within a brass helical spring. Each is constructed of 20-mm-diameter bar. For the brass spring G_b = 40 GPa and $(\tau_b)_{all}$ = 150 MPa, and for the steel spring G_s = 80 GPa and $(\tau_s)_{all}$ = 250 MPa. Each spring has the same number of coils n = 10, with ends constrained to deflect an identical amount. The brass spring has a mean radius of 100 mm, and the steel spring a mean radius of 70 mm. Calculate (a) the total permissible load the two springs can sustain jointly and (b) the ratio of the spring constants.

8.17 Determine the maximum normal stress and the maximum shear stress at a point B of the aluminum bar loaded as shown in Fig. 8.3a. Let a = 5 ft, d = 2 in., R = 40 lb, P = 20 kips, and T = 6 kip·in.

8.18 Redo Prob. 8.17 for the case in which the axial load is compressive, P = − 20 kips.

8.19 Find the required diameter d of the shaft shown in Fig. 8.3a for P = 0, R = 50π N, T = 120π N·m, L = 2 m, σ_{all} = 120 MPa, and τ_{all} = 70 MPa.

8.20 Determine the principal stresses and the maximum shearing stress at point D in the L-shaped bracket of the uniform thin-walled cross section shown in Fig. P8.20.

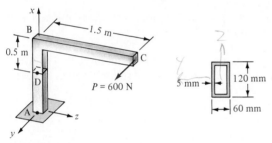

Figure P8.20

8.21 A steel shaft, 2 in. in diameter and rotating at 2400 rpm, is acted upon by a bending moment of 5 kip·in. Calculate the torque and the horsepower that can also be applied simultaneously to the shaft. Use τ_{all} = 9 ksi and σ_{all} = 14 ksi.

8.22 Using Fig. P8.22, calculate the maximum principal stress and the maximum shearing stress for a d = 5-mm-diameter shaft in equilibrium. The bearings at A and B are assumed to be frictionless.

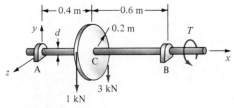

Figure P8.22

8.23 A solid 40-mm-diameter shaft is subjected to a bending moment of $M = 200\pi$ N·m. Determine the torque T that can also act if $\sigma_{all} = 120$ MPa and $\tau_{all} = 80$ MPa.

8.24 Determine the maximum shearing stress at point E on the surface at midspan of the 3-in.-diameter shaft of the assembly shown in Fig. P8.24. Bearings A and B are taken to be frictionless.

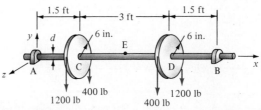

Figure P8.24

8.25 Redo Prob. 8.24 for the case in which the shaft is hollow, with a 3-in. outer diameter and a 2-in. inner diameter.

8.26 If the allowable normal stress at a point D in the simple beam of Fig. 8.4a is 10 MPa, determine the maximum permissible value of P.

8.27 An overhanging beam of rectangular cross section, 6 in. deep by 4 in. wide, is loaded as shown in Fig. P8.27. Determine the principal stresses at point D. Sketch the results on a properly oriented element.

8.28 Repeat Prob. 8.27 for a beam of S 8 × 23 cross section. (Refer to Table B.3.)

8.29 For a beam loaded as shown in Fig. P8.29, by taking into account the beam's own weight of 500 N/m, calculate the maximum normal and the maximum shearing stresses at point D of the cross section at midspan. Show the results on a properly oriented element.

Figure P8.27

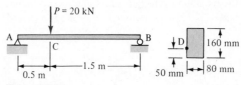

Figure P8.29

8.30 If the allowable shearing stress at a point B in the cantilever beam shown in Fig. 8.6a is 80 MPa, calculate the largest permissible value of P for $L = 2$ m.

8.31   Determine the value of α for which the bending stress in the cantilever beam, loaded as shown in Fig. P8.31, is a maximum. Use $b = 6$ in. and $h = 9$ in.

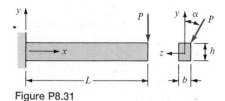

Figure P8.31

8.32   A cantilever beam is loaded as shown in Fig. P8.31. Given $L = 2$ m, $h = 2b = 100$ mm, $\alpha = 20°$, and $P = 500$ N, determine (*a*) the maximum normal stress and (*b*) the orientation of the neutral axis.

8.33 through 8.35   As depicted in cross section in Figs. P8.33 through P8.35, a beam is acted upon by a moment M with its vector forming an angle α with the horizontal axis. Calculate the stress at points A, B, and D.

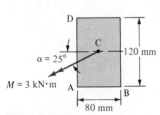

Figure P8.33

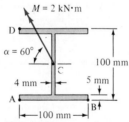

Figure P8.34

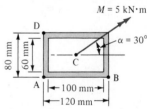

Figure P8.35

8.36 through 8.38   As shown in cross section in Figs. P8.36 through P8.38, a beam is subjected to a moment M with its vector forming an angle α with the horizontal axis. Calculate (*a*) the orientation of the neutral axis and (*b*) the maximum bending stress.

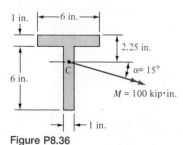

Figure P8.36

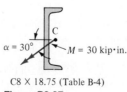

C8 X 18.75 (Table B-4)
Figure P8.37

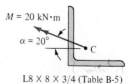

L8 X 8 X 3/4 (Table B-5)
Figure P8.38

8.39 Redo Prob. 8.37 for a C 4 × 5.4 channel (refer to Table B.4) with $M = 10$ kip·in.

8.40 Redo Prob. 8.38 for an L 5 × 3 × ¼ angle (refer to Table B.6) with $M = 15$ kip·in.

8.41 through 8.43 A beam of cross section shown (Figs. P8.41 through P8.43) is acted upon by a moment M with its vector forming an angle α with the horizontal axis. For $M = 2$ kN·m and $\alpha = 15°$, determine (a) the orientation of the neutral axis and (b) the maximum bending stress.

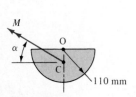

Figure P8.41

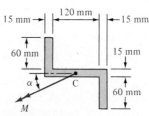

Figure P8.42

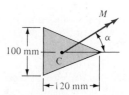

Figure P8.43

8.44 through 8.46 For the beam cross section shown in Figs. P8.41 through P8.43, determine the largest permissible value of the moment M for $\sigma_{all} = 100$ MPa and $\alpha = 0°$.

8.47 through 8.49 Redo Probs. 8.44 through 8.46 for $\sigma_{all} = 80$ MPa and $\alpha = 30°$.

8.50 As shown in Fig. P8.50, a simply supported beam, L meters long, has a Z section of uniform thickness for which $I_y = 2tc^3/3$, $I_z = 8tc^3/3$, and $I_{yz} = -tc^3$. Determine the maximum bending stress in the beam if it is subjected to a load P at midspan and $c = 10t$.

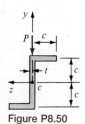

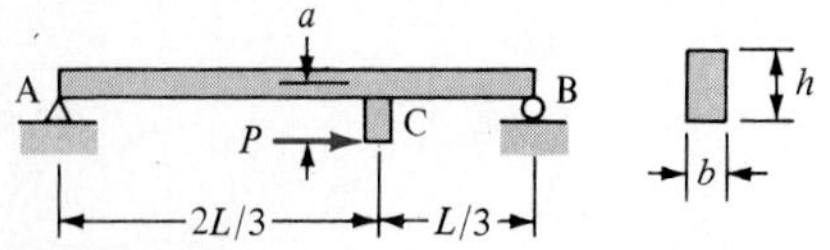

Figure P8.50

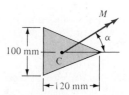

Figure P8.51

8.51 For the beam and loading shown in Fig. P8.51, determine the maximum tensile and compressive stresses if (a) $a = 2h$ and (b) $a = h/4$.

8.52 A closed-ended cylinder of 12-in. inside diameter and ½-in. wall thickness is subjected to an internal pressure of 500 psi and a 4-in. off-center axial load P. Determine the largest permissible value of P if $(\sigma_t)_{all} = 6$ ksi and $(\sigma_c)_{all} = 4.2$ ksi.

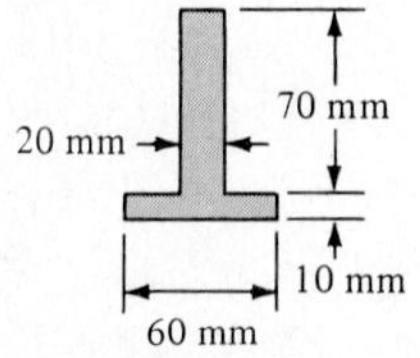

8.53 The link loaded as shown in Fig. 8.12a has a T cross section as shown in Fig. P8.53. Determine (a) the maximum stress and (b) the location of the neutral axis.

Section A-B
Figure P8.53

8.54 A W 410 × 60 beam supports an axial tensile load $P = 400$ kN, which does *not* pass through the centroid. What is the largest eccentricity (y_0) if $z_0 = 0$, $(\sigma_c)_{all} = 120$ MPa, $(\sigma_t)_{all} = 80$ MPa?

8.55 For the bracket of b by h rectangular cross section, loaded as shown in Fig. P8.55, determine (a) the maximum normal stress and (b) the location of the neutral axis. Use $P = 8$ kips, $b = 1$ in., and $h = 4$ in.

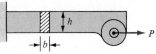

Figure P8.55

8.56 Determine the largest load P that the bracket of Fig. P8.55 can carry if $h = 6b = 6$ in. and $\sigma_{\text{all}} = 18$ ksi.

8.57 A typical cantilever steel beam ($E = 200$ GPa) is loaded as shown in Fig. P8.57. Given the normal strains $\varepsilon_A = 600 \mu$ and $\varepsilon_B = -200 \mu$ at the extreme fibers at points A and B, compute the magnitude of load P and distance d.

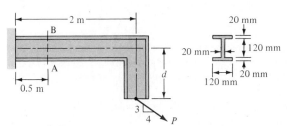

Figure P8.57

8.58 For the beam loaded as shown in Fig. P8.57, determine, for $P = 20$ kN and $d = 0.2$ m, (a) the maximum bending stress and (b) the location of the neutral axis.

8.59 Figure P8.59 shows a cast-iron C clamp. What is the largest permissible load P that may be applied if $(\sigma_t)_{\text{all}} = 35$ MPa, $(\sigma_c)_{\text{all}} = 50$ MPa, and $h = 2b = 100$ mm?

8.60 The C clamp in Fig. P8.59 is tightened until $P = 4$ kN. If $h = 1.6b = 80$ mm, calculate (a) the maximum normal stress and (b) the location of the neutral axis.

8.61 Figure P8.59 shows a steel C clamp ($\sigma_{\text{all}} = 140$ MPa). Given applied load $P = 2$ kN and $h = 60$ mm, calculate the required dimension b of the cross section.

8.62 A 50-mm wide by 100-mm deep bracket is acted upon by a load of $P = 20$ kN, as shown in Fig. P8.62. Determine the principal stresses and the maximum shearing stress at point A. Show the results on a properly oriented element.

8.63 Calculate the largest load P that can be applied to the bracket in Fig. P8.62 if the allowable normal stress at point B is 90 MPa.

8.64 Member ABC of the frame shown in Fig. P8.64 is made of a C 6 × 8.2 steel channel having $\sigma_{\text{yp}} = 36$ ksi. Refer to Table B.4 and determine the largest load P that can be supported by the member, based upon a factor of safety of $f_s = 1.2$. Neglect the effect of direct shear.

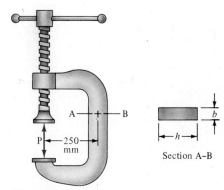

Figure P8.59

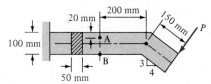

Figure P8.62

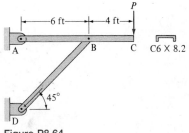

Figure P8.64

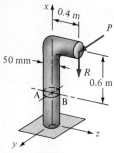

Figure P8.66

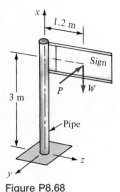

Figure P8.68

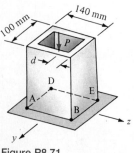

Figure P8.71

8.65 Determine the maximum normal stress in the member ABC of the structure shown in Fig. P8.64 for $P = 2$ kips.

8.66 Loads $P = 400\pi$ N and $R = 800\pi$ N are applied at the free end of the 50-mm-diameter post shown in Fig. P8.66. Calculate the principal stresses and the maximum shearing stress at (a) point A and (b) point B. Indicate the results on properly oriented elements.

8.67 Redo Prob. 8.66 for the case in which the vertically applied load R is zero.

8.68 A sign of weight $W = 1.5$ kN is supported by a 125-mm outer-diameter and 100-mm inner-diameter pipe ($\sigma_{yp} = 250$ MPa), as shown in Fig. P8.68. For a wind load of $P = 2$ kN on the sign, determine the factor of safety f_s against failure by permanent deformation.

8.69 The sign of Fig. P8.68 is supported by a 100-mm outer-diameter and 75-mm inner-diameter steel pipe. Given the weight of the sign $W = 500$ N and a wind load of $P = 800$ N, determine the maximum shearing stress in the pipe.

8.70 For the member loaded as shown in Fig. P8.70, calculate the maximum normal stress at (a) point A, (b) point B, and (c) point D.

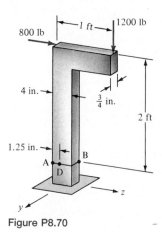

Figure P8.70

8.71 The tube shown in Fig. P8.71 is of constant 20-mm thickness. Given distance d of 30 mm, $(\sigma_t)_{all} = 80$ MPa, and $(\sigma_c)_{all}$ of 120 MPa, determine the maximum permissible load P.

8.72 The tube shown in Fig. P8.71 is of 20-mm uniform thickness. For $P = 250$ kN, $(\sigma_t)_{all} = 90$ MPa, and $(\sigma_c)_{all} = 130$ MPa, what is the minimum permissible distance d?

8.73 Determine the normal stress at A, B, D, and E of the tube seen in Fig. P8.71, which has a constant 10-mm thickness, for $d = 50$ mm and $P = 150$ kN.

8.74 Consider a rectangular cross section (Fig. P8.74a) with a compressive load P which is off-center in two directions. Verify, using Eq. (8.11), that the stress at A will be zero when P acts on line mn (Fig. P8.74b) represented by

$$\frac{y_0}{b/6} + \frac{z_0}{h/6} = 1 \qquad\qquad (P8.74)$$

The other three dashed lines shown can be established by symmetry. Interestingly, the shaded area in Fig. P8.74b is called the *core*, or *kern*, *of the cross section*. As long as P lies on or within this diamond-shaped figure, *every part* of the cross section is in compression.

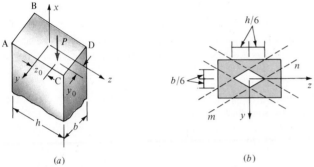

(a)

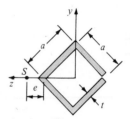

Figure P8.78

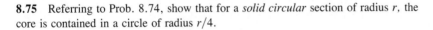

(b)

Figure P8.74   Core, or kern, of a rectangular section.

8.75 Referring to Prob. 8.74, show that for a *solid circular* section of radius r, the core is contained in a circle of radius $r/4$.

Secs. 8.8 to 8.10

8.76 A channel section of uniform thickness is loaded as shown in Fig. 8.17a. Given the dimensions $b = 120$ mm, $h = 180$ mm, $t = 5$ mm, and $V_y = 2$ kN, calculate (a) the distance e; (b) the shearing stress at D; and (c) the maximum shearing stress.

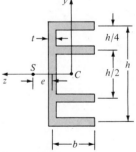

Figure P8.79

8.77 For a C 6 × 10.5 channel, calculate the maximum shearing stress caused by a 10 kips vertical shear force applied at the shear center S. (Refer to Table B.4.)

8.78 through 8.80 Determine the distance e that locates the shear center S of a thin-walled beam of constant thickness t of the cross section shown in Figs. P8.78 through P8.80.

8.81 and 8.82 A thin-walled beam of cross section shown in Figs. P8.79 and P8.80 is acted upon at S by a vertical shear force of 10 kN. Determine the maximum shearing stress if $b = 30$ mm, $h = 100$ mm, and $t = 2$ mm.

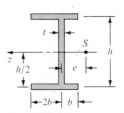

Figure P8.80

8.83 and 8.84   Determine the distance e that locates the shear center S of a thin-walled beam of cross section shown in Figs. P8.83 and P8.84.

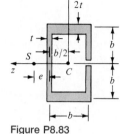

Figure P8.83

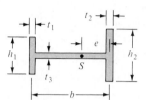

Figure P8.84

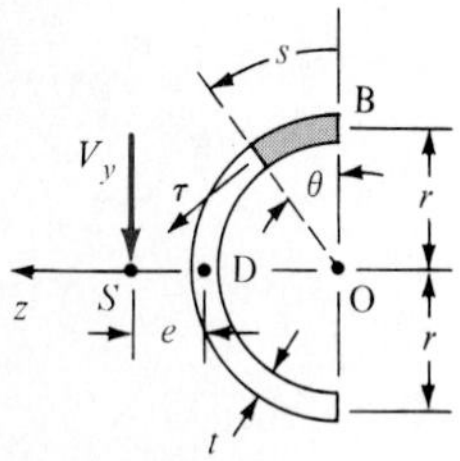

8.85   A thin-walled half tube is loaded as shown in Fig. P8.85. (*a*) Demonstrate that the location of shear center S is given by $e \approx 0.27r$. (*b*) Calculate the shear stress at D for $r = 2$ in., $t = \frac{1}{8}$ in., and $V_y = 100$ lb.

8.86   Figure P8.86 shows a $d = 50$-mm-diameter steel shaft ($\sigma_{yp} = 260$ MPa, $\tau_{yp} = 140$ MPa) subjected to a load R and $P = Q = 0$. If the factor of safety is $f_s = 2$, calculate the largest permissible value of R in accordance with (*a*) the maximum shear stress theory and (*b*) the von Mises theory.

8.87   Redo Prob. 8.86 for $P = 40R$ and $Q = 0$.

8.88   A thin-walled, closed-ended metal tube ($\sigma_u = 36$ ksi, $\sigma_u' = 74$ ksi) having outer and inner diameters of 8 and $7\frac{1}{2}$ in., respectively, is subjected to an internal pressure of 800 psi and a torque of 400 kip·in. Determine the factor of safety f_s according to the maximum principal stress theory.

8.89   A steel circular bar ($\sigma_{yp} = 210$ MPa) of 50-mm diameter is acted upon by combined moments M and axial compressive loads P at its ends. If $M = 1.2$ kN·m, calculate, on the basis of the von Mises theory, the maximum permissible value of P.

8.90   A 60-mm-wide, 80-mm-deep, and 1.2-m-long cantilever is subjected to a concentrated load of $P = 2$ kN at its free end. For $\sigma_{yp} = 240$ MPa, what is the factor of safety f_s, assuming failure occurs in accordance with the maximum energy of distortion theory?

8.91   Referring to Fig. P8.86, design a solid steel shaft ($\tau_{yp} = 60$ MPa) subjected to loads of $R = 600$ N, $Q = 200$ N, and $P = 0$. Apply the maximum shear stress theory using a factor of safety $f_s = 1.2$.

8.92   Design the cross section of a rectangular beam b meters wide by $2b$ meters deep, loaded as shown in Fig. P8.27, if $\sigma_{all} = 18$ ksi. Use the maximum principal stress theory of failure.

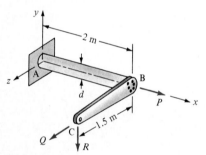

Figure P8.85

Figure P8.86

8.93   A 6-m-long steel shaft of allowable strength $\sigma_{all} = 120$ MPa carries a torque of 400 N·m and its own weight. Use $\rho = 7.86$ Mg/m³ as the mass per unit volume (see Table B.1) and assume that the shaft is supported by frictionless bearings at its ends. Calculate the required shaft diameter in accordance with the von Mises theory of failure.

8.94   As shown in Fig. P8.94, a solid shaft AB is to transmit 20 kW at 180 rpm from the motor and gear to pulley D, where 8 kW is taken off, and to pulley C where the remaining 12 kW is taken off. Assume the ratios of the pulley tensions to be $T_1/T_2 = 3$ and $T_3/T_4 = 3$. If $\tau_{all} = 45$ MPa, determine the required diameter of the shaft AB according to the maximum shear stress theory of failure.

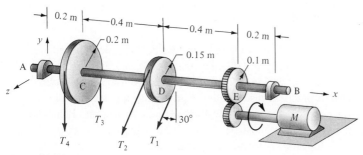

Figure P8.94

8.95   Repeat Prob. 8.94 for the case in which tensions T_1 and T_2 on pulley D are in the horizontal (z) direction.

8.96   Design the steel shaft ($\tau_{yp} = 60$ MPa) of the system shown in Fig. P8.22. Use the maximum shear stress theory of failure and a factor of safety $f_s = 1.2$.

8.97   For the shaft of the system shown in Fig. P8.24, determine the diameter d, applying the maximum principal stress theory of failure. Use $\sigma_{all} = 20$ ksi.

8.98   Redo Prob. 8.97, assuming that the tensions on the pulley C are in the horizontal (z) direction.

9

DEFLECTION OF BEAMS BY INTEGRATION

9.1 INTRODUCTION

As discussed in Sec. 7.6, the axis of a straight beam is bent under the action of applied forces into a curve defined as the *elastic,* or *deflection, curve.* We now treat the *small deflection* of beams subjected to transverse loading.

In numerous practical cases, beam requirements are described by a given load-carrying capacity and an allowable value of deflection. For example, beams in many machines must be stiff enough—that is, deflect within pre-scribed limits—to satisfy tolerances between various moving parts, and floor beams must resist excessive deflection that could lead to cracks in a plastered ceiling. In addition, a deflection study is essential to the analysis of statically indeterminate beams.

Our analysis of beam deflection will be based upon the fundamental assumptions of the beam theory, and only deflections caused by bending will be considered. Thus contributions of shear to deflection will be regarded as negligible, since for static bending problems, the shear deflection represents no more than a few percent of the total deflection (as we shall see in Secs. 11.5 and 11.6).

Two common procedures for determining elastic beam deflection are discussed in this chapter: the successive integration approaches and the su-perposition method. The special techniques, the so-called moment-area and finite-difference methods, will be presented in the chapter which follows. Castigliano's theorem, which will be discussed in Sec. 11.7, is based upon work and energy concepts and provides an alternative approach for calcula-tion of displacements at specified points along a beam axis. The inelastic deformations of beams will be considered in Chap. 12.

9.2 EQUATION OF THE ELASTIC CURVE

We observed in Sec. 7.8 that the *deflection v* of a beam is related to its curvature via the expression $1/\rho = d^2v/dx^2$, which is valid for a beam of *any* material. For a linearly elastic beam whose cross section is symmetrical about the plane (xy) of loading, we also have $1/\rho = M/EI$ (Sec. 7.9). Here ρ is the radius of curvature, and EI is the flexural rigidity. Both the bending moment M and the curvature $1/\rho$ of the beam may vary from section to section, as shown in Fig. 9.1. The basic differential *equation of the deflection curve* is therefore

$$\frac{d^2v}{dx^2} = \frac{M}{EI} \qquad (9.1)$$

This moment-curvature relation is known as the *Bernoulli-Euler law* of the technical bending theory.

Careful consideration should be given to the sign conventions associated with Eq. (9.1). Recall from Sec. 7.2 that applied forces and moments are assumed to be positive when their vectors are in a positive coordinate direction. Also, positively directed internal force vectors act on a positive face. Accordingly, the loads (M_0, P, w) and internal force resultants (V, M) depicted in Fig. 9.1 are all positive. Note that the positive sense of the curvature d^2v/dx^2, concave up, agrees with the curvature induced by the applied positive moments.

The y-directed deflection v is measured from the x axis to the elastic curve. An upward ($\uparrow$) deflection is therefore positive. The angle of rotation θ (measured in radians) is the angle between the x axis and the tangent to the curve at a point. The sense of θ vector, as also of the M vector, follows the right-hand rule. That is, θ is *positive* when *counterclockwise* ($\curvearrowleft$), as shown in the figure. For small displacements, the angle of rotation is approximately equal to the *slope* (dv/dx) of the elastic curve.

The internal moment M, internal shear force V, and applied loading intensity w are related by Eqs. (7.1) and (7.3). These, combined with Eq. (9.1), compose the useful sequence of expressions:

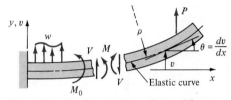

Figure 9.1   Positive loads and internal forces.

$$\text{Deflection} = v$$

$$\text{Slope} = \theta = \frac{dv}{dx} = v'$$

$$\text{Moment} = M = EI\frac{d\theta}{dx} = EIv'' \qquad (9.2)$$

$$\text{Shear} = V = -\frac{dM}{dx} = -(EIv'')'$$

$$\text{Load} = w = -\frac{dV}{dx} = (EIv'')''$$

For a beam of *constant* flexural rigidity EI, the differential equations given above become

$$EIv'' = M \qquad (9.3)$$

$$EIv''' = -V \qquad (9.4)$$

$$EIv'''' = w \qquad (9.5)$$

Deflection v can thus be found by successively integrating any one of Eqs. (9.3) through (9.5), as will be described in Sec. 9.4. The choice of equation depends upon mathematical convenience and individual preference.

9.3 BOUNDARY CONDITIONS

The restrictions imposed on the beam by its supports are known as *boundary conditions*. In order to solve a beam-deflection problem, it is necessary to prescribe the conditions at each support, in addition to using the differential equations. These conditions may be a given deflection and slope, or force and moment, or some combination.

We shall formulate a variety of the most frequently occurring situations. The boundary conditions which apply at the end ($x = a$) of a beam (see Fig. 9.2) are as follows:

The *fixed or clamped end* (Fig. 9.2a): In this case both the deflection and slope must vanish. Hence

$$v(a) = 0 \qquad \theta(a) = v'(a) = 0 \qquad (9.6)$$

The *simply supported* end (Fig. 9.2b): The end considered is free to rotate. Therefore the deflection as well as the bending moment must be zero. That is,

$$v(a) = 0 \qquad M(a) = EIv''(a) = 0 \qquad (9.7)$$

The *free end* (Fig. 9.2c): Such an end is free of the bending moment and shear force. Thus

$$M(a) = EIv''(a) = 0 \qquad V(a) = -(EIv'')'_{x=a} = 0 \qquad (9.8)$$

Other boundary conditions may be represented similarly. We observe from Eqs. (9.6) through (9.8) that the *force* (static) variables M, V and the *geometric* (kinematic) variables v, θ are zero for commonly encountered cases.

Figure 9.2 Boundary conditions: (a) fixed end; (b) simply supported end; and (c) free end.

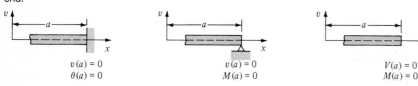

Many beams are subjected to concentrated loads, to reactions, or to the discontinuities in a distributed longitudinal loading. In such cases, the solution requires consideration of the continuity conditions of the elastic curve as well as of the boundary conditions. The former require that both the slope and the deflection be the same at any junction of two parts of a beam; the elastic curve must be *smooth*.

9.4 DIRECT-INTEGRATION METHODS

The approach to solving deflection problems beginning with Eq. (9.4) or (9.5) is called the *multiple-integration method*. This procedure is initiated by expressing the loading on the beam in terms of the shear V or the load intensity w. Then, in general, Eq. (9.5) is integrated four times to obtain v. That is,

$$EIv'''' = w$$

$$EIv''' = \int w \, dx + C_1$$

$$EIv'' = \int dx \int w \, dx + C_1 x + C_2 \qquad (9.9)$$

$$EIv' = \int dx \int dx \int w \, dx + \tfrac{1}{2}C_1 x^2 + C_2 x + C_3$$

$$EIv = \int dx \int dx \int dx \int w \, dx + \tfrac{1}{6}C_1 x^3 + \tfrac{1}{2}C_2 x^2 + C_3 x + C_4$$

The constants C_1, C_2, C_3, and C_4 are determined from the boundary conditions imposed on both force and geometric variables.

The first step in calculating the deflection using Eq. (9.3) is *expressing the bending amount M as a function of x* in terms of the applied loading. Next, the moment-curvature relation may be integrated to obtain the deflection as a function of position along the beam. Because two integrations are required, this is referred to as the *double-integration method*. Thus for a beam of uniform flexural rigidity EI,

$$EIv'' = M$$

$$EIv' = \int M \, dx + C_1 \qquad (9.10)$$

$$EIv = \int dx + \int M \, dx + C_1 x + C_2$$

where the constants C_1 and C_2 are found from the geometric boundary conditions. Clearly these coefficients do *not* correspond to those used in Eqs. (9.9).

The method of multiple integration is helpful if the expression for the load w (or the shear V) is simple to write or the bending moment difficult to represent. Otherwise, the double-integration approach is much easier and is preferred.

When there are abrupt changes in loading or in EI along the beam, there will be separate moment expressions for each segment of the beam in which discontinuity occurs. In this case, the solution is found by satisfying the continuity conditions at the common boundaries of the beam segments. However, this "matching" procedure may be avoided by representing all the loadings by a single equation (as will be seen in Sec. 9.5). The merit of the double-integration method becomes obvious particularly when the latter approach is employed.

Both statically *determinate* and *indeterminate* elastic-beam problems can be treated by using the successive integration procedures outlined above. It should be emphasized that in solving beam deflections, the three principles of solid mechanics set forth in Sec. 1.3 are used. The equations of static equilibrium provide the moment in $EIv'' = M$, which itself is a force-deformation relationship. The requirements of geometric compatibility complete the solution by fixing the constants in Eqs. (9.10). Equations (9.9) are interpreted similarly.

The following examples illustrate the analytical techniques.

EXAMPLE 9.1

A simply supported beam AB carries a uniform load w per unit length, as shown in Fig. 9.3a. (a) Determine the equation of the elastic curve using the double-integration method. (b) Redo the problem using the multiple-integration method. (c) Find the maximum deflection and the slopes θ_A and θ_B.

Solution The reactions are noted in Fig. 9.3a. The expression for the bending moment (from Fig. 7.7c) is

$$M = \tfrac{1}{2}wLx - \tfrac{1}{2}wx^2$$

(a) Substitution of the above into Eq. (9.3) gives

$$EIv'' = \tfrac{1}{2}wLx - \tfrac{1}{2}wx^2 \qquad (a)$$

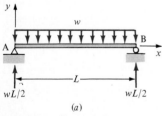

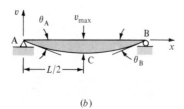

(a)

(b)

Figure 9.3 Example 9.1.

Integrating twice in x, we obtain

$$EIv' = \tfrac{1}{4}wLx^2 - \tfrac{1}{6}wx^3 + C_1 \qquad (b)$$
$$EIv = \tfrac{1}{12}wLx^3 - \tfrac{1}{24}wx^4 + C_1x + C_2$$

Boundary conditions $v(0) = v(L) = 0$ are applied to Eq. (b) to yield $C_2 = 0$ and

$$0 = \tfrac{1}{12}wL^4 - \tfrac{1}{24}wL^4 + C_1L$$

from which $C_1 = -wL^3/24$. Alternatively, $v'(L/2) = 0$ can also be used to find the constant C_1. Substituting the values of C_1 and C_2 into Eqs. (a) and (b), we obtain the expressions for the slope and the elastic curve, respectively:

$$v' = -\frac{w}{24EI}(L^3 - 6Lx^2 + 4x^3)$$
$$v = -\frac{w}{24EI}(L^3x - 2Lx^3 + x^4) \tag{9.11}$$

(b) Successive integrations of Eq. (9.5), with load intensity $-w$, lead to

$$EIv'''' = -w$$
$$EIv''' = -wx + C_1$$
$$EIv'' = -\tfrac{1}{2}wx^2 + C_1x + C_2 \tag{c}$$
$$EIv' = -\tfrac{1}{6}wx^3 + \tfrac{1}{2}C_1x^2 + C_2x + C_3$$
$$EIv = -\tfrac{1}{24}wx^4 + \tfrac{1}{6}C_1x^3 + \tfrac{1}{2}C_2x^2 + C_3x + C_4$$

The boundary conditions require that

$$v(0) = 0 \quad v(L) = 0 \quad M(0) = 0 \quad M(L) = 0$$

Insertion of the first and the third of these equations into Eqs. (c) yields $C_4 = C_2 = 0$. The remaining conditions then give

$$EIv''(L) = 0 = -\tfrac{1}{2}wL^2 + C_1L \qquad \text{or} \qquad C_1 = \tfrac{1}{2}wL$$

$$v(L) = 0 = -\tfrac{1}{24}wL^4 + \tfrac{1}{6}C_1L^3 + C_3L \qquad \text{or} \qquad C_3 = -\frac{wL^3}{24}$$

Carrying the values of C_1, C_2, C_3, and C_4 into the last two of Eqs. (c), we obtain Eqs. (9.11).

(c) Because of symmetry, the maximum deflection occurs at the middle of the span where the slope is zero (Fig. 9.3b). Setting $x = L/2$ in the second of Eqs. (9.11), we have

$$v_{\text{max}} = v_C = -\frac{5wL^4}{384EI} = \frac{5wL^4}{384EI} \downarrow \tag{9.12}$$

The largest slopes occur at the supports of the beam. Introducing $x = 0$ and $x = L$ into the first of Eqs. (9.11), we find that

$$\theta_A = v'(0) = -\frac{wL^3}{24EI} = \frac{wL^3}{24EI} \quad \diagdown \tag{9.13}$$

$$\theta_B = v'(L) = \frac{wL}{24EI} \quad \diagup \tag{9.13}$$

showing, as expected, that the angles of rotation at the ends are equal (Fig. 9.3b).

EXAMPLE 9.2

A force P and a moment M_0 are applied at the free end of a cantilever of uniform cross section, as shown in Fig. 9.4a. Determine the equation of the deflection curve and the deflection and slope at B. Calculate the maximum angle of rotation and the deflection for $P = 20$ kN, $M_0 = 5$ kN·m, $L = 2$ m, and $EI = 10$ MN·m².

Solution Referring to the free-body diagram of Fig. 9.4b, we find that the equilibrium condition $\Sigma M_O = 0$ yields

$$M = Px - M_0 - PL$$

Inserting the above into Eq. (9.3) and integrating twice, we obtain

$$EIv'' = Px - M_0 - PL$$

$$EIv' = \tfrac{1}{2}Px^2 - M_0 x - PLx + C_1$$

$$EIv = \tfrac{1}{6}Px^3 - \tfrac{1}{2}M_0 x^2 - \tfrac{1}{2}PLx^2 + C_1 x + C_2$$

But $v(0) = 0$ and $v'(0) = 0$; hence, $C_2 = 0$ and $C_1 = 0$. The equations of the slope and elastic curve are therefore

$$v' = -\frac{1}{2EI}[Px(2L - x) + 2M_0 x]$$

$$v = -\frac{1}{6EI}[Px^2(3L - x) + 3M_0 x^2] \tag{9.14}$$

The deflection and the slope at B are found upon substitution of $x = L$ into the foregoing expressions:

Figure 9.4 Example 9.2.

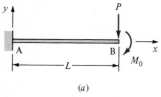

(a)

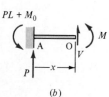

(b)

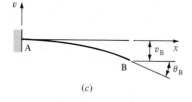

(c)

$$\theta_B = v'(L) = -\frac{L}{2EI}(PL + 2M_0)$$

$$v_B = v(L) = -\frac{L^2}{6EI}(2PL + 3M_0)$$

(9.15)

The results are indicated in Fig. 9.4c. The largest values of θ and v occur at the free end B of the beam:

$$\theta_{max} = -\frac{2(10^3)}{2(10 \times 10^6)}[20(2) + 2(5)] = -0.005 \text{ rad} = 0.005 \text{ rad} \quad \text{\reflectbox{$\searrow$}}$$

$$v_{max} = -\frac{2^2(10^3)}{6(10 \times 10^6)}[2(20)2 + 3(5)] = -6.3 \text{ mm} = 6.3 \text{ mm} \downarrow$$

as found by substituting the given data into Eqs. (9.15).

EXAMPLE 9.3

For the simply supported beam of Fig. 9.5a, determine the expression of the elastic curve and the deflection at point C.

Solution   The reactions are shown in Fig. 9.5a. The expressions for the moments in segments AC and CB of the beam are, from Example 7.4,

$$M_1 = \frac{Pb}{L}x \qquad\qquad (0 \le x \le a)$$

$$M_2 = \frac{Pb}{L}x - P(x - a) = Pa\left(1 - \frac{x}{L}\right) \qquad (a \le x \le L)$$

(d)

Double integration of Eqs. (d) results in the following:

For segment AC

$$EIv_1'' = \frac{Pb}{L}x$$

$$EIv_1' = \frac{Pb}{2L}x^2 + C_1$$

$$EIv_1 = \frac{Pb}{6L}x^3 + C_1x + C_2$$

For segment CB

$$EIv_2'' = Pa - \frac{Pa}{L}x$$

$$EIv_2' = Pax - \frac{Pa}{2L}x^2 + C_3$$

$$EIv_2 = \frac{1}{2}Pax^2 - \frac{Pa}{6L}x^3 + C_3x + C_4$$

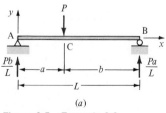

(a)

(b)

Figure 9.5   Example 9.3.

The boundary and the continuity conditions are applied to the above equations to yield

$$v_1(0) = 0: \quad 0 = C_2 \qquad v_2(L) = 0: \quad \frac{PaL^2}{3} + C_3 L + C_4$$

$$v_1(a) = v_2(a): \quad \frac{Pa^3 b}{6L} + C_1 a = \frac{Pa^3}{2} - \frac{Pa^4}{6L} + C_3 a + C_4$$

$$v_1'(a) = v_2'(a): \quad \frac{Pa^2 b}{2L} + C_1 = Pa^2 - \frac{Pa^3}{2L} + C_3$$

Solving these expressions, we obtain

$$C_1 = -\frac{Pb}{6L}(L^2 - b^2) \qquad C_2 = 0$$

$$C_3 = -\frac{Pa}{6L}(2L^2 + a^2) \qquad C_4 = \frac{Pa^3}{6}$$

With the foregoing values, the elastic curves for the left- and right-hand segments take the form:

$$v_1 = -\frac{Pbx}{6EIL}(L^2 - b^2 - x^2) \qquad (0 \leq x \leq a)$$

$$v_2 = -\frac{Pbx}{6EIL}(L^2 - b^2 - x^2) - \frac{P(x-a)^3}{6EI} \qquad (a \leq x \leq L) \tag{9.16}$$

Similarly, the slopes for the two parts of the beam are found to be

$$v_1' = -\frac{Pb}{6EIL}(L^2 - 3x^2 - b^2) \qquad (0 \leq x \leq a)$$

$$v_2' = -\frac{Pb}{6EIL}(L^2 - 3x^2 - b^2) - \frac{P(x-a)^2}{2EI} \qquad (a \leq x \leq L) \tag{9.17}$$

To determine the deflection and slope at point C, set $x = a$ in Eqs. (9.16) and (9.17). Thus

$$v_C = -\frac{Pba}{6EIL}(L^2 - a^2 - b^2)$$

$$\theta_C = -\frac{Pb}{6EIL}(L^2 - 3a^2 - b^2) \tag{9.18}$$

Note that inasmuch as $\theta_C \neq 0$, the maximum deflection of the span does *not* occur at C (Fig. 9.5b). For $a > b$, by equating to zero the slope v_1' from Eq. (9.17), it is found that the *largest* deflection occurs in the *longer* segment at $x_m = \sqrt{(L^2 - b^2)/3}$, a point very close to the center of the beam (see Table B.7).

In the special case in which force P acts *at the middle* of the span, we have $a = b = L/2$, and Eqs. (9.18) give

$$v_{max} = v_C = \frac{PL^3}{48EI} \downarrow \qquad \theta_C = 0 \tag{9.19}$$

The elastic curve is now symmetric about the beam center.

EXAMPLE 9.4

Determine the reactions of the beam of Fig. 9.6*a*, employing (*a*) the double-integration method and (*b*) the multiple-integration method.

Solution The free-body diagram of Fig. 9.6*b* reveals that the problem is statically indeterminate; there are three unknown reactions and only two independent equations of statics available. From the conditions $\Sigma F_y = 0$ and $\Sigma M_B = 0$, we have

$$R_A + R_B = \tfrac{4}{3}wL \tag{e}$$

$$R_A L - M_A = w\left(\frac{4L}{3}\right)\left(\frac{4L}{6} - \frac{L}{3}\right) = \frac{4}{9}wL^2 \tag{f}$$

The additional unknown necessitates the use of the deflection-curve equation.
(*a*) Referring to the free-body diagram of a segment AC of the beam (Fig. 9.6*c*), we find that condition $\Sigma M_O = 0$ gives

$$M = R_A x - M_A - \tfrac{1}{2}wx^2$$

Substitution of M into Eq. (9.3) and integrating twice leads to

$$EIv'' = R_A x - M_A - \tfrac{1}{2}wx^2$$
$$EIv' = \tfrac{1}{2}R_A x^2 - M_A x - \tfrac{1}{6}wx^3 + C_1$$
$$EIv = \tfrac{1}{6}R_A x^3 - \tfrac{1}{2}M_A x^2 - \tfrac{1}{24}wx^4 + C_1 x + C_2$$

The boundary conditions are

$$v(0) = 0 \qquad v'(0) = 0 \qquad v(L) = 0 \tag{g}$$

Application of the foregoing requirements to the preceding equations results in $C_1 = 0$, $C_2 = 0$, and

$$4R_A L - 12M_A = wL^2 \tag{h}$$

Simultaneous solution of Eqs. (*e*), (*f*), and (*h*) yields

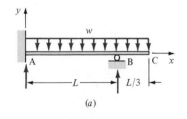

(*a*)

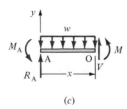

(*c*)

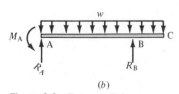

(*b*)

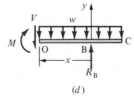

(*d*)

Figure 9.6 Example 9.4.

$$M_A = \tfrac{7}{72}wL^2 \; \circlearrowright \qquad R_A = \tfrac{13}{24}wL \uparrow \qquad R_B = \tfrac{19}{24}wL \uparrow \qquad (9.20)$$

Introducing these quantities into the expressions for deflection v, slope θ, and bending moment M, one obtains the complete solution for segment AB of the beam.

An alternate approach to analysis of this beam is as follows: We locate the origin of the coordinates at the right support (Fig. 9.6d) and express the bending moment within the span as

$$M = R_Bx - \tfrac{1}{2}w(x + L/3)^2$$

We then insert this relation into Eq. (9.3) and solve it as before.

(b) Successive integration of Eq. (9.5), with load intensity $-w$, gives

$$EIv'''' = -w$$
$$EIv''' = -wx + C_1$$
$$EIv'' = -\tfrac{1}{2}wx^2 + C_1x + C_2$$
$$EIv' = -\tfrac{1}{6}wx^3 + \tfrac{1}{2}C_1x^2 + C_2x + C_3$$
$$EIv = -\tfrac{1}{24}wx^4 + \tfrac{1}{6}C_1x^3 + \tfrac{1}{2}C_2x^2 + C_3x + C_4$$

Applying conditions (g), $M(0) = -M_A$, $V(0) = -R_A$ to the above, we obtain $C_3 = C_4 = 0$, $C_2 = -M_A$, $C_1 = R_A$. Finally, the requirement $v(L) = 0$ will lead to the same result given in Eq. (h).

EXAMPLE 9.5

An overhanging beam supports a concentrated force P, as seen in Fig. 9.7. Derive the equation of the elastic curve using (a) the second-order differential equation and (b) the third-order differential equation.

Solution The reactions are noted in Fig. 9.7.

(a) Equilibrium requirements yield the following bending moment at x:

$$M_1 = -\frac{Px}{2} \qquad (0 \le x \le L)$$

$$M_2 = Px \qquad \left(L \le x < \frac{2L}{3}\right)$$

Double integration of the foregoing leads to the following expressions:

For segment AB

$$EIv_1'' = -\frac{Px}{2}$$

$$EIv_1' = -\frac{Px^2}{4} + C_1$$

$$EIv_1 = -\frac{Px^3}{12} + C_1x + C_2$$

For segment BC

$$EIv_2'' = Px - \frac{3PL}{2}$$

$$EIv_2' = \frac{Px^2}{2} - \frac{3PLx}{2} + C_3$$

$$EIv_2 = \frac{Px^3}{6} - \frac{3PLx^2}{4} + C_3x + C_4$$

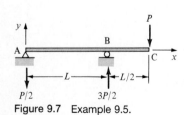

Figure 9.7 Example 9.5.

Employing the boundary and the continuity conditions in connection with the above, we obtain

$$v_1(0) = 0: \quad C_2 = 0 \qquad v_1(L) = 0: \quad C_1 = \frac{PL^2}{12}$$

$$v_1'(L) = v_2'(L): \quad C_3 = \frac{5PL^2}{6} \qquad v_2(L) = 0: \quad C_4 = -\frac{PL^3}{4}$$

The resulting elastic curves of the beam are

$$v_1 = \frac{Px}{12EI}(L^2 - x^2) \qquad\qquad (0 \le x \le L)$$

$$v_2 = \frac{P}{12EI}(-3L^3 + 10L^2x - 9Lx^2 + 2x^3) \qquad \left(L \le x \le \frac{3L}{2}\right)$$

(9.21)

The deflection at the free end of the beam is readily obtained by introducing $x = 3L/2$ into the second of Eqs. (9.21):

$$v_C = -\frac{PL^3}{8EI} = \frac{PL^3}{8EI} \downarrow \qquad\qquad (9.22)$$

(b) Summation of forces for the segments of the beam on either side of x yields

$$V_1 = \frac{P}{2} \qquad (0 \le x \le L)$$

$$V_2 = -P \qquad \left(L \le x \le \frac{3L}{2}\right)$$

Introducing the above into Eq. (9.4) and successive integrations give the following expressions:

For segment AB

$$EIv_1''' = -\frac{P}{2}$$

$$EIv_1'' = -\frac{Px}{2} + C_1$$

$$EIv_1' = -\frac{Px^2}{4} + C_1x + C_2$$

$$EIv_1 = -\frac{Px^3}{12} + \tfrac{1}{2}C_1x^2 + C_2x + C_3$$

For segment BC

$$EIv_2''' = P$$

$$EIv_2'' = Px + C_4 \qquad\qquad (i)$$

$$EIv_2' = \frac{Px^2}{2} + C_4x + C_5$$

$$EIv_2 = \frac{Px^3}{6} + \tfrac{1}{2}C_4x^2 + C_5x + C_6$$

The boundary and the continuity conditions yield

$$v_1''(0): \quad C_1 = 0 \qquad v_2''\left(\frac{3L}{2}\right) = 0: \quad C_4 = -\frac{3PL}{2}$$

$$v_1'(L) = v_2'(L): \quad C_5 = \frac{5PL^2}{6}$$

$$v_1(0) = 0: \quad C_3 = 0 \qquad v_1(L) = 0: \quad C_2 = \frac{PL^2}{12}$$

$$v_2(L) = 0: \quad C_6 = -\frac{PL^3}{4}$$

Substituting the values of C_1 through C_6 into Eqs. (i), we obtain the expressions for the elastic curves given by Eqs. (9.21).

EXAMPLE 9.6

A cantilever beam AB with a triangularly distributed load of maximum intensity w_0 is supported by a bearing block at A, as shown in Fig. 9.8a. Assume EI to be the flexural rigidity of the beam and $E_a I_a$, the axial rigidity of the block. Determine the equation of the deflection curve and the reactions using (a) the double-integration method and (b) the multiple-integration method. Consider the bearing block to be rigid.

Solution   The reactions are noted in Fig. 9.8a. The beam is clearly statically indeterminate to the first degree.
(a) The expression for the moment is (Fig. 9.8b)

$$M = R_A x - \frac{w_0 x^3}{6L}$$

From the above,

$$EIv'' = R_A x - \frac{w_0 x^3}{6L}$$

$$EIv' = \tfrac{1}{2}R_A x^2 - \frac{w_0 x^4}{24L} + C_1 \qquad (j)$$

$$EIv = \tfrac{1}{6}R_A x^3 - \frac{w_0 x^5}{120L} + C_1 x + C_2$$

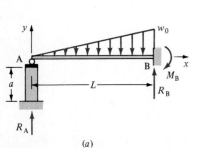

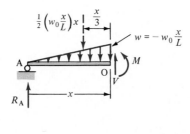

(a)   (b)

Figure 9.8   Example 9.6: cantilever beam supported by a bearing block.

We have $v(0) = -R_A a/A_a E_a$; hence $C_2 = -EIR_A a/A_a E_a$. Using the conditions at the fixed support,

$$v(L) = 0: \tfrac{1}{6}R_A L^3 - \frac{w_0 L^4}{120} + C_1 L - \frac{EIR_A a}{A_a E_a} = 0$$

$$v'(L) = 0: \tfrac{1}{2}R_A L^2 - \frac{w_0 L^3}{24} + C_1 = 0$$

we obtain

$$R_A = \frac{w_0 L^4 A_a E_a}{10(3aEI + L^3 A_a E_a)} \qquad (9.23a)$$

and $C_1 = w_0 L^3/24 - R_A L^2/2$. Then application of the equations of statics ($\Sigma F_y = 0$, $\Sigma M_B = 0$) to the free body of Fig. 9.8a leads to

$$R_B = \tfrac{1}{2}w_0 L - \frac{w_0 L^4 A_a E_a}{10(3aEI + L^3 A_a E_a)} \qquad (9.23b)$$

$$M_B = \tfrac{1}{6}w_0 L^2 - \frac{w_0 L^5 A_a E_a}{10(3aEI + L^3 A_a E_a)} \qquad (9.23c)$$

Introducing the notation

$$D_1 = w_0 L^4 A_a E_a \qquad D_2 = 3aEI + L^3 A_a E_a \qquad (9.24)$$

we can rewrite the reactions as follows:

$$R_A = \frac{D_1}{10D_2} \qquad R_B = \frac{w_0 L}{2} - \frac{D_1}{10D_2} \qquad M_B = \frac{w_0 L^2}{6} - \frac{D_1 L}{10D_2} \qquad (9.25)$$

These representations are convenient for computer implementation. Substitution of R_A, C_1, and C_2 into Eqs. (j) results in the expression for the elastic curve.

(b) The load intensity at an arbitrary section x (Fig. 9.8b) is expressed by $w = -w_0 x/L$. Therefore, direct integration of Eq. (9.5) yields

$$EIv'''' = -\frac{w_0 x}{L}$$

$$EIv''' = -\frac{w_0 x^2}{2L} + C_1$$

$$EIv'' = -\frac{w_0 x^3}{6L} + C_1 x + C_2 \qquad (k)$$

$$EIv' = -\frac{w_0 x^4}{24L} + \tfrac{1}{2}C_1 x^2 + C_2 x + C_3$$

$$EIv = -\frac{w_0 x^5}{120L} + \tfrac{1}{6}C_1 x^3 + \tfrac{1}{2}C_2 x^2 + C_3 x + C_4$$

Boundary conditions $v(0) = 0$ and $v''(0) = 0$ give $C_4 = 0$ and $C_2 = 0$, respectively, and

$$v(L) = 0: \quad -\frac{w_0 L^4}{120} + \tfrac{1}{6}C_1 L^3 + C_3 L = 0$$

$$v'(L) = 0: \quad -\frac{w_0 L^3}{24} + \tfrac{1}{2}C_1 L^2 + C_3 = 0$$

lead to $C_1 = w_0 L/10$ and $C_3 = -w_0 L^3/120$. Inserting the values of C_1 through C_4 into Eqs. (k), we obtain

$$v''' = \frac{w_0}{10EIL}(L^2 - 5x^2)$$

$$v'' = \frac{w_0}{30EIL}(3L^2 x - 5x^3)$$

$$v' = -\frac{w_0}{120EIL}(L^4 - 6L^2 x^2 + 5x^4)$$

$$v = -\frac{w_0}{120EIL}(L^4 x - 2L^2 x^3 + x^5)$$

The reactions may then be calculated readily from $R_A = EIv'''(0)$, $R_B = -EIv'''(L)$, and $M_B = EIv''(L)$, as follows:

$$R_A = \tfrac{1}{10}w_0 L \uparrow \qquad R_B = \tfrac{2}{5}w_0 L \uparrow \qquad M_B = \tfrac{1}{15}w_0 L^2 \,\rotatebox{180}{$\curvearrowright$} \qquad (9.26)$$

Note that $\Sigma F_y = 0$ and $\Sigma M_B = 0$ serve as a check for the calculations (Fig. 9.8a).

*9.5 USE OF SINGULARITY FUNCTIONS

The integration procedures of the preceding section become quite cumbersome for cases where several intervals and several sets of continuity conditions exist. The use of singularity functions may simplify the computations involved in solving problems of this kind.

In this section, we shall define and employ singularity functions to write one expression for the bending moment $M(x)$ that is valid throughout the entire beam length. The beam may be carrying concentrated forces and moments or discontinuous distributed loads. When the beam is subjected to *only* distributed loads that change abruptly in intensity, the distributed load $w(x)$ itself may be expressed in terms of singularity functions. Therefore the need for conditions of continuity is eliminated.

A singularity function of x is written as $\langle x - a \rangle^n$. Here n represents an integer and a is the value of x at which the function "begins."* We have, for $n \geq 0$, the definitions

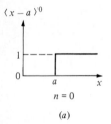

$\langle x - a \rangle^0$

$n = 0$

(a)

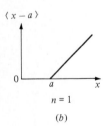

$\langle x - a \rangle$

$n = 1$

(b)

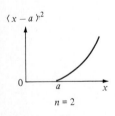

$\langle x - a \rangle^2$

$n = 2$

(c)

Figure 9.9 Graphs of singularity functions $\langle x - a \rangle^n$.

*In 1919 W. H. Macaulay first suggested the use of the pointed brackets for beam problems. See W. D. Pilkey, "Clebsch's Method for Beam Deflections," *Journal of Engineering Education*, January 1964, p. 170.

$$\langle x - a \rangle^n = \begin{cases} 0 & \text{when } x < a \\ (x - a)^n & \text{when } x \geq a \end{cases} \tag{9.27}$$

Three examples of singularity functions are shown in Fig. 9.9. As one may readily see, it follows from Eqs. (9.27) that

$$\langle x - a \rangle^0 = \begin{cases} 0 & \text{when } x < a \\ 1 & \text{when } x \geq a \end{cases}$$

$$\langle x - a \rangle = \begin{cases} 0 & \text{when } x < a \\ x - a & \text{when } x \geq a \end{cases} \tag{9.28}$$

and

$$\int \langle x - a \rangle^n \, dx = \frac{\langle x - a \rangle^{n+1}}{n + 1} \tag{9.29}$$

It is to be noted that the pointed brackets—*Macaulay notation*—have no particular mathematical significance: They serve only to abbreviate expressions. Clearly the units of the singularity functions are the same as the units of x^n (for example $\langle x - a \rangle^2$ has units of x^2).

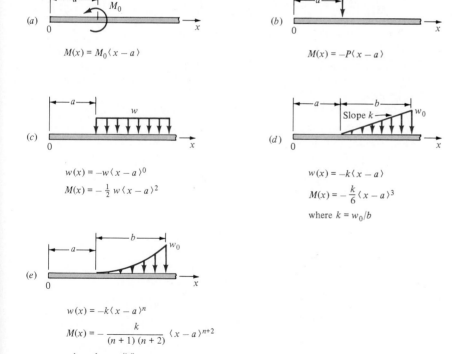

(a)

$M(x) = M_0 \langle x - a \rangle$

(b)

$M(x) = -P \langle x - a \rangle$

(c)

$w(x) = -w \langle x - a \rangle^0$

$M(x) = -\frac{1}{2} w \langle x - a \rangle^2$

(d)

$w(x) = -k \langle x - a \rangle$

$M(x) = -\frac{k}{6} \langle x - a \rangle^3$

where $k = w_0/b$

(e)

$w(x) = -k \langle x - a \rangle^n$

$M(x) = -\frac{k}{(n + 1)(n + 2)} \langle x - a \rangle^{n+2}$

where $k = w_0/b^n$

Figure 9.10 Common loadings and corresponding bending moments.

Figure 9.10 represents the bending moments associated with commonly encountered loadings. Note that each case can be verified by referring to Eqs. (9.27) and expressing the load intensity or moment at a point located to the right of the point of discontinuity. To derive the expression given for $M(x)$ in Fig. 9.10e, it is convenient to use Eqs. (9.9), $M(x) = \int dx \int w(x)\, dx$, with $w(x) = -k\langle x - a \rangle^n$. Here $k = w_0/b^n$ is the slope of the loading curve having the form of a general spandrel. We observe that, for $n = 0$ and $n = 1$, the equations for $w(x)$ and $M(x)$ corresponding to the loading of Fig. 9.10e reduce to the expressions for the cases of Fig. 9.10c and d, respectively.

The combined loads can be handled by superposition of the basic loads given in Fig. 9.10. Applications of the integration methods, together with singularity functions, are illustrated in the following sample problems.

EXAMPLE 9.7

Determine the deflection at the free end of the beam loaded as shown in Fig. 9.11a.

Solution Obviously, discontinuities at $x = L/4$, $x = L/2$, and $x = 3L/4$ exist in $M(x)$. Thus using singularity functions and the superposed distributed loads of Fig. 9.11b, we have

$$EI v'' = M(x) = -Px - M_0 \left\langle x - \frac{L}{4} \right\rangle^0 - \frac{w}{2} \left\langle x - \frac{L}{2} \right\rangle^2 + \frac{w}{2} \left\langle x - \frac{3L}{4} \right\rangle^2 \quad (a)$$

Note that, for $x < L/4$, the last three terms in the above expression vanish as defined by the first of Eqs. (9.27) and depicted in Fig. 9.9a and c. When $L/4 \le x < L/2$, one has $\langle x - L/4 \rangle^0 = 1$, while $\langle x - L/2 \rangle^2 = 0$ according to the first of Eqs. (9.28) and (9.27), respectively. The interval $L/2 \le x < 3L/4$ is described in a like manner.

Two integrations of Eq. (a) referring to Eq. (9.29) yield

$$EI v' = -\frac{Px^2}{2} - M_0 \left\langle x - \frac{L}{4} \right\rangle - \frac{w}{6} \left\langle x - \frac{L}{2} \right\rangle^3 + \frac{w}{6} \left\langle x - \frac{3L}{4} \right\rangle^3 + C_1$$

$$EI v = -\frac{Px^3}{6} - \frac{M_0}{2} \left\langle x - \frac{L}{4} \right\rangle^2 - \frac{w}{24} \left\langle x - \frac{L}{2} \right\rangle^4 \quad (b)$$

$$+ \frac{w}{24} \left\langle x - \frac{3L}{4} \right\rangle^4 + C_1 x + C_2$$

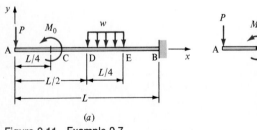

(a)

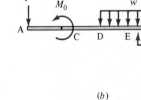

(b)

Figure 9.11 Example 9.7.

The geometric boundary conditions require that

$$EIv'(L) = 0: \quad 0 = -\frac{PL^2}{2} - M_0\left(\frac{3L}{4}\right) - \frac{w}{6}\left(\frac{L}{2}\right)^3 + \frac{w}{6}\left(\frac{L}{4}\right)^3 + C_1$$

$$EIv(L) = 0: \quad 0 = -\frac{PL^3}{6} - \frac{M_0}{2}\left(\frac{3L}{4}\right)^2 - \frac{w}{24}\left(\frac{L}{2}\right)^4 + \frac{w}{24}\left(\frac{L}{4}\right)^4 + C_1L + C_2$$

from which

$$C_1 = \frac{PL^2}{2} + \frac{3LM_0}{4} + \frac{7wL^3}{384}$$

$$C_2 = -\frac{PL^3}{3} - \frac{13M_0L^2}{32} - \frac{97wL^4}{6144}$$

Substitution of these values into Eqs. (b) leads to expressions for θ and v. At the free end of the beam ($x = 0$), the slope and deflection are thus

$$\theta_A = \frac{1}{EI}\left(\frac{PL^2}{2} + \frac{3LM_0}{4} + \frac{7wL^3}{384}\right)$$

$$v_A = -\frac{1}{EI}\left(\frac{PL^3}{3} + \frac{13M_0L^2}{32} + \frac{97wL^4}{6144}\right)$$

(9.30)

In a like manner, we can determine any other deflection and slope as required.

EXAMPLE 9.8

Redo Example 9.3, this time using singularity functions.

Solution   The moments M_1 and M_2 given in Example 9.3 may be represented in the form

$$M(x) = \frac{Pb}{L}x - P\langle x - a\rangle$$

Successive integrations then appear as

$$EIv'' = \frac{Pb}{L}x - P\langle x - a\rangle$$

$$EIv' = \frac{Pb}{2L}x^2 - \frac{P}{2}\langle x - a\rangle^2 + C_1$$

$$EIv = \frac{Pb}{6L}x^3 - \frac{P}{6}\langle x - a\rangle^3 + C_1x + C_2$$

From the boundary conditions $v(0) = 0$ and $v(L) = 0$, we have

$$C_1 = -\frac{Pb}{6L}(L^2 - b^2) \qquad C_2 = 0$$

These results agree with those found for C_1 and C_2 in Example 9.3. Note that the additional constants C_3 and C_4 are not needed now.

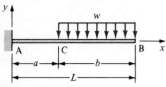

Figure 9.12   Example 9.9.

EXAMPLE 9.9

Determine the equation of the deflection curve for the cantilever beam of Fig. 9.12. Employ the fourth-order differential equation.

Solution   Using singularity functions, we find that after four integrations, Eq. (9.5) yields

$$Elv'''' = -w\langle x - a \rangle^0$$
$$Elv''' = -w\langle x - a \rangle + C_1$$
$$Elv'' = -\tfrac{1}{2}w\langle x - a \rangle^2 + C_1 x + C_2$$
$$Elv' = -\tfrac{1}{6}w\langle x - a \rangle^3 + \tfrac{1}{2}C_1 x^2 + C_2 x + C_3$$
$$Elv = -\tfrac{1}{24}w\langle x - a \rangle^4 + \tfrac{1}{6}C_1 x^3 + \tfrac{1}{2}C_2 x^2 + C_3 x + C_4$$

The boundary conditions are

$$v(0) = 0 \qquad v'(0) = 0 \qquad Elv'''(L) = 0 \qquad Elv''(L) = 0$$

From these we have

$$C_4 = 0 \qquad C_3 = 0 \qquad C_1 = wb \qquad C_2 = -\tfrac{1}{2}wb(L + a)$$

Thus the expressions for the slope and deflection are of the forms

$$Elv' = -\tfrac{1}{6}w\langle x - a \rangle^3 - \frac{wbx}{2}(L + a - x)$$
$$Elv = -\tfrac{1}{24}w\langle x - a \rangle^4 - \frac{wbx^2}{12}(3L + 3a - 2x)$$

(9.31)

For the deflection v_C at the free end ($x = L$), we have

$$v_C = -\frac{wb}{24EI}(2L^3 + 6L^2 a + b^3)$$

(9.32)

A negative value connotes a downward displacement.

9.6   APPLICATION OF THE METHOD OF SUPERPOSITION

The deflections of a simply loaded beam are found readily by applying the integration procedures and are often conveniently tabulated (Table B.7). For combined load configurations, the method of superposition may be used to good advantage to simplify the analysis. The method is valid when the displacements are linearly proportional to the applied loads. This will always be the case if Hooke's law holds for the structural material and if the deflections and slopes are small (see Sec. 3.2).

Consider, for example, the beam of Fig. 9.13a, replaced by the beams depicted in Fig. 9.13b and c. At point C, the beam experiences the deflections $(v_C)_w$ and $(v_C)_P$, owing to w and P, respectively. The total deflection v_C of point C is

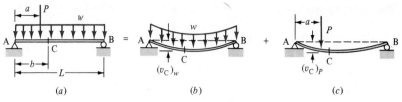

Figure 9.13 Deflections of a simply supported beam with two loads.

$$v_C = (v_C)_w + (v_C)_P$$

Note that, if $a = b = L/2$, we have, from Eqs. (9.12) and (9.19),

$$v_C = \frac{5wL^4}{384EI} + \frac{PL^3}{48EI} \downarrow$$

The expressions for the deflection curve and the angle of rotation of the beam can be found in the same way by combining Eqs. (9.11) and (9.18).

In situations in which the beam is *statically indeterminate,* the approach described in Sec. 5.4 is followed. That is, the redundant reactions are considered as unknown loads and the corresponding supports are removed or modified accordingly. Then superposition is used, the deformation diagrams are drawn, and expressions are written for the displacements caused by the individual loads (both known and unknown). Finally, these reactions are computed such that the displacements fulfill the *geometric* boundary conditions.

This procedure is illustrated for the case of a beam indeterminate to the first degree (Fig. 9.14a). Reaction R_B is selected as redundant and is treated as an unknown load by eliminating the support at B (Fig. 9.14b). Decomposition of the loads is shown in Fig. 9.14c. Solutions for each of the latter cases are (see Table B.7)

$$(v_B)_w = -\frac{wL^4}{8EI} \qquad (v_B)_R = \frac{R_B L^3}{3EI}$$

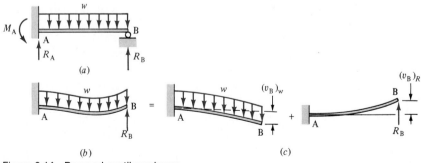

Figure 9.14 Propped cantilever beam.

The compatibility condition for the original beam requires that

$$v_B = -\frac{wL^4}{8EI} + \frac{R_B L^3}{3EI} = 0$$

from which

$$R_B = \tfrac{3}{8}wL \uparrow \tag{9.33a}$$

The remaining reactions are

$$R_A = \tfrac{5}{8}wL \uparrow \qquad M_A = \tfrac{1}{8}wL^2 \circlearrowleft \tag{9.33b}$$

as obtained from the equations of statics.

EXAMPLE 9.10

For the overhanging beam in Fig. 9.15a, determine the deflection at point C.

Solution A free-body diagram of segment AB is shown in Fig. 9.15b. It is observed that this segment carries a uniform load and resists a vertical load P and a moment $M_B = Pa$. The problem can thus be reduced to that of finding the deflection $(v_C)_M$ caused by the rotation of B (Fig. 9.15c) and the deflection $(v_C)_P$ due to the bending of segment BC as a cantilever.

The slope at B (see Table B.7), setting $M_B = Pa$, is

$$\theta_B = \frac{wL^3}{24EI} - \frac{PaL}{3EI}$$

Therefore

$$(v_C)_M = \theta_B a = \frac{wL^3 a}{24EI} - \frac{Pa^2 L}{3EI}$$

We have, from Table B.7, $(v_C)_P = -Pa^3/3EI$. Hence the total deflection at point C is

$$v_C = \frac{wL^3 a}{24EI} - \frac{Pa^2}{3EI}(L + a) \tag{9.34}$$

It is clear that the shape of the elastic curve for the beam depends upon the prescribed values of w, P, L, and a (Fig. 9.15c).

Figure 9.15 Example 9.10.

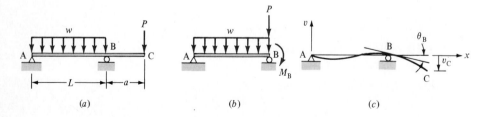

(a)   (b)   (c)

EXAMPLE 9.11

Determine the support reactions for the beam of Fig. 9.16a.

Solution The reactions are indicated in Fig. 9.16a. The couples M_A and M_B at supports are often referred to as the *fixed-end moments*. We shall select these as redundants and treat them as unknown loadings. The slopes at the ends due to P, M_A, and M_B are shown in Fig. 9.16b.

The conditions that the angles of rotation at both ends of the original beam must be zero yield (see Table B.7)

$$\theta_A = (\theta_A)_P - (\theta_A)_{M_A} - (\theta_A)_{M_B} = 0$$

or

$$\frac{M_A L}{3EI} + \frac{M_B L}{6EI} = \frac{Pab(L+b)}{6EIL}$$

and

$$\theta_B = (\theta_B)_P - (\theta_B)_{M_A} - (\theta_B)_{M_B} = 0$$

or

$$\frac{M_A L}{6EI} + \frac{M_B L}{3EI} = \frac{Pab(L+a)}{6EIL}$$

The solutions are

$$M_A = \frac{Pab^2}{L^2} \circlearrowright \qquad M_B = \frac{Pa^2 b}{L^2} \circlearrowright \qquad (9.35)$$

We can now use the equilibrium requirements to determine the remaining reactions:

$$R_A = \frac{Pb^2}{L^3}(L+2a) \uparrow \qquad R_B = \frac{Pa^2}{L^3}(L+2b) \uparrow \qquad (9.36)$$

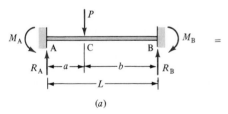

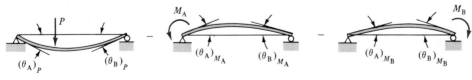

Figure 9.16 Example 9.11.

An alternative approach for solving this problem is to remove the fixed support at the right end and to consider the deformations caused by P, R_B, and M_B separately. In this case, we have $\theta_B = (\theta_B)_P + (\theta_B)_{R_B} + (\theta_B)_{M_B} = 0$ and $v_B = (v_B)_P + (v_B)_{R_B} + (v_B)_{M_B} = 0$, as required by geometric compatibility of deformations.

PROBLEMS

Secs. 9.1 to 9.4

9.1 through 9.6 A cantilever beam is loaded as shown in Figs. P9.1 through P9.6. Using the double-integration method, determine (a) the equation of the elastic curve; (b) the slope at the free end; and (c) the deflection at the free end.

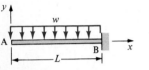

Figure P9.1

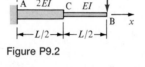

Figure P9.2

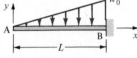

Figure P9.3

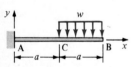

Figure P9.4

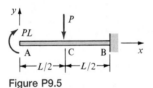

Figure P9.5

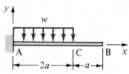

Figure P9.6

9.7 and 9.8 Solve Probs. 9.1 and 9.2, using the multiple-integration method.

9.9 and 9.10 Redo Probs. 9.3 and 9.4, using the multiple-integration method.

9.11 and 9.12 Solve Probs. 9.5 and 9.6, using the multiple-integration method.

9.13 A cantilever beam is loaded as shown in Fig. P9.13. Using the multiple-integration method, determine (a) the equation of the elastic curve; (b) the deflection at the free end; and (c) the reactions at the fixed support.

9.14 Determine the maximum slope and maximum deflection of the beam in Fig. P9.1. The beam is of S 100×14 rolled steel (see Table B.3); the remaining data are $w = 1.2$ kN/m, $L = 2$ m, and $E = 200$ GPa.

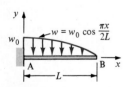

Figure P9.13

9.15 Determine the deflection at the center of the beam shown in Fig. P9.5 if the beam is a solid rod of diameter $d = 2$ in. and $P = 80$ lb, $L = 3$ ft, and $E = 10 \times 10^6$ psi.

9.16 through 9.21 A simple beam is loaded as shown in Figs. P9.16 through P9.21. Using the double-integration method, determine (*a*) the equation of the elastic curve; (*b*) the slope at the end A; and (*c*) the deflection at midspan.

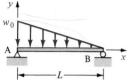

Figure P9.16

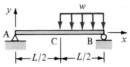

Figure P9.17

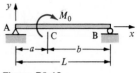

Figure P9.18

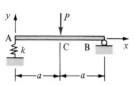

Figure P9.19

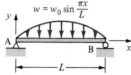

Figure P9.20

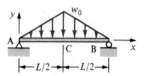

Figure P9.21

9.22 and 9.23 Redo Probs. 9.16 and 9.17, using the multiple-integration method.

9.24 and 9.25 Redo Probs. 9.18 and 9.19, using the multiple-integration method.

9.26 and 9.27 Solve Probs. 9.20 and 9.21, using the multiple-integration method.

9.28 Calculate the deflection at midspan of the beam shown in Fig. P9.17 if the beam is an S 200 × 34 rolled shape (see Table B.3), and $w = 15$ kN/m, $L = 3$ m, and $E = 210$ GPa.

9.29 Find the slope at A of the beam in Fig. P9.21 if the beam is a W 250 × 80 rolled shape (see Table B.2), and $w_0 = 50$ kN/m, $L = 4$ m, and $E = 200$ GPa.

9.30 through 9.35 Using a direct-integration method, determine, for the beam loaded as shown in Figs. P9.30 through P9.35, (*a*) the equation of the elastic curve; (*b*) the deflection at midspan; (*c*) the magnitude and location of the maximum deflection; and (*d*) the slope at the right end.

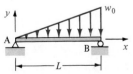

Figure P9.30

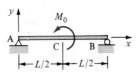

Figure P9.31

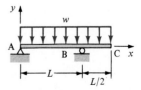

Figure P9.32

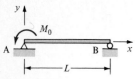

Figure P9.33

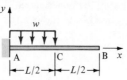

Figure P9.34

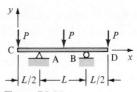

Figure P9.35

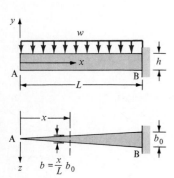

Figure P9.37

9.36 Calculate the slope at A of the beam shown in Fig. P9.31 if the beam is a W 12 × 72 rolled shape (see Table B.2) and $M_O = 400$ kips·in., $L = 15$ ft, and $E = 30 \times 10^6$ psi.

9.37 A cantilever of variable cross section supports a uniformly distributed load w (Fig. P9.37). Verify, using the last of Eqs. (9.2), that the expression for the deflection curve is

$$v = \frac{wL}{Eb_0 h^3} (-x^3 + 3L^2 x - 2L^3) \tag{P9.37}$$

Here h represents the depth of the beam and b_0 is the width at its fixed end.

9.38 The slope at the wall of the clamped beam shown in Fig. P9.38a is as seen in Fig. P9.38b. Find the reaction on the roller at A.

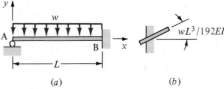

(a)   (b)

Figure P9.38

9.39 Before load w is applied to the beam shown in Fig. 9.6a, a small gap $\delta_B = \frac{1}{8}$ in. exists between the beam AC and the support at B. Determine the reaction at each support developed subsequent to the loading. Let $w = 1.2$ kips/ft, $L = 12$ ft, and $EI = 400(10^6)$ lb·in.^{2}.

9.40 Before load w_0 is applied to the beam shown in Fig. 9.8a, a small gap, $\delta_A = \frac{1}{16}$ in., exists between the beam AB and the support at A. Determine the reaction at each support developed subsequent to the loading. Assume that $w_0 = 2.4$ kips/ft, $L = 10$ ft, $E = 30 \times 10^6$ psi, $I = 65$ in.^{4}, and the bearing block is rigid.

9.41 through 9.49 A beam is loaded and supported as shown in Figs. P9.41 through P9.49. Using a direct-integration method, determine (a) all of the reactions and (b) the midspan deflection if $w = w_0/2 = 15$ kN/m, $P = 25$ kN, $M_0 = 10$ kN·m, $L = 4$ m, $E = 200$ GPa, and $I = 18 \times 10^6$ mm^4, as needed.

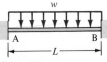

Figure P9.41

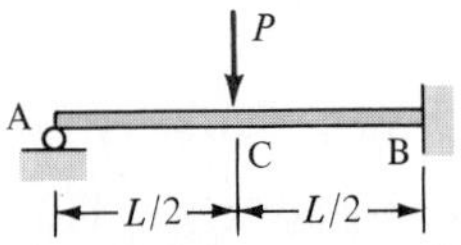

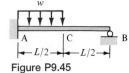

Figure P9.42

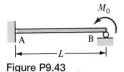

Figure P9.43

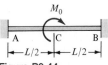

Figure P9.44

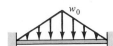

Figure P9.45

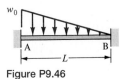

Figure P9.46

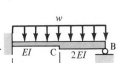

Figure P9.47

Figure P9.48

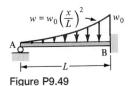

Figure P9.49

Sec. 9.5

9.50 Using singularity functions, solve Prob. 9.4.

9.51 Using singularity functions, solve Prob. 9.5.

9.52 Using singularity functions, solve Prob. 9.6.

9.53 through 9.55 For the beam loaded and supported as shown in Figs. P9.53 through P9.55, using singularity functions, determine (a) the equation of the elastic curve and (b) slope at end B if $P = 4$ kips, $a = 3$ ft, $L = 10$ ft, $E = 10 \times 10^6$ psi, and $I = 72$ in.^{4}.

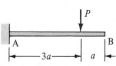

Figure P9.53

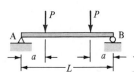

Figure P9.54

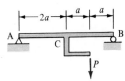

Figure P9.55

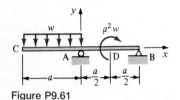

Figure P9.61

9.56 Using singularity functions, solve Prob. 9.17.

9.57 Using singularity functions, solve Prob. 9.18.

9.58 Using singularity functions, solve Prob. 9.19.

9.59 Using singularity functions, solve Prob. 9.21.

9.60 Using singularity functions, solve Prob. 9.31.

9.61 Employing singularity functions, determine the deflection at the free end of the beam shown in Fig. P9.61.

9.62 Using singularity functions, solve Prob. 9.42.

9.63 Using singularity functions, solve Prob. 9.44.

9.64 Using singularity functions, solve Prob. 9.45.

9.65 through 9.70 A beam is loaded and supported as shown in Figs. P9.65 through P9.70. Using singularity functions, determine (a) the reactions at support A and (b) the deflection at the midspan.

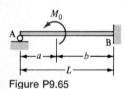

Figure P9.65

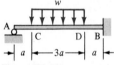

Figure P9.66

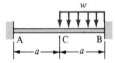

Figure P9.67

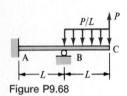

Figure P9.68

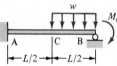

Figure P9.69

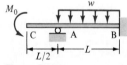

Figure P9.70

9.71 Before the loads are applied, a small gap δ_B exists between beam AC and support B, as seen in Fig. P9.71. Subsequent to loading, the gap closes and reactions develop at each support. Using singularity functions, determine the reactions.

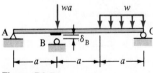

Figure P9.71

Sec. 9.6

9.72 Using the method of superposition, solve part c of Prob. 9.5.

9.73 Using the method of superposition, determine the slope and deflection at end C of the beam loaded as shown in Fig. 9.7.

9.74 Using the method of superposition, solve part c of Prob. 9.4.

9.75 Using the method of superposition, solve part b of Prob. 9.54.

9.76 Using the method of superposition, solve part b of Prob. 9.55.

9.77 through 9.79 A beam is loaded and supported as shown in Figs. P9.77 through P9.79. Using the method of superposition, determine the deflection at the free end.

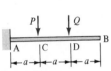

Figure P9.77

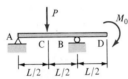

Figure P9.78

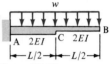

Figure P9.79

9.80 An S 250 × 38 rolled-shape beam (see Table B.3) is loaded as shown in Fig. P9.77. Using the method of superposition, calculate the slope at B. Let $E = 200$ GPa, $I = 51.4 \times 10^{-6}$ m^4, $a = 2$ m, and $P = 2Q = 10$ kN.

9.81 The beam AC in Fig. P9.78 is made of a W 150 × 24 rolled-steel shape (see Table B.2), and $M_0 = 10$ kN·m, $P = 4$ kN, $L = 3$ m, and $E = 210$ GPa. Using the method of superposition, calculate the deflection at C.

9.82 Redo Example 9.4, using the method of superposition.

9.83 Using the method of superposition, solve part a of Prob. 9.42.

9.84 Using the method of superposition, solve part a of Prob. 9.43.

9.85 Using the method of superposition, solve part a of Prob. 9.44.

9.86 Using the method of superposition, solve part a of Prob. 9.45.

9.87 Using the method of superposition, solve part a of Prob. 9.46.

9.88 Using the method of superposition, solve part a of Prob. 9.47.

9.89 Using the method of superposition, solve part a of Prob. 9.48.

9.90 Using the method of superposition, solve part a of Prob. 9.70.

9.91 through 9.96   A beam is loaded as shown in Figs. P9.91 through P9.96. Employ the method of superposition to determine the reaction at each support.

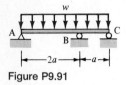

Figure P9.91

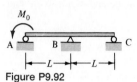

Figure P9.92

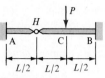

Figure P9.93

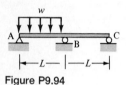

Figure P9.94

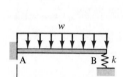

Figure P9.95

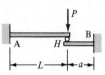

Figure P9.96

10

DEFLECTION OF BEAMS BY SPECIAL METHODS*

10.1 INTRODUCTION

In many applications in which beam displacements must be obtained, the loading varies arbitrarily and the beam is nonprismatic. Examples of such situations include those involving stepped beams and shafts of machines and tapered beams employed in aircraft and bridge structures. The use of moment-area or finite-difference methods enables the engineer to expand his or her ability to solve such problems. These special methods thus provide tools with which the engineer may be freer to undertake the solution of problems as they exist in practice.

The *moment-area procedures* (Secs. 10.2 and 10.3) interpret semigraphically the mathematical operations concerned with solving the governing differential equation of a beam. The method of *finite differences* (Sec. 10.4) is a general and very useful numerical method. This approach approximates differentials by finite differences, and as a result, it transforms the beam differential equation into a set of algebraic equations, written for every nodal point in the beam (Sec. 10.5).

The foregoing methods are merely alternative approaches for solving beam-deflection problems and carry with them the limitations of the technical theory of bending. Their application to statically determinate as well as indeterminate beams will be considered in this chapter.

The most general and commonly employed numerical technique is the *finite-element method*. In this powerful procedure, the structural member is discretized by a finite number of elements, connected at their corner points or nodes and along hypothetical interelement boundaries. The conditions of equilibrium and compatibility of displacements considered at each node and along the boundaries between elements lead to a set of algebraic equations.

*The material presented in this chapter is optional, and the entire chapter can be omitted without destroying the continuity of the text.

Thus an essential difference exists between the finite-element and finite-difference analyses. The former is relatively lengthy and is not treated in this book, but both approaches have clear application to digital computers (Sec. 10.6).

10.2 MOMENT-AREA METHOD

We now treat the displacements of elastic beams employing a semigraphical technique called the moment-area method. This approach utilizes the relationships between the derivatives of the deflection v and the properties of the area of bending-moment diagram. It usually offers a more rapid solution than integration methods when the deflection and slope at *only* one point of the beam are required. The moment-area method is particularly effective in dealing with beams of variable cross section with uniform or concentrated loading.

Two theorems form the basis of the moment-area approach. They are developed by considering the bending of a beam carrying an arbitrary loading (Fig. 10.1a). The M/EI diagram is assumed to be as shown in Fig. 10.1b. A sketch of the greatly exaggerated deflection curve is depicted in Fig. 10.1c. The change in the angle $d\theta$ of the tangents at the ends of an element of length dx and the bending moment are related through the second of Eqs. (9.2):

$$d\theta = \frac{M}{EI}\,dx \qquad (a)$$

Thus the *difference in slope* between any two points A and B for the beam of Fig. 10.1a can be expressed in the form

$$\theta_{BA} = \theta_B - \theta_A = \int_A^B \frac{M}{EI}\,dx \qquad (10.1a)$$

or

$$\theta_{BA} = \text{area of } \frac{M}{EI} \text{ diagram between A and B} \qquad (10.1b)$$

This is the *first moment-area theorem*: The change in angle θ_{BA} between the tangents to the elastic curve at two points A and B equals the area of the M/EI diagram between those points.

Observe that angle θ_{BA} and the area of the M/EI diagrams are of the *same sign*. That is, a positive (negative) area corresponds to a counterclockwise (clockwise) rotation of the tangent to the elastic curve as we proceed in the x direction. Accordingly, θ_{BA} shown in Fig. 10.1c is positive.

As the deflections of a beam are assumed small, we can see from Fig. 10.1c that the vertical distance dt owing to the effect of curvature of an element of length dx equals $x_1\,d\theta$, where $d\theta$ is given by Eq. (a). Therefore *vertical distance* AA′, the *tangential deviation* t_{AB} of point A from the tangent at B, is

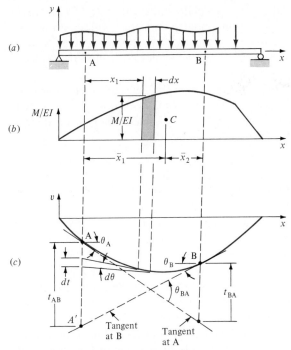

Figure 10.1   Moment-area method: (a) load diagram;
(b) M/EI diagram; and (c) elastic curve.

$$t_{AB} = \int_A^B x_1 \frac{M\,dx}{EI} \tag{10.2a}$$

or

$$t_{AB} = \left(\text{area of } \frac{M}{EI} \text{ diagram between A and B} \right) \bar{x}_1 \tag{10.2b}$$

In the above, $\bar{x}_1$ is the horizontal distance to the centroid C of the area from A. The foregoing represents the *second moment-area theorem*: The tangential deviation t_{AB} of point A with respect to the tangent at B equals the moment with respect to A of the area of the M/EI diagram between A and B.

Similarly, the tangential deviation t_{BA} of B from the tangent at A is equal to the moment with respect to B of the area of the M/EI diagram between A and B:

$$t_{BA} = \left(\text{area of } \frac{M}{EI} \text{ diagram between A and B} \right) \bar{x}_2 \tag{10.3}$$

Here $\bar{x}_2$ is the horizontal distance from point B to the centroid C of the area (Fig. 10.1b).

We observe from Eqs. (10.2) and (10.3) that the signs of t_{AB} and t_{BA} depend upon the sign of the bending moments. A point with a *positive* tangential deviation is located *above* the tangent drawn from the other point, as seen in Fig. 10.1c. Accordingly, a point with a negative tangential deviation lies below the corresponding tangent.

It should be noted that the direction of the deflections and slopes can usually be found by inspection. When this is the case, it is not necessary to follow the sign conventions described for the moment-area method: Only numerical values are used in the calculations.

10.3 APPLICATION OF THE MOMENT-AREA METHOD

In determining the deflections and slopes by the method of moment-area, a correctly constructed M/EI diagram and a sketch of the elastic curve are always necessary. To facilitate the application of the method, the areas and the centroidal distances of common shapes are furnished in a table inside the back cover of this book. The slopes of points on the beam with respect to one another can be determined using Eq. (10.1) and the deflection, applying Eq. (10.2) or (10.3).

The moment-area procedure is most conveniently applied to beams in which the direction of the tangent to the elastic curve at one or more points is known (for example, cantilever beams). Note that it may often be advantageous to draw the M/EI diagrams and make the formulations in terms of symbols, inserting numerical values in the final step of the solution.

For a statically determinate beam with several loads and for an indeterminate beam, the deflections obtained by the method of moment-areas are frequently best found by superposition. This requires a series of diagrams showing the moment owing to each load or reaction drawn on a separate sketch. In this way calculations can be simplified because the areas of the separate M/EI diagrams may be simple geometric forms. In the case of indeterminacy, each additional geometric compatibility condition is expressed by a moment-area equation to supplement the equations of equilibrium.

EXAMPLE 10.1

Compute the center deflection and the slope at the ends of the beam shown in Fig. 10.2a.

Solution Since the flexural rigidity EI is constant, the M/EI diagram has the same parabolic shape as the bending-moment diagram (Fig. 10.2b). (The area properties indicated in the figure are taken from the table on the inside back cover.) The elastic curve is shown in Fig. 10.2c, with the tangent drawn at A.

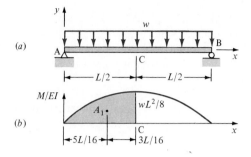

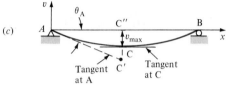

Figure 10.2   Example 10.1.

The beam and its loading are symmetric with respect to center C. Thus the tangent to the elastic curve at C is *horizontal*. Referring to the figure, it is observed that $\theta_C = 0$,

$$\theta_{CA} = 0 - \theta_A \quad \text{or} \quad \theta_A = -\theta_{CA}$$

and

$$A_1 = \frac{2}{3}\left(\frac{L}{2}\right)\left(\frac{wL^2}{8EI}\right) = \frac{wL^3}{24EI}$$

Then, using the first moment-area theorem, $\theta_{CA} = A_1$, or

$$\theta_A = -\frac{wL^3}{24EI} = \frac{wL^3}{24EI}\, \text{↖}$$

The end A of the beam rotates clockwise, as depicted in the figure, and $\theta_B = -\theta_A$.

Applying the second moment-area theorem, Eq. (10.3), we have

$$t_{CA} = A_1\left(\frac{3L}{16}\right) = \frac{wL^4}{128EI}$$

where $t_{CA} = CC'$ and $\theta_A L/2 = C'C''$ (Fig. 10.2c). The maximum deflection, $v_{max} = CC''$, is therefore

$$v_{max} = -\frac{wL^3}{24EI}\left(\frac{L}{2}\right) + \frac{wL^4}{128EI} = -\frac{5wL^4}{384EI} = \frac{5wL^4}{384EI}\, \downarrow$$

Alternatively, the moment of A_1 about point A, Eq. (10.2), readily yields the numerical value of v_{max}.

EXAMPLE 10.2

Determine the slope and deflection at B and C of the cantilever beam loaded as shown in Fig. 10.3a. The left half of the beam has an EI twice that of the right half. Calculate the maximum angle of rotation and maximum deflection for $P = 5$ kips, $L = 6$ ft, $E = 10 \times 10^6$ psi, and $I = 65$ in.[4].

Solution The M/EI diagram is divided conveniently into its component parts (rectangle and triangle) for which the areas

$$A_1 = -\frac{PL^2}{8EI} \qquad A_2 = -\frac{PL^2}{16EI} \qquad A_3 = -\frac{PL^2}{8EI}$$

and the centroids are readily determined (Fig. 10.3b). The elastic curve is concave downward throughout its length as the bending moments are negative (Fig. 10.3c). Since $\theta_A = 0$ and $v_A = 0$,

$$\theta_C = \theta_{CA} \qquad \theta_B = \theta_{BA} \qquad v_C = t_{CA} \qquad v_B = t_{BA}$$

Using the first moment-area theorem, we obtain

$$\theta_C = A_1 + A_2 = -\frac{3PL^2}{16EI}$$

$$\theta_B = A_1 + A_2 + A_3 = -\frac{5PL^2}{16EI} \qquad (10.4)$$

Applying the second moment-area theorem, we have

$$v_C = A_1\left(\frac{L}{4}\right) + A_2\left(\frac{L}{3}\right) = -\frac{5PL^3}{96EI}$$

$$v_B = A_1\left(\frac{3L}{4}\right) + A_2\left(\frac{5L}{6}\right) + A_3\left(\frac{L}{3}\right) = -\frac{3PL^3}{16EI} \qquad (10.5)$$

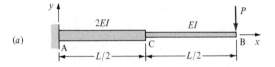

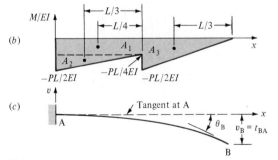

Figure 10.3 Example 10.2.

The largest values of θ and v occur at point B:

$$\theta_{max} = \theta_B = -\frac{5(5 \times 10^3)(72)^2}{16(10 \times 10^6)(65)} = -12.46 \times 10^{-3} \text{ rad}$$

$$= 12.46 \times 10^{-3} \text{ rad} \nwarrow$$

$$v_{max} = v_B = -\frac{3(5 \times 10^3)(72)^3}{16(10 \times 10^6)(65)} = -0.538 \text{ in.}$$

$$= 0.538 \text{ in.} \downarrow$$

In the above, the given data have been substituted into Eqs. (10.4) and (10.5).

EXAMPLE 10.3

Determine the slope and deflection at the free end B of the cantilever beam of Fig. 10.4a in terms of P, L, E, and I. Compute the maximum vertical displacement if the beam is a W 150 × 30 rolled shape (see Table B.2). Use $P = 50$ kN, $L = 3$ m, and $E = 210$ GPa.

Solution The reactions are noted in Fig. 10.4a. The flexural rigidity is constant, resulting in the M/EI diagram seen in Fig. 10.4b. The bending moment changes sign at $x = 0.4L$. The areas of the diagram are

$$A_1 = \frac{1}{2}\left(-0.4\frac{PL}{EI}\right)\frac{2L}{5} = -0.08\frac{PL^2}{EI}$$

$$A_2 = \frac{1}{2}\left(0.6\frac{PL}{EI}\right)\frac{3L}{5} = 0.18\frac{PL^2}{EI}$$

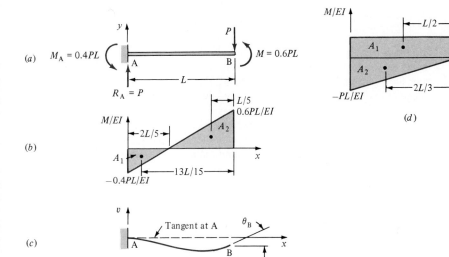

Figure 10.4 Example 10.3.

Corresponding to the positive moment, the elastic curve is concave up, and vice versa, and $\theta_A = 0$ (Fig. 10.4c). Therefore $\theta_{BA} = \theta_B$ and the first moment-area theorem yields

$$\theta_B = A_1 + A_2 = 0.1 \frac{PL^2}{EI} \quad \measuredangle \qquad (10.6)$$

Inasmuch as the tangent at A is a horizontal, the deflection at B is equal to the tangential deviation t_{BA}. Thus, employing the second moment-area theorem, we obtain

$$v_B = t_{BA} = A_1 \left(\frac{13L}{15}\right) + A_2 \left(\frac{L}{5}\right) = -\frac{1}{30}\frac{PL^3}{EI} = \frac{1}{30}\frac{PL^3}{EI} \downarrow \qquad (10.7)$$

After inserting the numerical values, we have

$$v_{max} = v_B = \frac{1}{30} \frac{50 \times 10^3 (3^3)}{6(17.2 \times 210 \times 10^3)} = 12.5 \times 10^{-3} \, \text{m} = 12.5 \, \text{mm} \downarrow$$

Alternative Solution. A solution may conveniently be found by summing the M/EI diagrams obtained from each of the loads acting separately. Referring to the *composite M/EI* diagram sketched this way (Fig. 10.4d), we have

$$A_1 = 0.6 \frac{PL^2}{EI} \qquad A_2 = -0.5 \frac{PL^2}{EI}$$

The moment-area theorems then readily result in

$$\theta_B = A_1 + A_2 = 0.1 \frac{PL^2}{EI} \quad \measuredangle$$

$$v_B = t_{BA} = A_1 \left(\frac{L}{2}\right) + A_2 \left(\frac{2L}{3}\right) = -\frac{1}{30}\frac{PL^3}{EI} = \frac{1}{30}\frac{PL^3}{EI} \downarrow$$

as before.

EXAMPLE 10.4

For the beam in Fig. 10.5a determine (*a*) the deflection of the free end; (*b*) the slope at the left end; and (*c*) the maximum deflection between supports.

Solution The reactions are indicated in Fig. 10.5a. Figure 10.5b represents the M/EI diagram. The deflection curve with a tangent drawn at A is shown in Fig. 10.5c, concave downward as the bending moments are negative. Obviously, the curve must pass through the support points at A and B. Referring to the table inside the back cover, we find that the areas of the triangle and parabolic spandrel are, respectively,

$$A_1 = -\frac{1}{2}\left(\frac{wL^2}{8EI}\right) L = -\frac{wL^3}{16EI}$$

$$A_2 = -\frac{1}{3}\left(\frac{L}{2}\right)\frac{wL^2}{8EI} = -\frac{wL^3}{48EI}$$

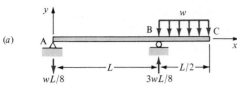

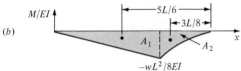

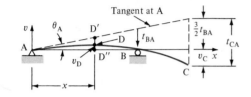

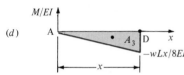

Figure 10.5   Example 10.4.

(a) Applying the second moment-area theorem, we find the tangential deviation at B to be

$$t_{BA} = A_1 \left(\frac{L}{3} \right) = -\frac{wL^4}{48EI} \qquad (a)$$

In a like manner,

$$t_{CA} = A_1 \left(\frac{5L}{6} \right) + A_2 \left(\frac{3L}{8} \right) = -\frac{23wL^4}{384EI}$$

Thus

$$v_C = t_{CA} - \frac{3}{2} t_{BA} = -\frac{11wL^4}{384EI} = \frac{11wL^4}{384EI} \downarrow \qquad (10.8)$$

(b) The angle between the tangent at A and the horizontal line is, from Fig. 10.5c and Eq. (a),

$$\theta_A = \frac{t_{BA}}{L} = \frac{wL^3}{48EI} \quad \measuredangle \qquad (10.9)$$

(c) The largest deflection between A and B occurs at point D, where $\theta_D = 0$ (Fig. 10.5c). The area of the M/EI diagram between A and D is (Fig. 10.5d)

$$A_3 = -\frac{1}{2}\left(\frac{wLx}{8EI}\right)x = -\frac{wLx^2}{16EI} \qquad (b)$$

Applying the first moment-area theorem with $\theta_D = 0$, we have $\theta_{DA} = 0 - \theta_A = A_3$. Substitution of Eqs. (10.9) and (b) into the foregoing yields

$$\frac{wL^3}{48EI} = \frac{wLx^2}{16EI}$$

from which

$$x = \frac{L}{\sqrt{3}}$$

It follows that

$$A_3 = -\frac{wL^3}{48EI}$$

Referring to Fig. 10.5c, we have $D''D' = x\theta_A$ and $DD' = t_{DA}$. From the second moment-area theorem, we have

$$t_{DA} = -A_3\left(\frac{x}{3}\right) = -\frac{wL^4}{144\sqrt{3}\ EI}$$

The maximum deflection v_D is, therefore,

$$v_D = \frac{L}{\sqrt{3}}\left(\frac{wL^3}{48EI}\right) - \frac{wL^4}{144\sqrt{3}\ EI} = \frac{wL^4}{72\sqrt{3}\ EI}\ \uparrow \qquad (10.10)$$

Alternatively, the numerical value of v_D may easily be found by obtaining the moment of the area A_3 about A (Fig. 10.5d):

$$|v_D| = \frac{wL^3}{48EI}\left(\frac{2}{3}\frac{L}{\sqrt{3}}\right) = \frac{wL^4}{72\sqrt{3}\ EI}$$

This is the same as given by Eq. (10.10).

EXAMPLE 10.5

Determine the reactions for the beam of Fig. 10.6a.

Solution The reactions noted in Fig. 10.6a show that the beam is indeterminate to the first degree. This may be reduced to determinacy by selecting R_B as redundant and removing it. We next consider the reaction at B as the unknown load, placing the loadings on the released beam (Fig. 10.6b), and drawing the corresponding composite M/EI diagram (Fig. 10.6c). The component triangular areas are

$$A_1 = \frac{R_B L^2}{2EI} \qquad A_2 = -\frac{Pa^2}{2EI}$$

One geometric condition relative to beam displacement is needed to deter-

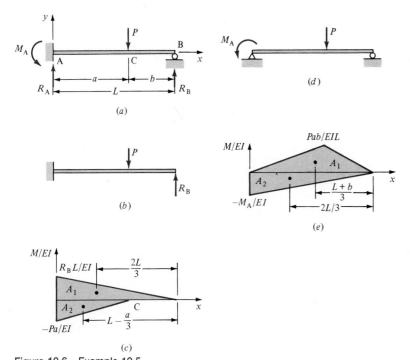

Figure 10.6   Example 10.5.

mine the redundant. Note that the slope at the fixed end and the deflection at the supported end are zero. Thus the tangent to the elastic curve at A passes through B, or $t_{BA} = 0$, and we have, from the second moment-area theorem,

$$\frac{R_B L^2}{2EI}\left(\frac{2L}{3}\right) - \frac{Pa^2}{2EI}\left(L - \frac{a}{3}\right) = 0$$

This provides the redundant reaction

$$R_B = \frac{Pa^2}{2L^3}(3L - a) \uparrow \qquad (10.11a)$$

The equations of equilibrium then result in the following:

$$R_A = \frac{Pb}{2L^3}(3L^2 - b^2) \uparrow \qquad (10.11b)$$

$$M_A = \frac{Pab}{2L^2}(L + b) \circlearrowleft \qquad (10.11c)$$

Since all reactions are known, the slope and deflection are found as needed by applying the usual moment-area procedure.

Alternative Solution.   Select the reactive couple at A as redundant, and replace the fixed end by a pin support (Fig. 10.6d). The corresponding composite M/EI

diagram owing to P and M_A is shown in Fig. 10.6e, where, from the table inside the back cover, we have

$$A_1 = \frac{Pab}{2EI} \qquad A_2 = -\frac{M_A L}{2EI}$$

Then the condition that the tangential deviation t_{BA} must be zero,

$$t_{BA} = A_1 \left(\frac{L+b}{3}\right) + A_2 \left(\frac{2L}{3}\right) = 0$$

yields the same result for M_A as given by Eq. (10.11c). The remaining two reactions follow from application of statics.

EXAMPLE 10.6

Determine the support reactions for the beam and loading of Fig. 10.7a.

Solution The beam is statically indeterminate to the second degree, thus requiring two supplementary equations. We choose the reactions at A as redundants and release the beam from that support (Fig. 10.7b). Considering the effects of known and unknown loads, we draw the composite M/EI diagram (Fig. 10.7c). The component areas are as follows:

$$A_1 = \frac{9R_A a^2}{2EI} \qquad A_2 = -\frac{3M_A a}{EI} \qquad A_3 = -\frac{4wa^3}{3EI}$$

The deflections and slopes of the elastic curve vanish at both supports. Hence the angle between tangents drawn at A to B is zero ($t_{BA} = 0$). The two moment-area theorems then yield

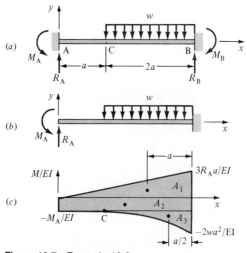

Figure 10.7 Example 10.6.

$$\theta_{BA} = A_1 + A_2 + A_3 = 27R_A a - 18M_A - 8wa^2 = 0$$

$$t_{BA} = A_1(a) + A_2 \left(\frac{3a}{2}\right) + A_3 \left(\frac{a}{2}\right) = 27R_A a - 27M_A - 4wa^2 = 0$$

Solving these equations simultaneously, we have

$$M_A = \tfrac{4}{9}wa^2 \; \curvearrowleft$$

$$R_A = \tfrac{16}{27}wa \; \uparrow$$
(10.12)

The other two reactions are

$$M_B = \tfrac{2}{3}wa^2 \; \curvearrowright$$

$$R_B = \tfrac{38}{27}wa \; \uparrow$$
(10.13)

as determined from the equations of equilibrium.

10.4 FINITE DIFFERENCES

In considering the fundamental relationships used in the finite-differences approach, we begin with a continuous function $v = f(x)$. Figure 10.8 depicts a graphical representation of the function at equidistant small values of x, h. In the derivations that follow, we will use the symbols Δ, ∇, and δ to denote forward, backward, and central differences, respectively, in terms of the h and a pivotal, or *nodal, point* n.

The first derivative of v at n on the curve may be approximated as follows:

$$\left(\frac{dv}{dx}\right)_n \approx \frac{\Delta v_n}{h} = \frac{1}{h}(v_{n+1} - v_n)$$
(a)

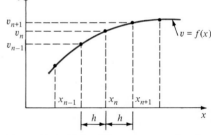

Figure 10.8 Finite-difference approximation.

Here

$$\Delta v_n = v_{n+1} - v_n \approx h \left(\frac{dv}{dx}\right)_n$$
(10.14)

is called the *first forward difference*. Similarly, the derivative can also be approximated:

$$\left(\frac{dv}{dx}\right)_n \approx \frac{\nabla v_n}{h} = \frac{1}{h}(v_n - v_{n-1})$$
(b)

in which

$$\nabla v_n = v_n - v_{n-1} \approx h \left(\frac{dv}{dx}\right)_n$$
(10.15)

is termed the *first backward difference*. The central differences involve the points symmetrically located with respect to n. Central differences often result in a more accurate approximation than forward and backward differences. For this case,

$$\left(\frac{dv}{dx}\right)_n \approx \frac{1}{2h}\,(v_{n+1} - v_{n-1}) \qquad (c)$$

In this expression,

$$\delta v_n = \frac{1}{2}\,(v_{n+1} - v_{n-1}) \approx h\left(\frac{dv}{dx}\right)_n \qquad (10.16)$$

is referred to as the *first central difference*. It is interesting to observe that the expression $\delta v_n = (\Delta v_n + \nabla v_n)/2$ relates the three first differences.

A procedure identical with that employed above will lead to higher-order derivatives. Hereafter only the central differences will be considered. The second derivative is expressed as

$$h^2\left(\frac{d^2v}{dx^2}\right)_n \approx \Delta(\nabla v_n) = \nabla(\Delta v_n) = \delta^2 v_n$$

Upon substituting Eqs. (10.14) and (10.15) into the foregoing, we have

$$\delta^2 v_n = \Delta v_n - \Delta v_{n-1} = (v_{n+1} - v_n) - (v_n - v_{n-1})$$

The *second central difference* is thus

$$\delta^2 v_n = v_{n+1} - 2v_n + v_{n-1} \approx h^2\left(\frac{d^2v}{dx^2}\right)_n \qquad (10.17)$$

The *third central difference* is readily determined as follows:

$$\delta^3 v_n = \delta(\delta^2 v_n) = \delta v_{n+1} - 2\delta v_n + \delta v_{n-1}$$
$$= \tfrac{1}{2}(v_{n+2} - v_n) - (v_{n+1} - v_{n-1}) + \tfrac{1}{2}(v_n - v_{n-2})$$

or

$$\delta^3 v_n = \frac{1}{2}\,(v_{n+2} - 2v_{n+1} + 2v_{n-1} - v_{n-2}) \approx h^3\left(\frac{d^3v}{dx^3}\right)_n \qquad (10.18)$$

In a like manner,

$$\delta^4 v_n = v_{n+2} - 4v_{n+1} + 6v_n - 4v_{n-1} + v_{n-2} \approx h^4\left(\frac{d^4v}{dx^4}\right)_n \qquad (10.19)$$

Having available the various derivatives in terms of (central) difference approximations, we can readily obtain the finite-difference equivalents of the beam equations, as described in the next section.

10.5   FINITE-DIFFERENCE METHOD

In this section consideration is given a numerical method for determining beam displacements. Known as the finite-difference method, it transforms the governing differential equation (and the equation of the boundary conditions) into a set of simultaneous finite-difference expressions at selected points along the beam. These points are located at the joints of beam segments. We describe the elastic curve of the beam through approximate values of the deflection v at these nodal points.

To understand the method, we first consider the deflection of a nonprismatic cantilever beam with depth varying arbitrarily and of constant width (Fig. 10.9a). We shall divide the beam into m segments of length $h = L/m$ and replace the variable loading by a load changing linearly between nodes (Fig. 10.9b). This is a simple finite-difference representation of the original beam and loading. Refinements which may serve to reduce the number of segments required to effect an accurate solution include replacement of the actual force distribution with a series of parabolically varying loadings. The optimum choice of segment size is dependent upon the load-averaging technique used. With a larger number of segments, it is usually the case that there will be a smaller error in the result.

We can now write the finite-difference equations at a nodal point n (Fig. 10.9b). Referring to Eq. (10.17), we find the difference equation corresponding to $EI(d^2v/dx^2) = M$ to be of the form

$$v_{n+1} - 2v_n + v_{n-1} = h^2 \left(\frac{M}{EI} \right)_n \qquad (10.20)$$

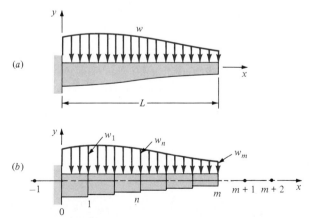

Figure 10.9   Nonprismatic cantilever beam with varying load: (a) actual structure and (b) finite-difference representation.

Here M_n and $(EI)_n$ represent the moment and the flexural rigidity, respectively, of the beam at n. Similarly, the equation $EI(d^4v/dx^4) = w$ is written

$$v_{n+2} - 4v_{n+1} + 6v_n - 4v_{n-1} + v_{n-2} = h^4 \left(\frac{w}{EI}\right)_n \qquad (10.21)$$

where w_n is the load intensity at n. Expressions identical with Eqs. (10.20) or (10.21) can be established at each remaining nodal point in the beam. There will then be m such equations. Thus the problem involves the solution of m unknowns, $v_1, \ldots, v_m$. The boundary conditions (Fig. 10.9b) are

$$v = 0 \qquad dv/dx = 0 \qquad\qquad (x = 0)$$
$$M = EI(d^2v/dx^2) = 0 \qquad V = EI(d^3v/dx^3) = 0 \qquad (x = L)$$

In finite-difference form, these become

$$v_0 = 0 \qquad v_1 = v_{-1}$$
$$v_{m+1} - 2v_m + v_{m-1} = 0 \qquad v_{m+2} - 2v_{m+1} + 2v_{m-1} - v_{m-2} = 0 \qquad (a)$$

When the fictitious points at $m = -1, m + 1, m + 2$ are introduced, there will be a sum of $m + 4$ unknown deflections (including $m = 0$). Equation (10.20) or (10.21), written for m nodal points within the beam, together with Eqs. (a), results in a total of $m + 4$ expressions for obtaining the $m + 4$ deflections. Rewriting such equations in matrix form makes the finite-difference approach more systematic and provides concise presentation.

The above described method is valid for *any* beam and loading. Upon evaluating the v's, we can relate them to the angular displacements along the beam axis using Eq. (10.16). At a nodal point n, for instance, the slope is given by

$$\theta_n = \left(\frac{dv}{dx}\right)_n = \frac{1}{2h}(v_{n+1} - v_{n-1}) \qquad (10.22)$$

a predictable result.

We note that the common boundary conditions (see Fig. 9.2) at a support designated by nodal point n may readily be written in finite-difference form, referring to Eqs. (9.6) through (9.9). For reference purposes, these are presented in Table 10.1.

Table 10.1   **Finite-Difference Boundary Conditions**

Edge	Fixed	Simply Supported	Free
At x_n	$v_n = 0$ $v_{n+1} = v_{n-1}$	$v_n = 0$ $v_{n+1} = -v_{n-1}$	$v_{n+1} - 2v_n + v_{n-1} = 0$ $v_{n+2} - 2v_{n+1} + 2v_{n-1} - v_{n-2} = 0$

It is apparent that the finite-difference approach has many advantages: simplicity and versatility; suitability for digital computer and programmable calculator use (or in case of coarse segmentation, for hand computation); and acceptable accuracy for most practical purposes.

The finite-difference method is best illustrated by reference to numerical examples.

EXAMPLE 10.7

Apply Eq. (10.21) to determine the midspan deflection and slope at support A of the beam loaded as shown in Fig. 10.10a.

Solution For simplicity, use $h = L/4$. In labeling nodal points, it is important to take into account any conditions of symmetry which may exist. This has been done in Fig. 10.10b. We observe that only half of the beam span need be considered. From the boundary conditions $v(0) = v''(0) = 0$ (Table 10.1) and symmetry, we find

$$v_0 = v_4 = 0 \qquad v_1 = -v_{-1} \qquad v_1 = v_3 \qquad (b)$$

Application of Eq. (10.21) at points 1 and 2 results in

$$v_3 - 4v_2 + 6v_1 - 4v_0 + v_{-1} = 0 \qquad (c)$$

$$v_4 - 4v_3 + 6v_2 - 4v_1 + v_0 = -\frac{wh^4}{EI}$$

Introducing Eqs. (b) into Eqs. (c), we have

$$6v_1 - 4v_2 = 0$$

$$-4v_1 + 3v_2 = -\frac{wh^4}{2EI}$$

Solving the above and setting $h = L/4$ yields $v_1 = -0.0039pL^4/EI$ and

$$v_2 = v_{\text{max}} = -0.0059 \frac{wL^4}{EI} = 0.0059 \frac{wL^4}{EI} \downarrow$$

From Eq. (10.22), we then obtain

$$\theta_A = \frac{1}{2h}(v_1 - v_{-1}) = -0.0156 \frac{wL^3}{EI} = 0.0156 \frac{wL^3}{EI} \;\;\diagdown .$$

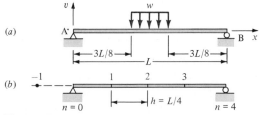

Figure 10.10 Example 10.7.

Note that the exact value of the maximum deflection is $0.0052wL^4/EI$. Thus even a coarse segmentation leads to a satisfactory solution in this case.

EXAMPLE 10.8

Using Eq. (10.20), compute the deflections at points 1 through 5 for beam and loading shown in Fig. 10.11.

Solution Application of Eq. (10.20) at points 1 through 5 gives, respectively,

$$v_2 - 2v_1 + v_0 = \frac{PL}{9EI}h^2$$

$$v_3 - 2v_2 + v_1 = \frac{2PL}{9EI}h^2$$

$$v_4 - 2v_3 + v_2 = \frac{PL}{6EI}h^2 \tag{d}$$

$$v_5 - 2v_4 + v_3 = \frac{PL}{9EI}h^2$$

$$v_6 - 2v_5 + v_4 = \frac{PL}{18EI}h^2$$

For this case, $h = L/6$ and the boundary conditions are $v_0 = v_6 = 0$. Then Eqs. (d) may be represented in matrix form:

$$\begin{bmatrix} -2 & 1 & 0 & 0 & 0 \\ 1 & -2 & 1 & 0 & 0 \\ 0 & 1 & -2 & 1 & 0 \\ 0 & 0 & 1 & -2 & 1 \\ 0 & 0 & 0 & 1 & -2 \end{bmatrix} \begin{Bmatrix} v_1 \\ v_2 \\ v_3 \\ v_4 \\ v_5 \end{Bmatrix} = \begin{Bmatrix} 2 \\ 4 \\ 3 \\ 2 \\ 1 \end{Bmatrix} C$$

Here $C = PL^3/648EI$. Solving the foregoing, we obtain $v_1 = -6.67C$, $v_2 = -11.33C$, $v_4 = -9.67C$, $v_5 = -5.33C$, and $v_3 = -12C$, or

$$v_3 = -0.01852\frac{PL^3}{EI} = 0.01852\frac{PL^3}{EI} \downarrow$$

Note that the exact value of the deflection at the center (see Table B.7) is $0.01775PL^3/EI$.

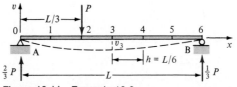

Figure 10.11 Example 10.8.

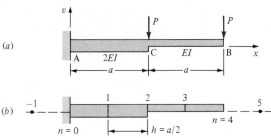

Figure 10.12   Example 10.9.

EXAMPLE 10.9

Determine the deflection of the free end of the cantilever beam loaded as shown in Fig. 10.12a. Use $h = a/2$ (Fig. 10.12b).

Solution   The boundary conditions $v(0) = 0$ and $v'(0) = 0$, referring to Table 10.1, yield

$$v_0 = 0 \qquad v_1 = v_{-1} \qquad (e)$$

We have $M_0 = -3Pa$, $M_1 = -2Pa$, $M_2 = Pa$, and $M_3 = -Pa/2$. When Eq. (10.20) is used at points 0, 1, 2, and 3, the following expressions are obtained:

$$v_1 - 2v_0 + v_{-1} = -\frac{3Pa}{2EI}h^2$$

$$v_2 - 2v_1 + v_0 = -\frac{Pa}{EI}h^2$$

$$v_3 - 2v_2 + v_1 = -\frac{Pa}{1.5EI}h^2 \qquad (f)$$

$$v_4 - 2v_3 + v_2 = -\frac{Pa}{2EI}h^2$$

Note that at node 2, the *average* flexural rigidity is used. From Eqs. (e) and (f), $v_1 = -3C/4$, $v_2 = -5C/2$, $v_3 = -59C/12$, and $v_4 = -47C/6$, where $C = (Pa/EI)h^2$. Thus, after setting $h = a/2$, we obtain

$$v_B = v_4 = -\frac{47Pa^3}{24EI} = \frac{47Pa^3}{24EI} \downarrow$$

This deflection is approximately 2.2 percent larger than the exact value.

EXAMPLE 10.10

Determine the reactions R_A, R_B, and R_C and the deflections at nodes 2 and 3 of the beam shown in Fig. 10.13a. Use Eq. (10.20) with $h = a$.

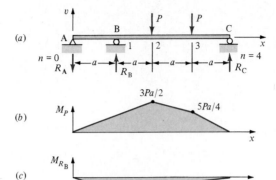

Figure 10.13 Example 10.10.

Solution The beam is statically indeterminate to the first degree and we choose the reaction R_B as redundant. The bending moment at any point of the beam may be obtained by superposing the moments caused separately by the load P (Fig. 10.13b) and by the reaction R_B (Fig. 10.13c). Applying Eq. (10.20) at points 1, 2, and 3, we obtain

$$v_2 - 2v_1 + v_0 = \left(\frac{3}{4}Pa - \frac{3}{4}R_Ba\right)\frac{a^2}{EI}$$

$$v_3 - 2v_2 + v_1 = \left(\frac{3}{2}Pa - \frac{1}{2}R_Ba\right)\frac{a^2}{EI} \qquad (g)$$

$$v_4 - 2v_3 + v_2 = \left(\frac{5}{4}Pa - \frac{1}{4}R_Ba\right)\frac{a^2}{EI}$$

The number of unknowns in the above is reduced from six to three through the use of the support conditions $v_0 = v_1 = v_4 = 0$. Solving Eqs. (g), we then have

$$R_B = \frac{13}{7}P \uparrow \qquad v_2 = \frac{9}{14}\frac{Pa^3}{EI} \downarrow \qquad v_3 = \frac{5}{7}\frac{Pa^3}{EI} \downarrow$$

Next, applying the conditions of statics, we determine $R_A = -9P/14$ and $R_C = 33P/42$. The exact values of the reactions are $R_B = 2P$ and $R_C = -R_A = 3P/4$. By means of finer segmentation, improved results are expected.

10.6 USE OF DIGITAL COMPUTERS

Numerical methods such as we have been discussing lead to systems of linear algebraic equations. The digital computer is often used to provide rapid solution of these simultaneous expressions, usually by means of matrix methods. Obviously, problem solving is greatly facilitated by use of computers and a number of special-purpose and general-purpose programs have been written to analyze both statically determinate indeterminate structures.*

For many types of slender structures, numerical solutions may be achieved only through application of computer programs. Computer software libraries contain programs that are particularly useful in finding the properties of complex sections, shear force, bending moment, deflection, and stress in beams, plates, and shells. It is now common practice to rely upon such programs.

A drawback in the use of the computer is that there is a tendency to lose the physical grasp of a problem. However, this is more than offset by many advantages of digital computation. To have the best of both worlds, the student should develop an ability to draw accurate sketches, formulate the basic equations symbolically, and systematically program for computer solution.

*See, for example, F. Fleming, *Structural Engineering Analysis on Personal Computers*, McGraw-Hill, New York, 1986.

PROBLEMS

Secs. 10.1 to 10.3

10.1 For the W 10×33 rolled-shape beam (see Table B.2) and loading shown in Fig. 10.5a, with $w = 6$ kips/ft, $L = 6$ ft, and $E = 30 \times 10^6$ psi, calculate (a) the deflection at point C; (b) the slope at point A; and (c) the maximum deflection between supports.

10.2 A beam carries a load of $P = 40$ kN and is supported as shown in Fig. 10.6a. Compute the support reactions and draw the bending-moment diagram for (a) $a = 3$ m and $b = 1$ m and (b) $a = b = 2$ m.

10.3 through 10.8 A cantilever beam is loaded as shown in Figs. P10.3 through P10.8. Use area moments to determine (a) the slope at the free end and (b) the deflection at the free end.

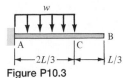

Figure P10.3

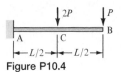

Figure P10.4

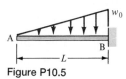

Figure P10.5

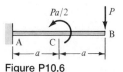

Figure P10.6

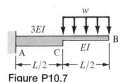

Figure P10.7

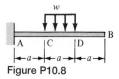

Figure P10.8

10.9 Use area moments to calculate the maximum slope and maximum deflection of the beam in Fig. P10.5 for an S 6 × 12.5 rolled shape (use Table B.3) and w_0 = 6 kips/ft, L = 8 ft, and E = 30 × 10^6 psi.

10.10 Use area moments to compute the slope and deflection at point D of the beam seen in Fig. P10.8 for a W 460 × 82 rolled-shape beam (use Table B.2), with w = 150 kN/m, a = 1 m, and E = 200 GPa.

10.11 through 10.16 A simple beam is loaded as shown in Figs. P10.11 through P10.16. Use area moments to determine (a) the slope at point C and (b) the deflection at point C.

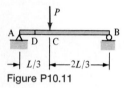

Figure P10.11

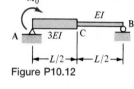

Figure P10.12

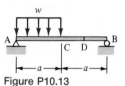

Figure P10.13

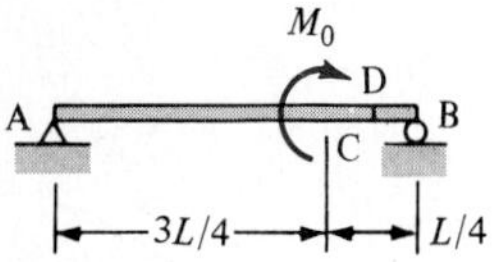

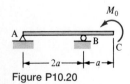

Figure P10.14

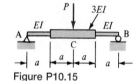

Figure P10.15

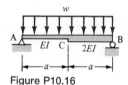

Figure P10.16

10.17 A beam is loaded and supported as shown in Fig. P10.11. Use moment areas to calculate the deflection at point D located $L/4$ from A for P = 4 kips, L = 9 ft, E = 10 × 10^6 psi, and I = 20 in.^{4}.

10.18 For the W 610 × 125 rolled-shape beam and loading shown in Fig. P10.13, let w = 120 kN/m, a = 4 m, and E = 200 GPa. Use Table B.2 and calculate the slope at point D located $a/2$ from B. Use the moment-area method.

10.19 A 4-m-long beam is subjected to a moment of M_0 = 50 kN·m (see Fig. P10.14). Use area moments to determine the value of EIv_D at point D located $L/8$ from B.

10.20 through 10.25 An overhanging beam is loaded as shown in Figs. P10.20 through P10.25. Use the area-moment approach to determine (a) the slope at point B and (b) the deflection at point C.

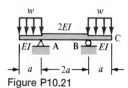

Figure P10.20

Figure P10.21

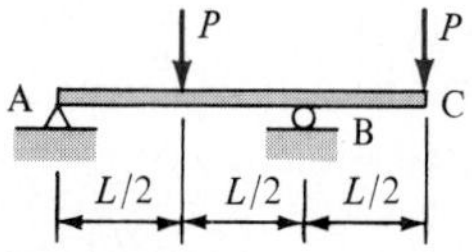

Figure P10.22

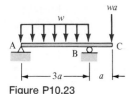

Figure P10.23

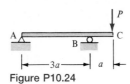

Figure P10.24

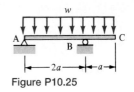

Figure P10.25

10.26 The 16-ft-long overhanging beam AC seen in Fig. P10.23 supports the loading shown, where $w = 400$ lb/ft. Use area moments to compute the value of $EI\theta_A$ at point A.

10.27 A 6-m-long overhanging beam AC supports a uniform load of intensity $w = 40$ kN/m (see Fig. P10.25). Use area moments to calculate the slope at point C. Use $E = 70$ GPa and $I = 20 \times 10^6$ mm^4.

10.28 The 12-ft overhanging beam AC seen in Fig. P10.20 supports a moment of $M_0 = 500$ kip·in. Use area moments to compute the maximum value of EIv_{max} between supports.

10.29 through 10.34 A beam is supported and loaded as shown in Figs. P10.29 through P10.34. Use area moments to determine the maximum deflection between the supports.

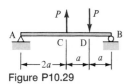

Figure P10.29

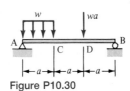

Figure P10.30

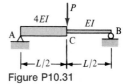

Figure P10.31

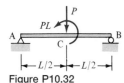

Figure P10.32

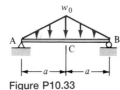

Figure P10.33

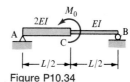

Figure P10.34

10.35 For the overhanging S 4 × 7.7 rolled-shape beam seen in Fig. P10.21, let $w = 200$ lb/ft, $a = 3$ ft, and $E = 30 \times 10^6$ psi. Use Table B.3 and area moments to calculate the slope at C.

10.36 The 4-m-long simple beam shown in Fig. P10.33 supports linearly varying loads of maximum intensity $w_0 = 60$ kN/m. Use area moments to compute the maximum value of $EI\theta_A$.

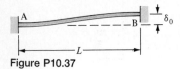

Figure P10.37

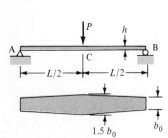

Figure P10.40

10.37 In the fixed beam AB shown in Fig. P10.37, the left-hand support has settled a distance δ_0 below the right-hand support. Use area moments to determine the reactions.

10.38 Using area moments, determine the free-end deflection v_A of the cantilever beam loaded as shown in Fig. P10.38. Assume that the beam has a rectangular cross section of constant width b.

Figure P10.38

10.39 Redo Prob. 10.38 for a beam of solid circular cross section and of the diameters h_0 and $2h_0$ at the ends A and B, respectively.

10.40 For the beam of variable cross section illustrated in Fig. P10.40, applying area moments, determine the maximum deflection v_C.

10.41 through 10.49 A beam is supported and loaded as shown in Figs. P10.41 through P10.49. Use area moments to determine the reactions.

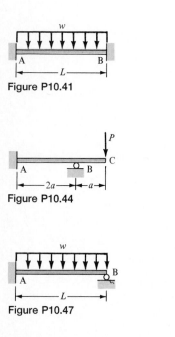

Figure P10.41

Figure P10.44

Figure P10.47

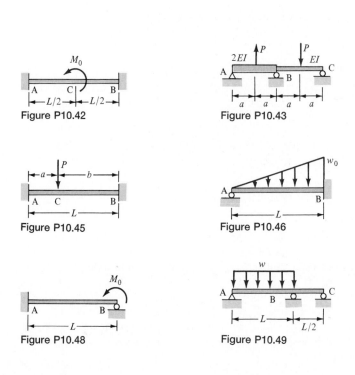

Figure P10.42

Figure P10.43

Figure P10.45

Figure P10.46

Figure P10.48

Figure P10.49

10.50   Before the load is applied to the beam shown in Fig. P10.50, a small gap δ_C exists between beam AB and the support at C. Following the application of load, the gap closes and reactions develop at each support. Using area moments, determine the reactions.

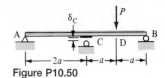

Figure P10.50

10.51   Rework Prob. 10.45 for the case in which end A of the beam is simply supported and $a = b = L/2$.

10.52   Rework Prob. 10.46 for the case in which end A of the beam is fixed.

Secs. 10.4 to 10.6

10.53 and 10.54   Use the finite-difference method, with $h = L/4$, to solve Prob. 10.4 and part b of Prob. 10.5.

10.55 and 10.56   Apply the finite-difference method to solve part b of Prob. 10.6 and Prob. 10.7. Use $h = L/4$ or $h = a/2$, as needed.

10.57 through 10.59   Use the finite-difference method to solve Probs. 10.12, 10.13, and 10.15. Let $h = L/4$ or $h = a/2$, as required.

10.60   Apply the finite-difference method to solve part b of Prob. 10.16. Use $h = a/4$.

10.61 and 10.62   Use the finite-difference method to solve part b of Probs. 10.20 and 10.22. Let $h = L/4$ or $h = a/2$, as needed.

10.63 and 10.64   Apply the finite-difference method to solve Prob. 10.24 and part b of Prob. 10.25. Use $h = L/4$ or $h = a/2$, as needed.

10.65 and 10.66   Use the finite-difference method to solve Probs. 10.29 and 10.30. Let $h = L/2$ or $h = a/2$, as needed.

10.67 and 10.68   Use the finite-difference method to solve Probs. 10.31 and 10.33. Let $h = L/6$ or $h = a/4$, as needed.

10.69 and 10.70   Apply the finite-difference method to solve Probs. 10.32 and 10.34. Use $h = L/6$.

10.71   Use the finite-difference method, with $h = L/4$, to solve Prob. 10.38.

10.72   Use the finite-difference method, with $h = L/4$, to solve Prob. 10.40.

10.73   Use the finite-difference method to determine the slope at point A and deflection at the midspan of the beam shown in Fig. P10.73. Let $h = L/8$.

10.74   Redo Prob. 10.73 for $h = L/6$.

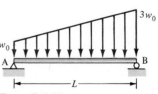

Figure P10.73

10.75 For the beam loaded and supported as shown in Fig. P10.41, use the finite-difference method, with $h = L/6$, to determine (*a*) the maximum deflection and slope and (*b*) the moment at the fixed support.

10.76 Redo Prob. 10.75 for the beam and loading shown in Fig. P10.46. Use $h = L/4$.

10.77 Repeat Prob. 10.75 for the beam and loading shown in Fig. P10.47. Use $h = L/4$.

10.78 and 10.79 Use the finite-difference method to solve Probs. 10.43 and 10.45. Let $h = a$ or $h = L/4$ and $a = b$, as needed.

10.80 and 10.81 Apply the finite-difference method to solve Probs. 10.44 and 10.48. Use $h = L/4$ or $h = a/2$, as needed.

10.82 Redo Example 10.10; this time choose the reaction R_A as redundant.

10.83 Use the finite-difference method, with $h = L/4$, to solve Prob. 10.49.

CHAPTER
11

ENERGY METHODS

11.1 INTRODUCTION

As an alternative to the equilibrium methods discussed in previous chapters, analysis of displacements and forces can be accomplished through the use of energy methods. These methods are based upon the concept of strain energy, which we introduced in Sec. 3.9. Application of energy techniques is particularly effective in cases involving members of variable cross sections, in complex problems dealing with elastic stability, and in dealing with multielement structures. In addition, strain-energy methods can greatly facilitate the determination of displacement of slender members under combined loads.

In this chapter we begin by developing the general expressions relative to strain energy, as applied to a linearly elastic body. Following this, in Secs. 11.3 through 11.5, we discuss the elastic strain energy stored in members subjected to axial loads, torsion, and bending moments. Then in Sec. 11.6 we use the principle of conservation of energy, and by equating the internal-strain energy to the external work, we determine the deflection for a number of members under a single load. Next, work and energy under several loads and resulting displacements are considered (Secs. 11.7 and 11.8), and the chapter concludes with a discussion of impact loading.

*11.2 STRAIN ENERGY FOR A GENERAL STATE OF STRESS

When a body is subjected to a three-dimensional stress, the total work done by each normal and shearing stress component is simply the sum of the expressions analogous to Eqs. (3.18) and (3.20), respectively. The strain-energy density, for a *general state of stress*, is thus

$$U_0 = \tfrac{1}{2}(\sigma_x \varepsilon_x + \sigma_y \varepsilon_y + \sigma_z \varepsilon_z + \tau_{xy} \gamma_{xy} + \tau_{yz} \gamma_{yz} + \tau_{xz} \gamma_{xz}) \quad (11.1)$$

Substituting Eqs. (3.13) into the above, we obtain

$$U_0 = \frac{1}{2E}[\sigma_x^2 + \sigma_y^2 + \sigma_z^2 - 2\nu(\sigma_x \sigma_y + \sigma_y \sigma_z + \sigma_x \sigma_z)]$$

$$+ \frac{1}{2G}(\tau_{xy}^2 + \tau_{yz}^2 + \tau_{xz}^2) \quad (11.2)$$

307

If the *principal* axes are used as coordinate axes, the shearing stresses vanish and Eq. (11.2) becomes

$$U_0 = \frac{1}{2E} [\sigma_1^2 + \sigma_2^2 + \sigma_3^2 - 2\nu(\sigma_1\sigma_2 + \sigma_2\sigma_3 + \sigma_1\sigma_3)] \quad (11.3)$$

where σ_1, σ_2, and σ_3 are the principal stresses.

The total strain energy stored in an elastic body can be obtained by integrating the strain-energy density over the entire volume:

$$U = \iiint U_0 \, dx \, dy \, dz = \int U_0 \, dV \quad (11.4)$$

Using this expression, we can evaluate the strain energy for members under combined loads. Note that the strain energy is a *nonlinear* (quadratic) function of loading or deformation. The principle of superposition is therefore *not* valid for strain energy (see Example 11.3).

Components of Strain Energy.   A new perspective relative to strain energy may be gained by separating the principal stresses at a point into two parts, as shown in Fig. 11.1a. The state of stress in Fig. 11.1b is associated with the *volume changes,* or so-called *dilatations.* On the other hand, the *shape changes,* or *distortions,* are caused by the set of stresses indicated in Fig. 11.1c. Thus the dilatational strain-energy density can be determined from Eq. (11.3) by setting $\sigma_1 = \sigma_2 = \sigma_3 = \sigma_m$:

$$U_{0v} = \frac{3(1 - 2\nu)}{2E} (\sigma_m^2) = \frac{1 - 2\nu}{6E} (\sigma_1 + \sigma_2 + \sigma_3)^2 \quad (11.5)$$

The distortional strain-energy density is readily found by subtracting the above from Eq. (11.3):

$$U_{0d} = \frac{1}{12G} [(\sigma_1 - \sigma_2)^2 + (\sigma_2 - \sigma_3)^2 + (\sigma_3 - \sigma_1)^2] \quad (11.6)$$

Here G and E are connected by Eq. (3.14).

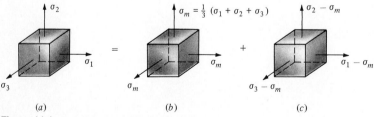

Figure 11.1 Resolution of principal stresses into dilatational and distortional stresses.

Test results indicate that the dilatational strain energy is ineffective in causing failure by yielding. The energy of distortion is assumed to be completely responsible for the material failure by inelastic action (recall Sec. 8.9). The volumetric and distortional stresses and associated strains also play an important role in the plastic behavior of the materials.

EXAMPLE 11.1

An element of a mild-steel member is subjected to a simple tension. Determine the strain-energy density and its components. Use $\nu = \frac{1}{4}$.

Solution   The state of stress in the element (see Fig. 11.1) is

$$\sigma_1 = \sigma \qquad \sigma_2 = \sigma_3 = 0 \tag{a}$$

Substitution of Eqs. (a) into Eqs. (11.3), (11.5), and (11.6) yields

$$U_0 = \frac{\sigma^2}{2E}$$

$$U_{0v} = \frac{(1 - 2\nu)}{6E}\sigma^2 = \frac{\sigma^2}{12E} \tag{11.7}$$

$$U_{0d} = \frac{(1 + \nu)}{3E}\sigma^2 = \frac{5\sigma^2}{12E}$$

We observe from the above expressions that $U_0 = U_{0v} + U_{0d}$ and $5U_{0v} = U_{0d}$. Therefore in changing the shape of a unit volume element under uniaxial stressing, *five times more* energy is absorbed than in changing the volume.

11.3 STRAIN ENERGY UNDER AXIAL LOADING

The state of stress at any given transverse section through a nonprismatic bar under axial force P_x (see Fig. 5.1b) is $\sigma_x = P_x/A_x$, where A_x represents the area of the section. In such a situation, substituting for σ_x into Eq. (3.19), we have

$$U = \int \frac{P_x^2 dV}{2A_x E}$$

As P_x and A_x can be functions of x alone, we set $dV = A\,dx$. The general expression for the strain energy is thus

$$U = \int \frac{P_x^2\,dx}{2A_x E} \tag{11.8}$$

which is integrated over the bar length.

In the case of a *prismatic bar,* subjected at its ends to equal and opposite forces of magnitude P, Eq. (11.8) appears as

$$U = \frac{P^2 L}{2AE} \tag{11.9a}$$

Recalling from Sec. 5.2 that $\delta = PL/AE$, we have

$$U = \frac{AE\delta^2}{2L} \tag{11.9b}$$

where δ is the deflection of the bar.

EXAMPLE 11.2

Two bars of the same elastic material and equal lengths L, but of different shapes, are subjected to the same axial loads P. Determine the strain energy stored (a) in the bar of constant cross-sectional area A of diameter d (Fig. 11.2a) and (b) in the bar of variable cross section having the ratio n of the two diameters (Fig. 11.2b).

Solution The normal stresses σ_x are assumed to be uniformly distributed in any given transverse section.
(a) As $\sigma_x = P/A$ and the bar volume $V = AL$, the total strain energy for the uniform bar, applying Eq. (11.9a), is

$$U_1 = \frac{P^2 L}{2AE} \tag{a}$$

(b) Similarly, the strain energy of the second bar is

$$U_n = \frac{P^2(L/3)}{2AE} + \frac{P^2(2L/3)}{2E(n^2 A)} = \frac{P^2 L}{6AE}\left(1 + \frac{2}{n^2}\right)$$

or

$$U_n = \frac{n^2 + 2}{3n^2}\frac{P^2 L}{2AE} \tag{11.10}$$

Note that, for $n = 1$, the foregoing yields the same result as obtained in part a, and for $n > 1$, $U_n < U_1$. For example, when $n = 2$ and $n = 3$, one has $U_2 = U_1/2$ and $U_3 = 11U_1/27$, respectively. It is clear that both bars have the same maximum stress $\sigma_{max} = P/A$. Thus for a given allowable stress, the strain-energy-absorbing capacity decreases as the volume of the bar increases.

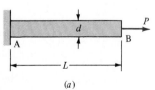

(a)

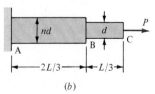

(b)

Figure 11.2 Example 11.2.

11.4   STRAIN ENERGY IN CIRCULAR SHAFTS

In the torsion of an elastic shaft of circular cross section, the only nonvanishing stress component is τ_{xy} (see Sec. 6.3). Thus if we introduce the torsion formula into Eq. (3.21), the general expression for the strain energy for a shaft becomes

$$U = \int \frac{T_x^2 \rho^2}{2GJ_x^2}\, dV$$

Here $dV = dA\, dx$, dA representing an element of cross-sectional area. Inasmuch as T_x and J_x may be a function of x alone,

$$U = \int_0^L \frac{T_x^2}{2GJ_x^2} \left(\int \rho^2\, dA \right) dx$$

where integration is performed over the shaft length. By definition, the term in the parentheses is the polar moment of inertia J_x of the cross section at a location x along the shaft axis. The strain energy is therefore

$$U = \int_0^L \frac{T_x^2\, dx}{2GJ_x} \qquad (11.11)$$

In the case of a prismatic shaft subjected at its ends to equal and opposite torques of magnitude T, Eq. (11.11) becomes

$$U = \frac{T^2 L}{2GJ} \qquad (11.12a)$$

An alternative form of the strain energy can be found by introducing $T = GJ\phi/L$ (recall Sec. 6.5) into Eq. (11.12a). By so doing, we obtain

$$U = \frac{GJ\phi^2}{2L} \qquad (11.12b)$$

where ϕ is the total angle of twist of the shaft.

EXAMPLE 11.3

Three identical shafts of torsional rigidity GJ and length L are subjected to torques T as shown in Fig. 11.3. Determine the strain energy stored in each member.

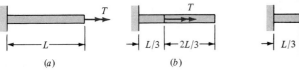
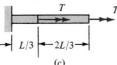

(a)   (b)   (c)

Figure 11.3   Example 11.3.

Solution Equation (11.12a) is applied to determine the strain energy in each case. Thus for the shafts of Fig. 11.3a, b, and c we write, respectively,

$$U_a = \frac{T^2 L}{2GJ}$$

$$U_b = \frac{T^2(L/3)}{2GJ} = \frac{T^2 L}{6GJ}$$

and

$$U_c = \frac{T^2(2L/3)}{2GJ} + \frac{(2T)^2(L/3)}{2GJ} = \frac{T^2 L}{GJ}$$

It is significant that $U_c \neq U_a + U_b$: The strain energy due to the two loads acting simultaneously is *not* equal to the sum of the strain energies owing to the loads acting alone.

11.5 STRAIN ENERGY IN BEAMS

In the case of pure bending, in accordance with the flexure formula for beams, the normal stress $\sigma_x = -My/I$. This relation is substituted into Eq. (3.19) to yield

$$U = \int \frac{M^2 y^2}{2EI^2} dV$$

where $dV = dA\,dx$. Noting that both M and I are functions of x alone, we have

$$U = \int_0^L \frac{M^2}{2EI^2} \left(\int y^2 \, dA \right) dx$$

Recalling that the integral in parentheses defines the moment of inertia, we find that the bending strain energy is

$$U = \int_0^L \frac{M^2 \, dx}{2EI} \tag{11.13}$$

where integration is carried over the length of the beam.

As described in Sec. 7.12, the transverse shear force V produces shearing stress τ_{xy} at every point in the beam. Introducing τ_{xy}, defined by Eq. (7.27), into Eq. (3.21), we have

$$U = \int \frac{V^2 Q^2}{2GI^2 b^2} dV \int_0^L \frac{V^2}{2GI^2} \left(\int \frac{Q^2}{b^2} dA \right) dx \tag{a}$$

Let us denote

$$\alpha = \frac{A}{I^2} \int \frac{Q^2}{b^2} dA \tag{11.14}$$

which is called the *form factor for shear*. The strain energy for shear in a beam may therefore be written in the form

$$U = \int_0^L \frac{\alpha V^2}{2AG} \, dx \qquad (11.15)$$

Integration extends over the length of the member. It is seen from Eq. (11.14) that the form factor is a dimensionless quantity specific to a given section geometry.

EXAMPLE 11.4

Determine the total strain energy of a cantilever beam with a rectangular section (Fig. 11.4) and compare the values of the bending and shear contributions.

Solution   The first moment of the cross-sectional area, from Eq. (7.30), is $Q = (b/2)[(h/2)^2 - y_1^2]$. As $A/I^2 = 144/bh^5$, Eq. (11.14) leads to

$$\alpha = \frac{144}{bh^5} \int_{-h/2}^{h/2} \frac{1}{4} \left(\frac{h^2}{4} - y_1^2 \right)^2 b \, dy_1 = \frac{6}{5}$$

The form factor for other cross sections is similarly obtained.

The bending moment at any section is $M = -Px$, and the shear force equals $V = P$ (Fig. 11.4). Upon introducing these, together with $\alpha = \frac{6}{5}$, into Eqs. (11.13) and (11.15) and then integrating, we have

$$U_b = \int_0^L \frac{P^2 x^2}{2EI} \, dx = \frac{P^2 L^3}{6EI} \qquad (b)$$

$$U_s = \int_0^L \frac{6}{5} \frac{V^2}{2AG} \, dx = \frac{3P^2 L}{5AG} \qquad (c)$$

Noting that $I/A = b^2/12$, we determine the total strain energy stored in the cantilever to be

$$U = U_b + U_s = \frac{P^2 L^3}{6EI} \left[1 + \frac{3E}{10G} \left(\frac{h}{L} \right)^2 \right] \qquad (11.16)$$

Observe that the ratio of the two contributions is equal to

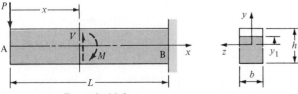

Figure 11.4   Example 11.4.

$$\frac{U_s}{U_b} = \frac{3E}{10G} \left(\frac{h}{L}\right)^2 = \frac{3}{5}(1 + \nu)\left(\frac{h}{L}\right)^2$$

If, for example, $L = 10h$ and $\nu = \frac{1}{3}$, the above quotient is only $\frac{1}{125}$; the strain energy due to the shear is less than 1 percent. It is customary in engineering practice to neglect the effect of shear in evaluating the strain energy in beams of ordinary proportion.

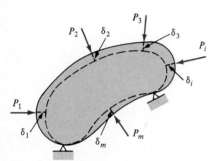

Figure 11.5 Displacements of an elastic body under several loads.

11.6 DISPLACEMENTS BY THE WORK-ENERGY METHOD

The strain energy of an elastic body subjected to several loads may be expressed in terms of the loads and resulting displacements. Suppose that all loads are applied statically, and let the final values of load and displacement be denoted by P_k ($k = 1, 2, \ldots, m$) and δ_k (see Fig. 11.5). Based upon linearity, force and displacement increase proportionately during the loading process, and thus the total work W is $\frac{1}{2}\Sigma P_k \, \delta_k$. This work is equal to the strain energy U gained by the body, provided that no energy is dissipated. Thus

$$U = W = \frac{1}{2}\sum_{k=1}^{m} P_k \, \delta_k \tag{11.17}$$

In other words, the work done by the loads acting on the body manifests itself as elastic strain energy.

In the case of a member or structure subjected to a *single concentrated* load P, Eq. (11.17) may be written as

$$U = \tfrac{1}{2}P\delta \tag{11.18}$$

where δ is the displacement through which the force P moves. It can similarly be demonstrated that

$$U = \tfrac{1}{2}M\theta \tag{11.19}$$

and

$$U = \tfrac{1}{2}T\phi \tag{11.20}$$

Here M (or T) and θ (or ϕ) are, respectively, the moment (or torque) and the associated slope (or angle of twist) at a point of a structure or member.

The preceding relationships, Eqs. (11.17) through (11.20), offer a simple approach for obtaining the displacement, provided that the strain energy can be determined. This is called the *work-energy method*. In the next section we shall present a more general method, which may be employed to obtain the displacement at a given point of a structure even when the structure is subjected simultaneously to various loads.

EXAMPLE 11.5

Determine the deflection v_A at the free end A of the cantilever beam shown in Fig. 11.4. Take into account the effects of both the bending and shear stresses.

Solution   The strain energy of the beam was determined in Example 11.4. Equating the expression found for U to the work of the load, we have

$$\frac{P^2L^3}{6EI}\left[1 + \frac{3E}{10G}\left(\frac{h}{L}\right)^2\right] = \frac{1}{2}Pv_A$$

The foregoing yields

$$v_A = \frac{PL^3}{3EI}\left[1 + \frac{3E}{10G}\left(\frac{h}{L}\right)^2\right] \qquad (11.21)$$

When the effect of shear is neglected, note that the relative error is the same as that obtained previously in Example 11.4. As it has already shown, this is less than 1 percent for a beam with ratio $h/L = 10$.

EXAMPLE 11.6

Obtain the vertical deflection of joint B for the truss of Fig. 11.6. The cross-sectional areas are $A_{AB} = 2A_{BC} = 0.004$ m^2. Use $E = 70$ GPa.

Solution   The axial forces in the bars are

$$F_{AB} = -30 \text{ kN} \qquad F_{BC} = 50 \text{ kN}$$

as found from the equilibrium of joint B. Thus the expression for the system strain energy is, referring to Fig. 11.6,

$$U = \frac{F_{AB}^2 L_{AB}}{2A_{AB}E} + \frac{F_{BC}^2 L_{BC}}{2A_{BC}E} = \frac{10^6}{2(70 \times 10^9)}\left[\frac{(-30^2)(1.5)}{0.004} + \frac{(50^2)(2.5)}{0.002}\right]$$

or

$$U = 24.7 \text{ J}$$

The work done by load P is $W = P\delta_V/2 = 20(10^3)\delta_V$. Equating U and W, we have

$$\delta_V = 1.24 \text{ mm} \downarrow$$

This result is identical with that obtained from the displacement diagram in Example 5.2.

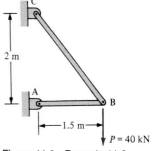

Figure 11.6   Example 11.6.

*11.7   DISPLACEMENTS BY CASTIGLIANO'S THEOREM

There is widespread application of Castigliano's theorem to the analysis of structural displacements. The theorem, proposed in 1879 by A. Castigliano

(1847–1884), is restricted to structures that behave linearly and for which the principle of superposition applies.

We refer again to Fig. 11.5, which shows a linearly elastic body subjected to loads P_k ($k = 1, 2, \ldots, m$). The strain energy U of the body equal to the work done by the applied forces (Sec. 11.6) is

$$U = \frac{1}{2} \sum_{k=1}^{m} P_k \, \delta_k$$

Let us now permit a *single load*, say P_i, to be *increased* a small amount dP_i, while the *other applied forces* P_k *remain unchanged*. The increase in strain energy is then $dU = (\partial U / \partial P_i) dP_i$, where $\partial U / \partial P_i$ represents the rate of change of the strain energy with respect to P_i. Hence the total energy is

$$U' = U + \left(\frac{\partial U}{\partial P_i}\right) dP_i \qquad (a)$$

Alternatively, an expression for U' may be derived by reversing the order of loading. Suppose that dP_i is applied first, followed by the forces P_k. In this case, application of dP_i causes a small displacement $d\delta_i$. The work, $dP_i \, d\delta_i / 2$, corresponding to this load increment can be neglected because it is of second order. Also, the work done during the application of the forces P_k is unaffected by the presence of dP_i. However, the latter force dP_i does perform work in moving an amount δ_i. Here δ_i represents the displacement caused by the application of P_k. The total strain energy attributable to the work done by this sequence of loads is thus

$$U' = U + dP_i \cdot \delta_i \qquad (b)$$

Equating the two preceding expressions for U', we have *Castigliano's theorem*:

$$\delta_i = \frac{\partial U}{\partial P_i} \qquad (11.22)$$

This expression states that for a linear structure, *the partial derivative of the strain energy with respect to an applied load is equal to the displacement at the point of application and in the direction of that load.*

Castigliano's theorem can similarly be demonstrated to be valid for applied moments M (or torques T) and the resulting slope θ (or angle of twist ϕ) of the structure. Therefore

$$\theta_i = \frac{\partial U}{\partial M_i} \qquad (11.23)$$

and

$$\phi_i = \frac{\partial U}{\partial T_i} \qquad (11.24)$$

In applying Castigliano's theorem, we must express the strain energy in terms of the external forces. In the case of a *beam,* for instance, recall from Sec. 11.5 that $U = \int M^2 \, dx/2EI$. To obtain the deflection v_i corresponding to load P_i, it is usually much simpler to differentiate under the integral sign:

$$v_i = \frac{\partial U}{\partial P_i} = \int_0^L \frac{M}{EI} \frac{\partial M}{\partial P_i} \, dx \qquad (11.25)$$

In a like manner, an expression may be written for the angle of rotation:

$$\theta_i = \frac{\partial U}{\partial M_i} = \int_0^L \frac{M}{EI} \frac{\partial M}{\partial M_i} \, dx \qquad (11.26)$$

For slender beams, as was seen in Sec. 11.6, the contribution of the shear force V to the displacement is negligible.

Similarly, in the case of a truss consisting of n members of length L_j, axial rigidity $A_j E_j$, and internal force F_j, the strain energy can be found from Eq. (11.9a) as

$$U = \sum_{j=1}^{n} \frac{F_j^2 L_j}{2A_j E_j}$$

The displacement δ_i of the point of application of load P_i is then

$$\delta_i = \frac{\partial U}{\partial P_i} = \sum_{j=1}^{n} \frac{F_j L_j}{A_j E_j} \frac{\partial F_j}{\partial P_i} \qquad (11.27)$$

When it is necessary to obtain the displacement at a point where there is *no* corresponding load, the problem is treated as follows. We place a *fictitious load* at the point in question in the direction of the desired displacement. We then determine the displacement by applying Castigliano's theorem, setting the fictitious load *equal to zero* in the expression obtained.

EXAMPLE 11.7

Calculate the vertical and horizontal deflections of point B of the truss shown in Fig. 11.7a. Let $A_{AB} = 2A_{BC} = 0.004$ m^2 and $E = 70$ GPa.

Solution A fictitious load Q, indicated by the dashed line, is introduced at point B (Fig. 11.7a). The axial forces in the bars are, from consideration of equilibrium of the joint (Fig. 11.7b),

$$F_{AB} = Q - 0.75P \qquad F_{BC} = 1.25P \qquad (c)$$

The displacements at B are, according to Eq. (11.27),

$$\delta_V = \frac{F_{AB} L_{AB}}{A_{AB} E} \frac{\partial F_{AB}}{\partial P} + \frac{F_{BC} L_{BC}}{A_{BC} E} \frac{\partial F_{BC}}{\partial P}$$

and

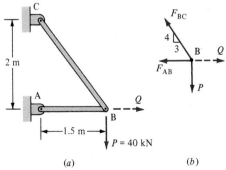

Figure 11.7   Example 11.7.

$$\delta_H = \frac{F_{AB}L_{AB}}{A_{AB}E}\frac{\partial F_{AB}}{\partial Q} + \frac{F_{BC}L_{BC}}{A_{BC}E}\frac{\partial F_{BC}}{\partial Q}$$

Substituting the given data and Eq. (c), differentiating, and setting $Q = 0$, we obtain

$$\delta_V = 1.24 \text{ mm} \downarrow \qquad \delta_H = -0.16 \text{ mm} = 0.16 \text{ mm} \leftarrow$$

These results are the same as those found previously (see Examples 5.2 and 11.6). Here the minus sign means that δ_H is in the direction opposite to that assumed for force Q (that is, leftward, as shown).

EXAMPLE 11.8

For the cantilever beam and loading shown in Fig. 11.8, determine the deflection at A.

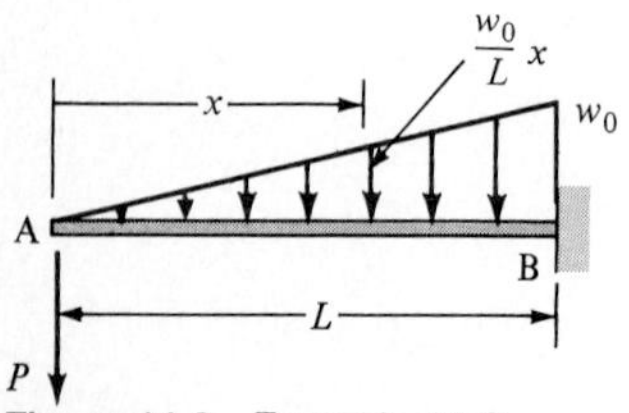

Solution At a distance x from A, one has

$$M = -Px - \frac{wx^3}{6L} \quad \text{and} \quad \frac{\partial M}{\partial P} = -x$$

Substitution of the above expression into Eq. (11.25) yields

$$v_A = \frac{1}{EI}\int_0^L \left(Px^2 + \frac{wx^4}{6L}\right) dx = \frac{PL^3}{3EI} + \frac{wL^4}{30EI} \downarrow \qquad (d)$$

Figure 11.8 Example 11.8.

EXAMPLE 11.9

A half-ring is fixed at one end and loaded by a force P at the other end, as depicted in Fig. 11.9a. Determine the horizontal displacement δ_H of the free end B. Assume that the radius of curvature R is large compared with the depth of the ring, so deformations due to shear and normal forces can be neglected.

Solution A free-body diagram of a portion of the ring subtended by angle θ is shown in Fig. 11.9b, where the internal forces and moment act as indicated. Referring to the figure, we find

$$M = PR \sin \theta \qquad dx = R \, d\theta$$

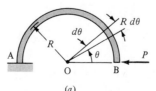

(a)

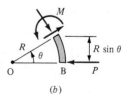

(b)

Figure 11.9 Example 11.9.

Substituting the above expressions into Eq. (11.25), we have

$$\delta_H = \int_0^L \frac{M}{EI} \frac{\partial M}{\partial P} \, dx = \frac{PR^3}{EI} \int_0^\pi \sin^2 \theta \, d\theta = \frac{PR^3}{EI} \left. \left(\frac{\theta}{2} - \frac{\sin 2\theta}{4} \right) \right|_0^\pi$$

from which

$$\delta_H = \frac{\pi PR^3}{2EI} \leftarrow \qquad (11.28)$$

EXAMPLE 11.10

For the simply supported beam shown in Fig. 11.10a, determine the deflection v_D at point D and the slope θ_A at the left end A caused by load P.

Solution   Referring to Fig. 11.10a, we write

$$M_1 = \frac{Pb}{L} x \quad \text{and} \quad \frac{\partial M_1}{\partial P} = \frac{bx}{L} \qquad (0 \le x \le a)$$

$$M_2 = \frac{Pa}{L} x' \quad \text{and} \quad \frac{\partial M_2}{\partial P} = \frac{ax'}{L} \qquad (0 \le x' \le b)$$

Insertion of these expressions into Eq. (11.25) yields

$$v_D = \frac{1}{EI} \int_0^a \frac{Pb}{L} x \left(\frac{bx}{L} \right) dx + \frac{1}{EI} \int_0^b \frac{Pa}{L} x' \left(\frac{ax'}{L} \right) dx' = \frac{Pa^2b^2}{3EIL} \downarrow \qquad (11.29)$$

As the slope is sought, a fictitious couple C is introduced at point A (Fig. 11.10b). From the conditions of equilibrium, the reactions are found to be

$$R_A = \frac{Pb}{L} - \frac{C}{L} \qquad R_B = \frac{Pa}{L} + \frac{C}{L}$$

The appropriate moment equations are then

$$M_1 = \left(\frac{Pb}{L} - \frac{C}{L} \right) x + C \qquad M_2 = \left(\frac{Pa}{L} + \frac{C}{L} \right) x'$$

Thus

$$\frac{\partial M_1}{\partial C} = 1 - \frac{x}{L} \qquad \frac{\partial M_2}{\partial C} = \frac{x'}{L}$$

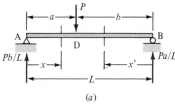

(a)

(b)

Figure 11.10   Example 11.10.

Substituting the foregoing into Eq. (11.26) and setting $C = 0$, we have

$$\theta_A = \frac{P}{EIL} \left[\int_0^a bx \left(1 - \frac{x}{L} \right) dx + \int_0^b ax' \left(\frac{x'}{L} \right) dx' \right]$$

$$= \frac{Pab}{6EIL^2} (-2a^2 + 3aL + 2b^2) \quad \text{↖} \tag{11.30}$$

The positive sign of θ_A connotes that the angle of rotation has the same sense as the couple C (that is, clockwise, as shown).

*11.8 UNIT-LOAD METHOD

In this section we will use Castigliano's theorem to derive the unit-load method, which is widely employed for finding displacements of structures. It is also known as the dummy-load method, method of virtual work, and the Maxwell-Mohr method. The procedure requires the use of a force or a unit couple in conjunction with actual loads.

We saw in the preceding section that the deflection at a point in a beam corresponding to a load P_i, expressed in terms of the moment produced by external forces, is

$$v_i = \frac{\partial U}{\partial P_i} = \int \frac{M}{EI} \frac{\partial M}{\partial P_i} dx \tag{a}$$

For a *linearly elastic structure*, the moment is linearly proportional to the loads, and consequently we are justified writing $M = mP$, m denoting a constant. It follows that

$$m = \frac{\partial M}{\partial P_i} \tag{b}$$

This represents the change in the bending moment M per unit change in P_i. That is, m is the bending moment in an arbitrary section due to a *unit* load. The deflection is then obtained in the following convenient form:

$$v_i = \int \frac{Mm}{EI} dx \tag{11.31}$$

An analogous equation may be written for the angle of rotation:

$$\theta_i = \int \frac{Mm'}{EI} dx \tag{11.32}$$

where $m' = \partial M / \partial M_i$ is the change in bending moment caused by a unit value of the moment M_i. Expressions (11.31) and (11.32) are the unit-load equations for the case when only flexural deformations are considered. Since both M and m (or m') usually vary along the beam length, both must be expressed by appropriate functions.

Similar derivations can be made for the effects of axial, shearing, and torsional deformations. In the case of a *truss,* Eq. (11.27) becomes

$$\delta_i = \sum_{j=1}^{n} \frac{f_i F_j L_j}{A_j E_j} \tag{11.33}$$

Here $f_i = \partial F_j/\partial P_i$ represents the change in axial forces due to a unit value of the load P_i.

The following examples help in understanding the unit-load method and illustrate its efficiency.

EXAMPLE 11.11

Determine the deflection at the free end B of a cantilever beam loaded as shown in Fig. 11.11a.

Solution   The unit load 1 lb is applied at B, whose deflection is sought (Fig. 11.11b). Referring to Fig. 11.11a and b, we obtain the expressions for the moments:

$$M_1 = 0 \qquad\qquad (0 \le x \le a)$$

$$M_2 = -P(x - a) \qquad (a \le x \le 2a)$$

and

$$m = -x \qquad\qquad (0 \le x \le 2a)$$

The deflection at B is then, from Eq. (11.31),

$$v_B = \int_0^a \frac{M_1 m}{EI}\, dx + \int_a^{2a} \frac{M_2 m}{EI}\, dx$$

$$= \frac{P}{EI} \int_0^a (0)(x)\, dx + \frac{P}{EI} \int_a^{2a} (x - a)(x)\, dx$$

Integrating, we obtain

$$v_B = \frac{5Pa^3}{6EI} \downarrow \tag{11.34}$$

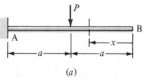

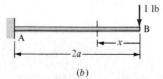

(a)   *(b)*

Figure 11.11   Example 11.11: (a) actual loading and (b) unit loading.

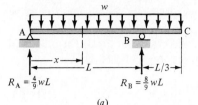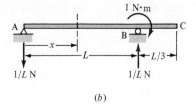

$R_A = \frac{4}{9}wL$ $\qquad\qquad R_B = \frac{8}{9}wL$ $\qquad\qquad$ $1/L$ N $\qquad\qquad$ $1/L$ N

(a) $\qquad\qquad\qquad\qquad$ (b)

Figure 11.12 Example 11.12: (a) actual loading and (b) unit loading.

EXAMPLE 11.12

Determine the slope of the elastic curve at the right support of the uniformly loaded beam shown in Fig. 11.12a.

Solution As a slope is sought, a unit couple moment 1 N·m is introduced at point B (Fig. 11.12b). The reactions, as obtained from equations of statics, are noted in the figure. Expressions for the moments corresponding to actual load and the unit load are, respectively,

$$M_1 = R_A x - \tfrac{1}{2}wx^2 \qquad (0 \le x \le L)$$

$$M_2 = R_A x - \tfrac{1}{2}wx^2 + R_B(x - L) \qquad \left(L \le x \le \frac{4L}{3}\right)$$

and

$$m_1' = -\frac{x}{L} \qquad (0 \le x \le L)$$

$$m_2' = 0 \qquad \left(L \le x \le \frac{4L}{3}\right)$$

Equation (11.32) is now applied to find the slope:

$$\theta_B = \int_0^L \frac{M_1 m_1'}{EI}\,dx + \int_L^{4L/3} \frac{M_2 m_2'}{EI}\,dx$$

$$= \frac{1}{EI}\int_0^L \left(R_A x - \frac{wx^2}{2}\right)\left(-\frac{x}{L}\right)dx = -\frac{1}{EI}\left(\frac{R_A L^2}{3} - \frac{wL^3}{8}\right)$$

Setting $R_A = 4wL/9$, we have

$$\theta_B = -\frac{5wL^3}{216EI} = \frac{5wL^3}{216EI} \qquad (11.35)$$

*11.9 STATICALLY INDETERMINATE STRUCTURES

Castigliano's theorem, or the unit-load method, is now applied as a supplement to the equations of statics in the solution of support reactions of a statically indeterminate elastic structure. Consider, for example, a structure indeterminate to the *first* degree. In this case, we select one of the reactions as the redundant (or unknown load), say R, by removing the corresponding support. All external forces, including both loads and reactions, must produce displacements compatible with the original supports. We first express the strain energy in terms of R and the given loads. Equation (11.22) may now be applied at the removed support and equated to the prescribed displacement:

$$\frac{\partial U}{\partial R} = \delta = 0 \qquad (a)$$

The expression of the type given by Eq. (a) is solved for the redundant reaction. The remaining reactions are found from the equations of statics.

In the case of a statically indeterminate structure with n redundant reactions R_n, after the strain energy is expressed in terms of these redundants and the given loads, we have

$$\frac{\partial U}{\partial R_n} = \delta_n = 0 \qquad (b)$$

The conditions of equilibrium, together with conditions of the type given by Eq. (b), constitute a set sufficient for the solution of all the reactions. The procedure is relatively straightforward.

Analytical techniques are illustrated in the examples that follow.

EXAMPLE 11.13

Determine the support reactions for the beam of Fig. 11.13a, considering (a) reaction R_A as redundant and (b) reaction M_B as redundant. Use Castigliano's theorem.

Solution   The free-body diagram of Fig. 11.13b indicates that the problem is statically indeterminate to the first degree.
(a) The expressions for the moments are

$$M_1 = R_A x \qquad \left(0 \le x \le \frac{L}{2}\right)$$

$$M_2 = R_A x - P\left(x - \frac{L}{2}\right) \qquad \left(\frac{L}{2} \le x \le L\right)$$

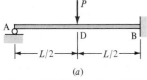

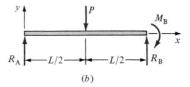

(a)   (b)

Figure 11.13   Example 11.13.

Note that the remaining unknowns, R_B and M_B, do *not* appear in the above. The deflection v_A at A must be zero. Equation (11.25) is therefore

$$v_A = \frac{1}{EI}\left\{\int_0^{L/2}(R_A x)x\,dx + \int_{L/2}^L\left[R_A x - P\left(x - \frac{L}{2}\right)\right]x\,dx\right\} = 0$$

from which

$$R_A = \tfrac{5}{16}P \uparrow \tag{11.36a}$$

The conditions of equilibrium of forces then yield

$$R_B = \tfrac{11}{16}P \uparrow \tag{11.36b}$$

$$M_B = \tfrac{3}{16}PL \;\rotatebox{0}{$\circlearrowright$} \tag{11.36c}$$

(b) Now R_A (or R_B) must be *expressed in terms of P and M_B*. Using the free-body diagram, we have $R_A = P/2 - M_B/L$. The moments are

$$M_1 = \left(\frac{P}{2} - \frac{M_B}{L}\right)x \qquad\qquad \left(0 \le x \le \frac{L}{2}\right)$$

$$M_2 = \left(\frac{P}{2} - \frac{M_B}{L}\right)x - P\left(x - \frac{L}{2}\right) \qquad \left(\frac{L}{2} \le x \le L\right)$$

As the slope at B is zero, Eq. (11.26) becomes

$$\theta_B = \frac{1}{EI}\int_0^{L/2}\left(\frac{P}{2} - \frac{M_B}{L}\right)(x)\left(-\frac{x}{L}\right)dx$$

$$+ \frac{1}{EI}\int_{L/2}^L\left[\left(\frac{P}{L} - \frac{M_B}{L}\right)x - P\left(x - \frac{L}{2}\right)\right]\left(-\frac{x}{L}\right)dx = 0$$

The foregoing, after integrating, yields $M_B = 3PL/16$, as before. The other reaction components (R_A and R_B) can be found from the conditions of equilibrium. It is observed that the approach in part *a* is slightly simpler than the approach in part *b*.

EXAMPLE 11.14

A truss constructed of three bars of equal axial rigidity AE supports load P at joint D, as shown in Fig. 11.14*a*. Employing Castigliano's theorem, determine the force in each bar.

Solution The truss is statically indeterminate to the first degree. We shall select R, the reaction at B, as redundant. This is equal to the force in bar BD. From symmetry, we observe that the forces in AD and CD are each equal to F. Therefore,

$$F = \tfrac{5}{8}(P - R) \tag{c}$$

as found from the equilibrium of forces at joint D (Fig. 11.14*b*).

We can now substitute the data given and Eq. (*c*) into Eq. (11.27) to obtain

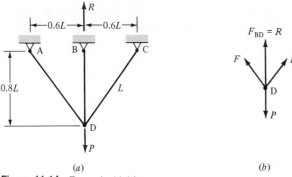

(a)   (b)

Figure 11.14 Example 11.14.

$$\delta_B = 2\frac{FL}{AE}\frac{\partial F}{\partial R} + \frac{R(0.8L)}{AE}\frac{\partial R}{\partial R}$$

$$= -\frac{25}{32AE}(P - R)L + \frac{0.8RL}{AE}$$

As the deflection at support B must be zero, setting $\delta_B = 0$ in the foregoing yields

$$R = 0.494P \uparrow \qquad\qquad (11.37)$$

The forces in the members are thus $F_{AD} = F_{CD} = 0.316P$ and $F_{BD} = 0.494P$.

EXAMPLE 11.15

A uniformly loaded beam, fixed at one end and simply supported at the other end, is shown in Fig. 11.15a. Apply the unit-load method to determine the reaction at B.

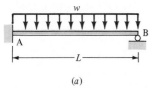

(a)

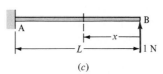

(c)

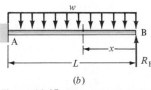

(b)

Figure 11.15 Example 11.15: (a) actual loading; (b) choice of redundant reaction; and (c) unit loading.

Solution The situation described is indeterminate to the first degree. Let the reaction R_A be taken as redundant by removing the right-hand support (Fig. 11.15b). Referring to this figure, we write

$$M = R_B x - \tfrac{1}{2}wx^2 \qquad (0 \le x \le L)$$

The corresponding unit-load diagram is seen in Fig. 11.15c and

$$m = x \qquad (0 \le x \le L)$$

We then have, from Eq. (11.32),

$$v_B = \int_0^L \frac{Mm}{EI}\,dx = \frac{1}{EI}\int_0^L \left(R_B x - \frac{1}{2}wx^2\right) x\,dx$$

$$= \frac{R_B L^3}{3EI} - \frac{wL^4}{8EI}$$

But, since point B does not move, $v_B = 0$. Thus

$$R_B = \tfrac{3}{8}wL \uparrow \qquad\qquad (c)$$

The remaining reactions may now be found by applying the equations of statics.

---

*11.10 IMPACT LOADS

A moving body striking a structure delivers a suddenly applied dynamic force that is called an *impact load*. Examples include the load resulting from the blow of a hammer, from the collision of two automobiles, from the rapid movement of a railroad train over a bridge, and cyclical loads caused by rotating machinery.

Impact loading causes elastic bodies to vibrate until equilibrium is re-established. Our concern here will be only with the influence of impact forces upon the maximum stress and deformation within the body. Impact analysis will be predicated upon the following idealizing assumptions, which lead to conservative structure or machine design:

1. The displacement is proportional to the loads, static or dynamic.
2. The material behaves elastically, and the stress-strain diagram obtained from a static test of the material is also valid under impact.
3. The inertia of the structure resisting impact may be neglected.
4. No energy is dissipated because of local deformations at the point of impact or at the structural supports.

As an illustration of an elastic system subjected to impact, consider the freestanding spring of Fig. 11.16a, on which is dropped a body of mass m from a height h. The body has initial zero velocity. At the instant of maximum dynamic deflection δ_{max} of the spring, the change in kinetic energy of the system is zero. The total work consists of the work done by gravity on the body of weight W and the resisting work done by the spring. Thus from the principle of conservation of energy, we have

$$W(h + \delta_{max}) - \tfrac{1}{2}k\delta^2_{max} = 0 \qquad (a)$$

where k is called the *spring constant*. Note that the body is assumed to remain in contact with the spring.

The static deflection δ_{st} corresponding to a static force equal to the weight of the body is W/k. We thus obtain the following general value of *maximum dynamic deflection* from Eq. (*a*):

$$\delta_{max} = \delta_{st} + \sqrt{\delta^2_{st} + 2\delta_{st}h} = \left(1 + \sqrt{1 + \frac{2h}{\delta_{st}}}\right)\delta_{st} = K\delta_{st} \quad (11.38)$$

where

$$K = \frac{\delta_{max}}{\delta_{st}} = 1 + \sqrt{1 + \frac{2h}{\delta_{st}}} \qquad (11.39)$$

is referred to as the *impact factor*. Multiplication of this factor by the static load yields an *equivalent static load,* that is, the *maximum dynamic load,*

$$P_{max} = KW \qquad (11.40)$$

To determine the maximum stress and maximum deflection caused by a maximum dynamic load, P_{max} may be used in the relationships derived for static loading.

Two extreme cases are of particular interest. If the *static deflection is very small* compared with height h, the foregoing becomes

$$\delta_{max} \approx \sqrt{2h\delta_{st}} \qquad (11.41)$$

Clearly this approximate formula gives displacements that are always less than those calculated from Eq. (11.38). When the *load is suddenly applied* (that is, $h = 0$), Eq. (11.38) reduces to

$$\delta_{max} = 2\delta_{st} \qquad (11.42)$$

Following a procedure similar to that developed above, expressions may be derived for the case of a mass m *moving horizontally* with a velocity v and stopped suddenly by an *elastic* body (Fig. 11.16b).* Now the kinetic energy $(Wv^2/2g = m^2/2)$ replaces $W(h + \delta_{max})$, the work done by W, in Eq. (*a*). Here g is the gravitational acceleration. It follows that

$$\delta_{max} = \delta_{st}\sqrt{\frac{v^2}{g\delta_{st}}} \qquad P_{max} = m\sqrt{\frac{v^2g}{\delta_{st}}} \qquad (11.43)$$

*For details of this case, see J. H. Faupel and F. E. Fisher, *Engineering Design,* 2d ed., Wiley, New York, 1986.

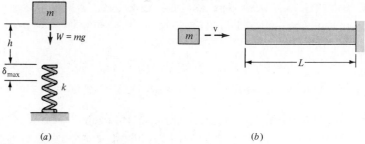

Figure 11.16   Impact loads.

in which δ_{st} is the static deflection produced by horizontal force W.

Note that if the body which is hit is a prismatic bar of length L and axial rigidity AE (Fig. 11.16b), from Eq. (11.43) we have $\delta_{st} = mgL/AE$. Equations (11.43) then become

$$\delta_{max} = \sqrt{\frac{mv^2L}{AE}} \qquad P_{max} = \sqrt{\frac{mv^2AE}{L}} \tag{11.44}$$

The corresponding maximum compressive stress, assumed to be uniform throughout the member, equals

$$\sigma_{max} = \frac{P_{max}}{A} = \sqrt{\frac{mv^2E}{AL}} \tag{11.45}$$

This expression indicates that the stress can be *reduced* by increasing the volume or decreasing the modulus of elasticity E of the bar.

EXAMPLE 11.16

A block of weight $W = 50$ N is dropped from a height h onto the free end of a cantilever steel beam ($E = 200$ GPa) of length $L = 1.6$ m (Fig. 11.17). The beam is of rectangular cross section: $b = 40$ mm wide and $d = 80$ mm deep. Determine the maximum deflection v_{max} and maximum stress for (a) $h = 40$ mm and (b) $h = 250$ mm.

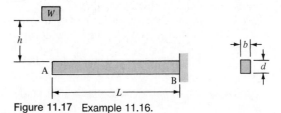

Figure 11.17   Example 11.16.

Solution The centroidal moment of inertia of the section is

$$I = \tfrac{1}{12}(0.04)(0.08)^3 = 1.7 \times 10^{-6}\,\text{m}^4$$

The static deflection v_{st} of the beam is

$$v_{st} = \frac{WL^3}{3EI} = \frac{50(1.6)^3}{3(200 \times 10^9)(1.7 \times 10^{-6})} = 0.2 \times 10^{-3}\,\text{m} = 0.2\,\text{mm} \downarrow$$

The maximum static stress, occurring at the fixed end B, is obtained from the flexure formula:

$$\sigma_{st} = \frac{Mc}{I} = \frac{50(1.6)(0.04)}{1.7 \times 10^{-6}} = 1.9\,\text{MPa}$$

(*a*) The impact factor for this system is obtained from Eq. (11.39):

$$K = 1 + \sqrt{1 + \frac{2(0.04)}{0.2 \times 10^{-3}}} = 21$$

We thus have

$$v_{max} = 21(0.2 \times 10^{-3}) = 4.2 \times 10^{-3}\,\text{m} = 4.2\,\text{mm} \downarrow$$

$$\sigma_{max} = 21(1.9) = 39.9\,\text{MPa}$$

(*b*) The impact factor now becomes

$$K = 1 + \sqrt{1 + \frac{2(0.25)}{0.2 \times 10^{-3}}} = 51$$

Hence

$$v_{max} = 51(0.2 \times 10^{-3}) = 10.2 \times 10^{-3}\,\text{m} = 10.2\,\text{mm}$$

$$\sigma_{max} = 51(1.9) = 96.9\,\text{MPa}$$

Comparing the results, we see that dynamic loading considerably increases deflection and stress. Note, however, that the *actual* quantities could be *less* than that calculated because of our simplifying assumptions 3 and 4.

PROBLEMS

Secs. 11.1 to 11.5

11.1 Determine the strain-energy density and its components at the point in the loaded structure described in Example 4.2. Use $E = 200$ GPa and $v = \tfrac{1}{3}$.

11.2 Calculate the strain-energy density and its components at the point in the loaded wooden structure described in Example 4.5. Use $E = 30 \times 10^6$ psi and $v = 0.3$.

11.3 Determine the strain-energy density and its components at the point in the loaded body of Prob. 4.41. Use $E = 70$ GPa and $v = \tfrac{1}{4}$.

11.4 Compute the strain-energy densities for the states of stress shown in Fig. 4.10a. Then, equating the results found, verify that $G = E/2(1 + \nu)$.

11.5 Determine the strain-energy density and its components for the triaxial state of stress shown in Fig. 4.10b. The following data apply: $\sigma_1 = 30$ MPa, $\sigma_2 = 40$ MPa, $\sigma_3 = 100$ MPa, $E = 110$ GPa, and $\nu = 0.3$.

11.6 A $1\frac{3}{4}$-in.-square steel machine part 4 ft long is to resist an axial energy load of 1.5 in.·kips. Assuming that $E = 30 \times 10^6$ psi and given a factor of safety of 2, calculate (a) the required proportional limit of the steel and (b) the corresponding modulus of resilience for the steel.

11.7 A 1.2-m-long stepped bar of diameters $d = 20$ mm and $nd = 30$ mm is subjected to tensile force P (Fig. 11.2b). For $\sigma_{yp} = 220$ MPa and $E = 200$ GPa, calculate the largest amount of energy that can be stored in the bar without causing any yielding.

11.8 Redo Prob. 11.7 for $d = nd = 25$ mm.

11.9 A cold-rolled bronze rod of cross-sectional area 12 in.^{2} is subjected to axial compression loading. If the allowable axial compressive stress is 14 ksi and $E = 16 \times 10^6$ psi, calculate the minimum length of the rod required for a strain energy of 1.3 in.·kips stored in the rod.

11.10 An aluminum bar AB is of yield strength 38 ksi (see Fig. 11.2a). A strain energy of 150 in.·lb must be stored in the bar when axial load P is applied. Given $L = 6$ ft and $E = 10 \times 10^6$ psi, calculate diameter d so that the factor of safety with respect to yielding is 4.

11.11 A solid tapered bar AB of length L is fixed at one end and subjected to axial force P at the other end (Fig. P11.11). Verify that the strain energy of the bar is

$$U = \frac{P^2 L}{\pi E d^2} \qquad \text{(P11.11)}$$

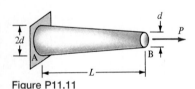

Figure P11.11

where d is the diameter at the free end.

11.12 A stepped shaft ABC ($G = 42$ GPa) is subjected to torques as shown in Fig. P11.12. For $T_B = 5$ kN·m and $T_C = 2$ kN·m, what is the strain energy of the shaft?

11.13 Rework Prob. 11.12 for the case in which the segment AB of the shaft is hollow, the inside diameter is 30 mm, and $T_B = 1$ kN·m and $T_C = 4$ kN·m.

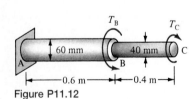

Figure P11.12

11.14 The steel shaft ABC ($G = 80$ GPa) of Fig. P11.12 is subjected to torques $T_B = 0$ and T_C. Calculate the strain energy of the shaft when the maximum shearing stress is 50 MPa.

11.15 Determine the strain energy of the aluminum shaft ($G = 28$ GPa) shown in Fig. P11.12 when $T_B = 0$ and the angle of twist ϕ_C is 0.05 rad.

11.16 through 11.21 A beam is supported and loaded as shown in Figs. P11.16 through P11.21. Determine the strain energy of the beam caused by bending.

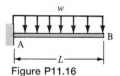

Figure P11.16

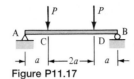

Figure P11.17

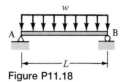

Figure P11.18

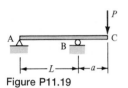

Figure P11.19

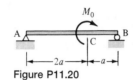

Figure P11.20

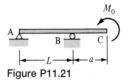

Figure P11.21

11.22 and 11.23 A beam of rectangular cross section is supported and loaded as shown in Figs. 11.16 and 11.17. Determine the strain energy of the beam caused by the shear deformation.

11.24 and 11.25 A beam of rectangular cross section is supported and loaded as shown in Figs. 11.18 and 11.19. Determine the strain energy of the beam caused by the shear deformation.

Sec. 11.6

11.26 Use the work-energy method to determine the deflection at end C of the bar ABC seen in Fig. 11.2b. Use $n = 2$.

11.27 Calculate the angle of twist for shaft ABC (shown in Fig. P11.12) due to torque $T_C = 1.2$ kN·m applied at the free end and $T_B = 0$. Employ the work-energy method. Use $G = 28$ GPa.

11.28 A beam of rectangular cross section is supported and loaded as shown in Fig. P11.17. Using the work-energy method, determine the deflection at point C owing to the bending and shear. (Note that due to symmetry, deformations at C and D are equal.)

11.29 Rework Prob. 11.28 for a beam of rectangular cross section which is supported and loaded as shown in Fig. P11.19.

11.30 A beam is supported and loaded as shown in Fig. P11.20. Employ the work-energy method to determine the slope at point C caused by bending.

11.31 Rework Prob. 11.30 for a beam supported and loaded as shown in Fig. P11.21.

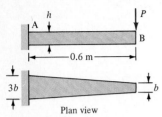

Figure P11.32

11.32 Using the work-energy method, determine the deflection due to bending at point B of the cantilever beam shown in Fig. P11.32. Let $P = 120$ N, $E = 200$ GPa, and $h = 2b = 60$ mm.

11.33 For the truss and loading shown in Fig. P11.33, obtain the vertical deflection at point C. Use the work-energy method and assume that all members are of equal axial rigidity AE.

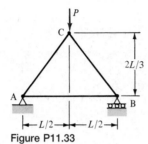

Figure P11.33

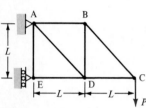

Figure P11.34

11.34 Rework Prob. 11.33 for the truss and loading shown in Fig. P11.34.

Secs. 11.7 to 11.9

11.35 For the beam loaded as shown in Fig. P11.35, determine (a) the vertical deflection v_B of the free end and (b) the slope θ_B of the free end. Use Castigliano's theorem.

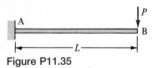

Figure P11.35

11.36 A cantilever beam is loaded as shown in Fig. P11.36. Use Castigliano's theorem to determine the deflection v_A of the free end.

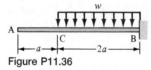

Figure P11.36

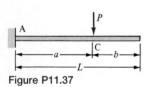

Figure P11.37

11.37 Determine the vertical deflection v_B of the free end of the cantilever beam loaded as shown in Fig. P11.37. Employ Castigliano's theorem.

11.38 A pin-ended truss in which all members have the same cross-sectional area A and elastic modulus E is loaded as depicted in Fig. P11.38. Using Castigliano's theorem, determine the vertical displacement δ_V of point D.

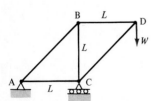

Figure P11.38

11.39 For the pin-ended truss loaded as shown in Fig. P11.38, determine the horizontal deflection δ_H of point D. Employ Castigliano's theorem. Assume that all members have the same cross-sectional area A and elastic modulus E.

11.40 For the beam and loading shown in Fig. P11.16, use Castigliano's theorem to determine the slope θ_B at point B.

11.41 For the beam and loading shown in Fig. P11.17, use Castigliano's theorem to determine the slope θ_B at point B.

11.42 through 11.44 For the beam and loading shown in Figs. P11.19 through P11.21, use Castigliano's theorem to determine (*a*) the deflection at point C and (*b*) the slope at point C.

11.45 For the truss loaded as shown in Fig. P11.33, determine the horizontal deflection at point C. Use Castigliano's theorem.

11.46 For the truss loaded as shown in Fig. P11.34, determine the vertical deflection at point D. Employ Castigliano's theorem.

11.47 through 11.49 A beam is loaded and supported as shown in Figs. P11.47 through P11.49. Use Castigliano's theorem to determine the deflection at point C.

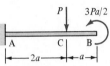

Figure P11.47

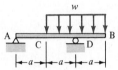

Figure P11.48

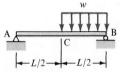

Figure P11.49

11.50 Employ Castigliano's theorem to determine the slope at point B of the curved bar of Fig. P11.50.

11.51 Use Castigliano's theorem to determine the vertical component of the deflection at point B of the curved bar of Fig. P11.50.

Figure P11.50

11.52 A uniform bar of flexural rigidity *EI* is fixed at one end and loaded at the other end as shown in Fig. P11.52. Use Castigliano's theorem to obtain (*a*) the vertical deflection at point D and (*b*) the slope at point D.

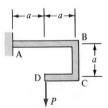

Figure P11.52

11.53 A load *P* is supported at joint B of a structure consisting of three bars of equal axial rigidity *AE* (Fig. P11.53). Use Castigliano's theorem to determine the force in each bar.

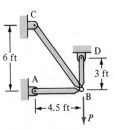

Figure P11.53

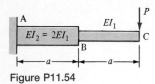

Figure P11.54

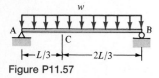

Figure P11.57

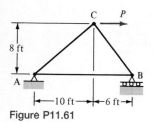

Figure P11.61

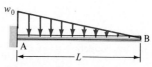

Figure P11.62

11.54 Apply Castigliano's theorem to determine the deflection of point C of the stepped cantilever beam shown in Fig. P11.54.

11.55 Determine the deflection of free end B of the uniformly loaded cantilever beam shown in Fig. P11.16. Apply the unit-load method.

11.56 Redo Prob. 11.35, using the unit-load method.

11.57 Determine the deflection of the point C of the uniformly loaded beam shown in Fig. P11.57. Use the unit-load method.

11.58 Determine the midspan deflection of a cantilever beam loaded as shown in Fig. P11.58. Employ the unit-load method.

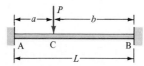

Figure P11.58

11.59 Redo Prob. 11.50, using the unit-load method.

11.60 Rework Prob. 11.51, employing the unit-load method.

11.61 Applying the unit-load method, determine the horizontal displacement of point C of the truss shown in Fig. P11.61. Let $P = 40$ kips, $E = 30 \times 10^6$ psi, and the cross-sectional area of each bar $A = 1.2$ in.^{2}.

11.62 A plane frame is loaded by a moment M_0 at B, as depicted in Fig. P11.62. Determine the reaction R at the roller. Apply the unit-load method.

11.63 Using the unit-load method, find the support reactions R_A and M_A for the beam of Fig. P11.63.

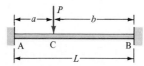

Figure P11.63

11.64 through 11.66 A beam is supported and loaded as shown in Figs. P10.41 through P10.43. Use Castigliano's theorem to determine the reactions.

11.67 through 11.69 A beam is supported and loaded as shown in Figs. P10.44 through P10.46. Use Castigliano's theorem to determine the reactions.

11.70 through 11.72 A beam is supported and loaded as shown in Figs. P10.47 through P10.49. Use Castigliano's theorem to determine the reactions.

Sec 11.10

11.73 A railroad car of mass 8 Mg is traveling at 1.1 m/s when it strikes a bumper of spring stiffness $k = 600$ kN/m. Determine the maximum deflection of the spring.

11.74 Rework Example 11.16a for the case in which the cantilever beam is supported at A by a spring of stiffness $k = 180$ kN/m.

11.75 A 20-kg block falls from a height $h = 50$ mm onto the midspan of a simply supported 2-m-long beam of rectangular cross section 40 mm wide and 60 mm deep. For $E = 70$ GPa, calculate (a) the maximum deflection and (b) the maximum stress.

11.76 Redo Prob. 11.75 for a beam supported at midspan by a spring of stiffness $k = 200$ kN/m.

11.77 Redo Prob. 11.75 for a beam supported at one end by a spring of stiffness $k = 180$ kN/m.

11.78 A 50-lb collar W is dropped from the position shown in Fig. P11.78 onto the end C of the stepped circular bar. If the maximum stress in the bar is not to exceed 30 ksi and if $E = 10 \times 10^6$ psi, determine the distance h.

11.79 A sliding collar of weight $W = 20$ lb is dropped $h = 1.2$ ft onto the end C of the stepped circular bar (Fig. P11.78). If $E = 10 \times 10^6$ psi, determine (a) the maximum deflection at end C and (b) the maximum stress in the bar.

11.80 Collar W, illustrated in Fig. P11.80, is dropped $h = 1.2$ m onto the end B of the circular bar of diameter $d = 25$ mm. Calculate the weight of the collar so that the largest stress in the bar does not exceed 200 MPa. The data are $E = 200$ GPa and $L = 5$ m.

11.81 A weight $W = 25$ lb falls a distance $h = 4.5$ ft when it comes into contact with end B of the steel rod ($E = 30 \times 10^6$ psi) of length $L = 6$ ft, as shown in Fig. P11.80. What diameter d should the rod have if the maximum stress is not to exceed 20 ksi?

11.82 A 40-kg diver jumps from the position shown in Fig. P11.82 onto the end C of a diving board. For $a = 5L = 2.5$ m, $h = 100$ mm, and $E = 14$ GPa, determine (a) the maximum deflection at C and (b) the maximum stress in the board.

11.83 Redo Prob. 11.82 for the case in which $a = L = 2$ m.

11.84 A circular shaft of cross-sectional area A and length L has a flywheel (of weight W and radius of gyration r) at one end and rotates at a given speed (ω rad/s). The shaft is suddenly stopped at the other end. Verify that the maximum angle of twist ϕ_{max} and maximum shearing stress τ_{max} produced by the impact are

$$\phi_{max} = \sqrt{\frac{2L}{GJ} U_k} \qquad \tau_{max} = \sqrt{\frac{4G}{AL} U_k} \tag{P11.84}$$

where U_k is the total kinetic energy ($W\omega^2 r^2/2g$) of the flywheel.

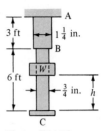

Figure P11.78

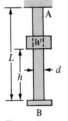

Figure P11.80

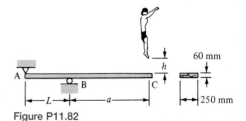

Figure P11.82

CHAPTER
12

INELASTIC BEHAVIOR

12.1 INTRODUCTION

The preceding chapters have been concerned with stress and deformation within the elastic range. Our consideration will now be extended to include inelastic behavior. As before, we must meet the requirements of equilibrium and geometric compatibility in our analysis. The only change we have to deal with lies in the stress-strain relationship, since in passing from the elastic to the plastic range, substantially different characteristics are observed. But when the stress-strain curve is known (Sec. 12.2), the three principles of analysis originally set forth in Sec. 1.3 suffice to describe the inelastic action occurring in variously loaded members.

The permanent deformation occurring in the plastic range of a material can assume considerable proportions. This inelastic deformation depends not only on the final state of stress, but also on the state of stress existing from the start of the loading process. We shall deal here with the maximum load that can be gradually applied to a structure before the onset of collapse. Design purposes dictate that this load, designated the *ultimate,* or *limit, load,* be known in order to determine the factor of safety against failure.

This chapter presents some of the fundamental concepts of plastic design involving an ideal elastoplastic material, and both statically determinate and indeterminate situations are discussed. It should be emphasized, however, that the plastic behavior considered here is applicable only to ductile materials, and in those cases in which material fatigue is a significant factor, inelastic analysis cannot be applied to machine design.

12.2 STRESS-STRAIN DIAGRAM
FOR ELASTOPLASTIC MATERIALS

The stress-strain relationship of a ductile material is frequently approximated by a diagram similar to that shown in Fig. 12.1a. In this text, a material having such an idealized *stress-strain diagram* is designated as *elastoplastic.* As before, the proportional limit and yield limit are assumed to be the same. The yield stress and yield strain are denoted σ_{yp} and ε_{yp}, respectively.

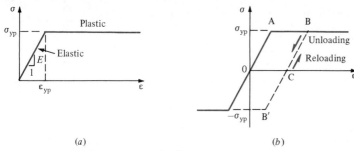

Figure 12.1   Idealized stress-strain diagram.

The *elastic* portion of the diagram is a straight line whose slope is E, the modulus of elasticity of the material. The *plastic* portion is a horizontal straight line beginning at the yield point. This is the region of considerable yielding under constant stress. Mild steel exhibits behavior of this form, both for tension and compression (see Fig. 3.8a). Note that, for steel, the plastic strains are roughly 15 to 20 times larger than the yield strain.

Stress-strain relations for many materials can be more accurately portrayed by elastic–linearly hardening diagrams, or so-called bilinear diagrams. Such diagrams are made up of two lines having different slopes (E and E_t) within the elastic and plastic ranges (see Sec. 3.7), respectively. This notwithstanding, the material will be treated here as elastoplastic in order to simplify the calculations. Thus the effect of strain hardening will not be considered. If we replace σ with τ and ε with γ, the stress-strain diagram of Fig. 12.1 also applies for shear.

When a mild-steel specimen that is stressed into the plastic range is unloaded at B of Fig. 12.1b, its response is along line BC. Notice particularly that this line is *parallel to* the *initial path* OA of the diagram. Then a permanent set or residual strain OC is left in the material. On reloading in tension, the material remains elastic from C to B; after that it again becomes plastic. If reloaded at C in compression, the material would remain elastic alone line CB$'$ which is the continuation of the straight line BC.

EXAMPLE 12.1

A steel bar of length $L = 0.4$ m and cross-sectional area $A = 50$ mm^2 is axially loaded until it is elongated 6 mm and unloaded. Compute the plastic deformation for $E = 200$ GPa and $\sigma_{yp} = 280$ MPa.

Solution   The stress-strain relationship can be represented by Fig. 12.1. At the onset of yielding,

$$\varepsilon_{yp} = \frac{\sigma_A}{E} = \frac{280(10^6)}{200(10^9)} = 1400 \ \mu$$

The maximum strain is

$$\varepsilon_{max} = \frac{\delta_B}{L} = \frac{6}{400} = 15,000 \ \mu$$

Upon load removal, the residual strain equals

$$\varepsilon_p = \varepsilon_C = 15,000 - 1400 = 13,600 \ \mu$$

The permanent *uniform* axial deformation is therefore $\delta_p = \varepsilon_p L = 0.0136(400) = 5.44$ mm.

---

*12.3 DUCTILITY AND DESIGN

The capacity of a material to undergo large strains while resisting a load is known as *ductility*. Ductile materials, in contrast with brittle materials, give warning of the onset of failure by excessive deformation. We shall discuss here the influence of ductility on the design of axially loaded members when stress concentrations are present.

Recall from Sec. 5.7 that the determination of stress-concentration factors is based upon a linear stress-strain relationship. For elastoplastic materials slowly and steadily loaded beyond the yield point, these factors decrease to a value approaching unity because of the redistribution of stress around a discontinuity. To illustrate this inelastic action, consider the behavior of a mild-steel bar that contains a hole and is subjected to a gradually increasing load P (Fig. 12.2).

When σ_{max} reaches the yield stress σ_{yp}, the stress distribution in the material is of the form of curve mn, and yielding impends at A. Some fibers are stressed in the plastic range, but enough others remain elastic, and the member can support additional load. Observe that the area under stress-distribution curve is equal to the load P. This area must increase as load P increases, and a *contained plastic flow* occurs in the material. It follows that, with the increase in the value of P, the stress-distribution curve assumes forms such as those depicted by line mp and finally mq. Hence the effect of abrupt change in geometry is nullified and $\sigma_{max} = \sigma_{avg}$, or $K = 1$; prior to necking, a nearly uniform stress distribution across the net section takes

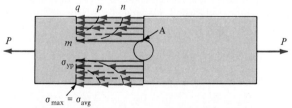

Figure 12.2 Redistribution of stress in a flat bar of mild steel.

place. This is referred to as a *fully plastic condition,* and the load at which it occurs is called the *plastic load.*

The foregoing discussion indicates that the stress-concentration factor is of no significance in design of a ductile material under static loading. However, for repeated or impact loading, even a ductile material may fail as a result of propagation of cracks originating at the points of high stress. Thus the presence of stress concentration in the case of dynamic loading must *not* be ignored, regardless of whether the material is brittle or ductile.

The effect of ductility upon the strength of shafts and beams is similar to that of axially loaded bars. That is, localized inelastic deformations enable these members to support high stress concentrations. Interestingly, material ductility introduces a certain element of forgiveness in analysis while producing acceptable design results—for example, rivets can carry equal loads in a riveted connection and angles for truss structures can be selected using the net section while disregarding the nonuniformity of stress on the net section.

12.4 PLASTIC DEFORMATIONS AND RESIDUAL STRESSES

The stress that remains in a structural member upon removal of external loads is called *residual stress.* This type of stress is always attributable to *nonuniform plastic deformation.* The presence of residual stress may be very harmful or, if properly controlled, may result in substantial benefit.

The magnitude and distribution of residual stresses can be obtained by superposition of the stresses due to loading and the *reverse,* or *rebound, stresses* due to unloading. (The strains corresponding to the latter are the reverse, or rebound, strains.) The reverse stress pattern is assumed to be *fully elastic* and hence can be obtained using Hooke's law. That is, the linear relationship between σ and ε remains valid (see line BB' in Fig. 12.1b) as long as the rebound stress does *not* decrease by more than $2\sigma_{yp}$. It is important to note that this superposition approach is not valid if the residual stresses thereby determined exceed the yield strength.

The nonuniform deformations that may be induced in a material by plastic torsion and plastic bending will be discussed in the two sections which follow. We are here concerned with a restrained or *statically indeterminate* structure that is *axially loaded* beyond the elastic range. In this case, some members of the structure experience different plastic deformation and these members retain stress following the release of load.

The value P_{yp} of the axial load at the onset of yielding in a statically determined ductile bar of cross-sectional area A is $\sigma_{yp}A$. This also equals the *plastic, limit,* or *ultimate* load P_u of the bar. For a statically indeterminate structure, however, after one member yields, additional load is applied until the remaining members also reach their yield limits. At this time *unrestricted* or *uncontained plastic flow* occurs and the limit load P_u is reached. Thus the ultimate load is the load at which yielding begins in *all* materials.

The following examples illustrate how plastic deformations and residual stresses are produced and how the limit load is determined in axially loaded structures.

EXAMPLE 12.2

Reconsider the steel bar of Example 5.6, assuming that $A_a = 2A_b = 800$ mm^2, $a = 0.6$ m, and $b = 0.9$ m. The material is elastoplastic with $E = 210$ GPa and $\sigma_{yp} = 420$ MPa (Fig. 12.3a). (a) Determine the displacement δ_b of the step as a function of applied load. (b) If the load is increased from zero to 480 kN and then reduced again to zero, calculate the residual stress in segment AC.

Solution Introducing the data into Eq. (5.10), we find that the axial forces in segments AC and CB of the bar are

$$R_A = \frac{P}{1 + (0.6 \times 400/0.9 \times 800)} = \frac{3P}{4} \qquad R_B = \frac{P}{4}$$

The normal stresses are therefore

$$\sigma_a = -\frac{3P}{4(800 \times 10^{-6})} = -937.5P$$

$$\sigma_b = \frac{P}{4(400 \times 10^{-6})} = 625P$$

where the minus sign indicates compression.

(a) Since $|\sigma_a| = |\sigma_b|$, the load P_{yp} at the onset of yielding is obtained from $|\sigma_a| = 420$ MPa. At this load, AC just yields, and the strain has the value of $\varepsilon_{yp} = \sigma_{yp}/E$. It follows that

$$P_{yp} = \frac{\sigma_{yp}}{937.5} = 448 \text{ kN} \qquad |\delta_b| = |\delta_a| = \varepsilon_{yp}a = 1.2 \text{ mm} \qquad (a)$$

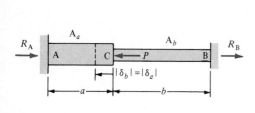

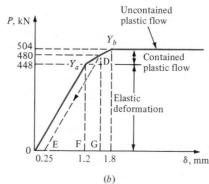

(a)

(b)

Figure 12.3 Example 12.2.

As the load P further increases, segment AC becomes *plastic*, and the stress is uniformly equal to σ_{yp}, thus resisting a compressive force of $R_A = \sigma_{yp} A_a = 336$ kN. On the other hand, at the start of yielding, segment CB carries a tensile load of $R_B = \sigma_{yp} A_b = 168$ kN and the strain just reaches $\varepsilon_{yp} = \sigma_{yp}/E$. Therefore

$$P_u = R_A + R_B = 504 \text{ kN} \qquad \varepsilon_b = \varepsilon_{yp} b = 1.8 \text{ mm} \qquad (b)$$

This value of load is the *ultimate load* of the stepped bar.

The load-deflection diagram of the bar is shown in Fig. 12.3b. Quantities given by Eqs. (a) and (b) locate the points Y_a and Y_b corresponding to the onset of yielding in segments AC and CB, respectively. We observe the *three stages of loading*: the range of linear elastic response; the range of contained plastic flow, in which AC yields and CB continues to deform elastically; and the condition of uncontained plastic flow beyond point Y_b. The last stage corresponds to the limit load of the structure.

(b) Load $P = 480$ kN locates point D in Fig. 12.3b; line DE is the elastic rebound of slope $448/1.2 = 373.3$; point E is the residual deflection, or permanent set. As segment CB is still in the elastic range, its elongation is

$$\delta_{OG} = \frac{(P - R_A)b}{A_b E} = \frac{(480 - 336)10^3 (900)}{400 \times 10^{-6}(210 \times 10^9)} = 1.54 \text{ mm}$$

Line segment EG in the figure represents the deformation of the bar during the unloading phase. From triangle EDG, we therefore obtain

$$\delta'_{EG} = -\frac{480}{373.3} = -1.29 \text{ mm}$$

The *residual deflection* in the bar is thus

$$\delta_{OE} = 1.54 - 1.29 = 0.25 \text{ mm}$$

The reverse strain in segment AC is $\varepsilon'_a = \delta'_{EG}/a = -1.29/600 = -2.15 \times 10^{-3}$ and the corresponding rebound stress equals $\sigma'_a = \varepsilon'_a E = -451.5$ MPa. This segment reaches the plastic range with $\sigma_a = 420$ MPa, and the *residual stress* is found as follows:

$$(\sigma_a)_{res} = \sigma_a - \sigma'_a = 420 - 451.5 = -31.5 \text{ MPa}$$

Segment CB of the bar is treated in a like manner.

EXAMPLE 12.3

As shown in Fig. 12.4a, a steel bar of 1.2 in.^{2} cross-sectional area is placed between two aluminum bars, each of 0.8 in.^{2} cross-sectional area. The ends of the bars are attached to a rigid support on one side and a rigid plate on the other. Let $E_s = 30 \times 10^6$ psi, $(\sigma_s)_{yp} = 35$ ksi, $E_a = 10 \times 10^6$ psi, and $(\sigma_a)_{yp} = 45$ ksi. If applied load P is increased from zero to P_u and removed, determine the residual stresses.

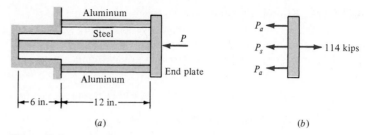

Figure 12.4   Example 12.3.

Solution   At ultimate load P_u *both* materials yield. Either material yielding by itself will not result in failure because the other material is still in the elastic range. Thus

$$P_a = 0.8(45 \times 10^3) = 36 \text{ kips}$$

$$P_s = 1.2(35 \times 10^3) = 42 \text{ kips}$$

and

$$P_u = 2(36) + 42 = 114 \text{ kips}$$

Application of an equal and opposite load of this amount, equivalent to a release of load (Fig. 12.4b), causes each member to rebound elastically. It follows from geometry that $\delta_a = \delta_s$ and

$$\frac{P'_a(12)}{0.8(10)} = \frac{P'_s(18)}{1.2(30)} \quad \text{or} \quad P'_s = 3P'_a \tag{c}$$

From statics,

$$2P'_a + P'_s = 114 \text{ kips} \tag{d}$$

Solving Eqs. (c) and (d), we obtain

$$P'_a = 22.8 \text{ kips} \qquad P'_s = 68.4 \text{ kips}$$

Superposition of the initial forces at P_u and the elastic rebound forces due to release of P_u leads to the residual forces:

$$(P_a)_{\text{res}} = 22.8 - 36 = -13.2 \text{ kips}$$

$$(P_s)_{\text{res}} = 68.4 - 42 = 26.4 \text{ kips}$$

The corresponding residual stresses are

$$(\sigma_a)_{\text{res}} = -\frac{13.2 \times 10^3}{0.8} = -16.5 \text{ ksi}$$

$$(\sigma_s)_{\text{res}} = \frac{26.4 \times 10^3}{1.2} = 22 \text{ ksi}$$

Note that subsequent to this prestressing process, the structure remains elastic as long as the value of $P_u = 114$ kips is not exceeded.

12.5   PLASTIC TORSION OF CIRCULAR SHAFTS

This section deals with the torsion of circular bars of ductile materials stressed into the plastic range. For such shafts, the first two basic assumptions associated with small deformations of circular bars in torsion (see Sec. 6.2) still apply. That is, the circular cross sections remain plane and their radii remain straight. As before, this means that strains vary linearly from the axis.

The relationships given in Secs. 6.3 and 6.5 are applicable as long as the shear strain in the bar does not exceed the yield strain γ_{yp}. With an increase in the applied torque, we expect yielding to occur on the boundary and to move progressively toward the interior. The cross-sectional stress distribution will appear as shown in Fig. 12.5a, b, and c.

At the onset of yield (Fig. 12.5a), the torque T_{yp}, from Eq. (6.2), is

$$T_{yp} = \frac{J}{c} \tau_{yp} = \frac{\pi c^3}{2} \tau_{yp} \tag{12.1}$$

This is referred to as the *maximum elastic torque*, or *yield torque*. When the twist is increased further, a *plastic portion* develops in the bar around an *elastic core* of radius ρ_0 (Fig. 12.5b). Applying Eq. (6.2), we find that the torque resisted by the elastic core is

$$T_1 = \frac{\pi \rho_0^3}{2} \tau_{yp} \tag{12.2a}$$

The outer part is subjected to constant yield stress τ_{yp} and resists the torque:

$$T_2 = \int \rho(\tau_{yp}\, dA) = \tau_{yp} \int_{\rho_0}^{c} \rho(2\pi\rho\, d\rho) = \frac{2\pi}{3}(c^3 - \rho_0^3)\tau_{yp} \tag{12.2b}$$

The total torque T, the sum of T_1 and T_2, may now be written in the form

$$T = \frac{\pi c^3}{6}\left(4 - \frac{\rho_0^3}{c^3}\right)\tau_{yp} = \frac{4}{3}T_{yp}\left(1 - \frac{1}{4}\frac{\rho_0^3}{c^3}\right) \tag{12.2c}$$

If twisting becomes very large, the region of yielding will approach the

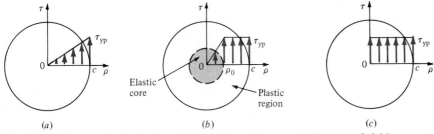

Figure 12.5   Stress distribution in a shaft as torque is increased: (a) onset of yield; (b) partially plastic; and (c) fully plastic.

middle of the shaft and hence ρ will approach zero (Fig. 12.5c). The corresponding torque T_u is the *plastic*, or *ultimate*, *shaft torque*, and its value from the above equation equals

$$T_u = \tfrac{2}{3}\pi c^3 \tau_{yp} = \tfrac{4}{3}T_{yp} \tag{12.3}$$

We thus observe that only one-third of the torque-carrying capacity remains after τ_{yp} is reached at the outermost fibers of a shaft.

The radius of the elastic core ρ_0 (Fig. 12.5b) is obtained from Eqs. (6.8) and (6.3), substituting $d\phi/dx = \phi/L$ and $\gamma_{max} = G\tau_{max} = \gamma_{yp}$:

$$\rho_0 = \frac{L\gamma_{yp}}{\phi} \tag{a}$$

Here L is the shaft length. If we denote the angle of twist at the onset of yielding (that is, when $\rho = c$) by ϕ_{yp}, the above leads to

$$c = \frac{L\gamma_{yp}}{\phi_{yp}} \tag{b}$$

Equations (a) and (b) result in

$$\frac{\rho_0}{c} = \frac{\phi_{yp}}{\phi} \tag{12.4}$$

Then Eq. (12.2c) may be written in the following useful form:

$$T_u = \frac{4}{3}T_{yp}\left(1 - \frac{1}{4}\frac{\phi_{yp}^3}{\phi^3}\right) \tag{12.5}$$

which is valid for $\phi > \phi_{yp}$. For $\phi < \phi_{yp}$, linear relation (6.10) applies.

A plot of Eqs. (6.10) and (12.5), shown in Fig. 12.6, reveals that after yield torque T_{yp} is reached, T and ϕ are connected by a *nonlinear* relationship. As T approaches T_u, the angle of twist grows without limit. Note, however, that the value of T_u is approached very rapidly (for example, $T_u = 1.32T_{yp}$ when $\phi = 3\phi_{yp}$).

If a shaft is strained beyond the elastic limit (point A in Fig. 12.6) and the applied torque is then removed, rebound is assumed to follow Hooke's law. Hence, once a portion of a shaft has yielded, residual stresses and residual rotations (ϕ_B) will develop. This process and the application of the foregoing relationships are illustrated in the next example. Statically indeterminate, inelastic torsion problems are treated analogously to those of axial load, as was shown in the preceding section.

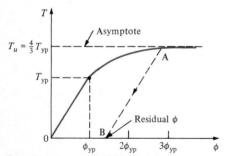

Figure 12.6 Torque-angle of twist relationship for a circular shaft.

EXAMPLE 12.4

A solid circular shaft 60 mm in diameter and 1.5 m long is subjected to a torque of 8 kN·m. Assume that $\tau_{yp} = 150$ MPa and $G = 80$ GPa. Determine (a) the radius of the elastic core; (*b*) the angle of twist of the shaft; (*c*) the residual stresses and the residual rotation remaining when the shaft is unloaded.

Solution   We have $c = 30$ mm and $J = \pi(0.03)^4/2 = 1272 \times 10^{-9}$ m^4.
(*a*) The yield torque, using Eq. (12.1), is

$$T_{yp} = \frac{J\tau_{yp}}{c} = \frac{1272 \times 10^{-9}(150 \times 10^6)}{0.03} = 6.36 \text{ kN·m}$$

Equation (12.2), together with the values of T and T_{yp}, leads to

$$\left(\frac{\rho_0}{c}\right)^3 = 4 - \frac{3T}{T_{yp}} = 4 - \frac{3(8 \times 10^3)}{6.36 \times 10^3} = 0.226$$

from which $\rho_0 = 0.609(0.03) = 18.3$ mm. The elastic-plastic stress distribution in the loaded shaft is depicted in Fig. 12.7*a*.
(*b*) The angle of twist at the onset of yielding, from Eq. (6.10), is

$$\phi_{yp} = \frac{T_{yp}L}{GJ} = \frac{6.36 \times 10^3(1.5)}{1272 \times 10^{-9}(80 \times 10^9)} = 0.0938 \text{ rad}$$

Substitution of the value obtained for ϕ_{yp} into Eq. (12.4) results in

$$\phi = \frac{c\phi_{yp}}{\rho_0} = \frac{30(0.0938)}{18.3} = 0.1538 \text{ rad} = 8.81°$$

(*c*) The removal of the torque causes elastic stresses as shown in Fig. 12.7*b*, and Eq. (6.2) results in

$$\tau'_{max} = \frac{Tc}{J} = \frac{8 \times 10^3(30 \times 10^{-3})}{1272(10^{-9})} = 188.7 \text{ MPa}$$

Upon superposition of the two distributions of stress, one obtains the residual stresses shown in Fig. 12.7*c*. The elastic rebound rotation is found from Eq. (6.10) as

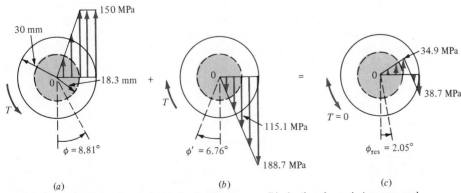

(*a*)   (*b*)   (*c*)

Figure 12.7   Example 12.4: (*a*) partially plastic stresses; (*b*) elastic rebound stresses; and (*c*) residual stresses.

$$\phi' = \frac{TL}{GJ} = \frac{8 \times 10^3(1.5)}{1272 \times 10^{-9}(80 \times 10^9)} = 0.1179 \text{ rad} = 6.76°$$

Thus, residual rotation equals $\phi_{\text{res}} = 8.81° - 6.76° = 2.05°$.

12.6   PLASTIC BENDING

Consideration is now given to *plastic bending* of an elastoplastic beam. The basic assumptions of bending theory, given in Sec. 7.7, remain valid even if the material behaves inelastically. Plane sections through a beam, normal to its axis, remain plane subsequent to bending. Hence a linear distribution of strain from the top to the bottom of the beam exists, and Eq. (7.7b) applies:

$$\varepsilon_x = -\frac{y}{\rho} \tag{7.7b}$$

where ρ denotes the radius of curvature of the deflection curve. However, the magnitude of the stress depends upon the stress-strain relationship, which varies according to the material. As the elastic flexure formula is based upon Hooke's law, it does not apply when the stresses exceed the proportional limit.

We shall focus our attention upon the analysis of a *rectangular beam* in pure bending subjected to a positive bending moment M (Fig. 12.8a). The beam is assumed to be made of an elastoplastic material having the stress-strain diagram shown in Fig. 12.8b. At the neutral axis, the location of which is yet to be determined, the strain equals zero. The bending moment at the onset of yield, M_{yp}, is obtained from the flexure formula:

$$M_{\text{yp}} = \frac{I}{c}\sigma_{\text{yp}} = S\sigma_{\text{yp}} = \frac{1}{6}bh^2\sigma_{\text{yp}} \tag{12.6}$$

This is termed the *yield moment*. Here S is the elastic modulus of the cross section of width b and height h. The stress distribution corresponding to M_{yp} is depicted in Fig. 12.9a. As the bending moment is increased, plastic zones develop in the member. The distance from the neutral surface to the point at which the elastic core ends is designated y_0 (Fig. 12.9b).

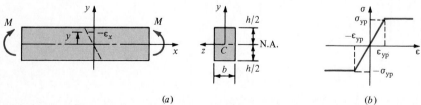

(a)   (b)

Figure 12.8   Plastic bending of an elastoplastic beam.

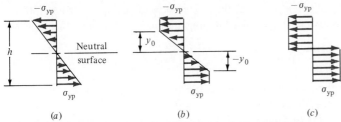

Figure 12.9 Stress distribution in a rectangular beam as bending moment is increased: (a) elastic; (b) partially plastic; and (c) fully plastic.

An examination of Fig. 12.9b shows that the bending stress varies as follows:

$$\sigma_x = -\frac{y}{y_0}\sigma_{yp} \qquad (-y_0 \le y \le y_0) \qquad (a)$$

Also $\sigma_x = -\sigma_{yp}$ for $y_0 \le y \le h/2$ and $\sigma_x = \sigma_{yp}$ for $-y_0 \le y \le -h/2$. As the beam is in pure bending, the neutral axis is located by using the equilibrium condition:

$$\sum F_x = \int \sigma_x \, dA = 0 \qquad (b)$$

or

$$\int_{-h/2}^{-y_0} \sigma_{yp}b \, dy + \int_{-y_0}^{y_0} -\frac{y}{y_0}\sigma_{yp}b \, dy + \int_{y_0}^{h/2} -\sigma_{yp}b \, dy = 0$$

Canceling the first and the third integrals, we have

$$\frac{\sigma_{yp}}{y_0}\int yb \, dy = 0$$

This means that the *neutral and the centroidal axes* of the cross section *coincide,* as in the case of an entirely elastic distribution of stress.

Similarly, from equilibrium of moments about the neutral axis,

$$\sum M_z = -\int \sigma_x y \, dA = M \qquad (c)$$

or

$$-\int_{-h/2}^{-y_0} \sigma_{yp}yb \, dy + \int_{-y_0}^{y_0} \frac{y}{y_0}\sigma_{yp}yb \, dy + \int_{y_0}^{h/2} \sigma_{yp}yb \, dy = M$$

The foregoing, after integration and substitution of Eq. (12.6), results in

$$M = \frac{3}{2}M_{yp}\left[1 - \frac{1}{3}\left(\frac{y_0}{h/2}\right)^2\right] \qquad (12.7)$$

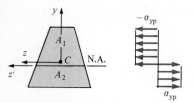

Figure 12.10   Plastic beam of trapezoidal cross section.

For the case in which $y_0 = h/2$, the above reduces to Eq. (12.6) and $M = M_{yp}$. As y_0 approaches zero, the bending moment approaches the limiting value:

$$M_p = M_u = \tfrac{3}{2}M_{yp} = \tfrac{1}{4}bh^2\sigma_{yp} \tag{12.8}$$

This value of the bending moment, corresponding to the stress distribution in Fig. 12.9c, is called the *plastic* or *ultimate moment*.

If instead of a rectangle we have a shape of cross section for which a *single plane* (*y*) *of symmetry* exists (Fig. 12.10), then Eq. (b) becomes $(-\sigma_{yp}A_1) + (\sigma_{yp}A_2) = 0$ or $A_1 = A_2$. We conclude therefore that the *neutral axis for plastic bending* (z') *divides the cross section into two equal areas.* Note that this axis is generally in a location *different* from that for the centroidal axis (z) of the section.

Application of the preceding analysis leads to similar relationships for other cross-sectional geometries. In general, for *any* cross section, the plastic moment for a beam can be written in the form

$$M_p = \sigma_{yp}Z \tag{12.9}$$

where Z is the *plastic section modulus*. The ratio of the plastic moment to the yield moment for a beam depends upon the geometric form of the cross section and is termed the *shape factor*:

$$f = \frac{M_p}{M_{yp}} = \frac{Z}{S} \tag{12.10}$$

For a rectangular beam, $S = bh^2/6$ and $Z = bh^2/4$. We thus verify the result given in Eq. (12.8) that $f = \tfrac{3}{2}$. It is observed that the plastic modulus represents the *sum of the first moments of the areas* (around the neutral axis) of the cross section above and below the neutral axis (see Fig. 12.8a):

$$Z = \frac{bh}{2}\left(\frac{h}{4} + \frac{h}{4}\right) = \frac{1}{4}bh^2 \tag{d}$$

In a like manner, the plastic modulus of other sections may readily be computed, as will be shown in Example 12.6.*

It is noted that when a beam is bent inelastically, some plastic deformation results, and the beam does not return to its initial configuration upon release of load. There will be some *residual stresses* in the beam (see Example 12.5). These stresses are determined by applying the principle of superposition in a way identical to that described in Secs. 12.4 and 12.5 for axial loading and torsion, respectively. The unloading phase may be treated by assuming the member to be fully elastic again, with Eq. (7.12) applying.

---

*The *Manual of Steel Construction* lists the values of Z for standard shapes of wide-flange beams. See *Manual of Steel Construction*, 8th ed., American Institute of Steel Construction, Chicago, 1980.

EXAMPLE 12.5

An elastoplastic beam of rectangular cross section 40 mm by 100 mm (Fig. 12.11) is subject to a bending moment of $M = 20$ kN·m. Determine, for $\sigma_{yp} = 240$ MPa and $E = 200$ GPa, (a) the thickness of the elastic core; (b) the radius of curvature of the neutral surface; and (c) the residual stresses following removal of the bending moment.

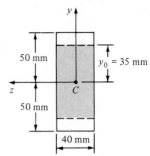

Figure 12.11 Example 12.5.

Solution We have $h = 0.1$ m and $b = 0.04$ m.

(a) Applying Eq. (12.6), we obtain

$$M_{yp} = \tfrac{1}{6}bh^2\sigma_{yp} = \tfrac{1}{6}(0.04)(0.1)^2(240 \times 10^6) = 16 \text{ kN·m}$$

Equation (12.7) then becomes

$$20 = \frac{3}{2}(16)\left(1 - \frac{4}{3}\frac{y_0^2}{0.01}\right)$$

from which

$$y_0 = 35 \times 10^{-3} \text{ m} = 35 \text{ mm} \qquad \text{and} \qquad 2y_0 = 70 \text{ mm}$$

The elastic core is shaded in Fig. 12.11.

(b) The yield strain ε_{yp} in the positive (upper) half of the beam is

$$\varepsilon_{yp} = -\frac{\sigma_{yp}}{E} = -\frac{240 \times 10^6}{200 \times 10^9} = -1200 \ \mu$$

Then, for $\varepsilon_x = \varepsilon_{yp}$, Eq. (7.7b) leads to

$$\rho = -\frac{y_0}{\varepsilon_{yp}} = \frac{35 \times 10^{-3}}{1.2 \times 10^{-3}} = 29.2 \text{ m}$$

(c) The stress distribution associated with moment $M = 20$ kN·m is shown in Fig. 12.12a. The release of moment M causes elastic stresses and Eq. (7.13) applies (Fig. 12.12b):

$$\sigma'_{max} = \frac{Mc}{I} = \frac{20 \times 10^3(0.05)}{0.04(0.1)^3/12} = 300 \text{ MPa}$$

Superimposing the two stress distributions, we can determine the residual stresses (Fig. 12.12c). It is observed that both tensile and compressive residual stresses remain in the beam.

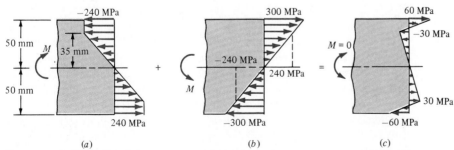

Figure 12.12 Example 12.5: stress distribution in a rectangular beam: (a) elastic-plastic; (b) elastic rebound; and (c) residual.

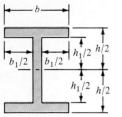

Figure 12.13 Example 12.6.

EXAMPLE 12.6

Determine the shape factor of an elastoplastic I beam in pure bending (see Fig. 12.13).

Solution The elastic section modulus is

$$S = \frac{I}{c} = \frac{1}{6h}(bh^3 - b_1 h_1^3)$$

and

$$M_{yp} = S\sigma_{yp} = \frac{1}{6h}(bh^3 - b_1 h_1^3)\sigma_{yp} \tag{12.11}$$

The plastic modulus is calculated by taking the first moment of one flange and the portion of the web above the neutral axis and multiplying by 2. From this we obtain

$$Z = \tfrac{1}{4}(bh^2 - b_1 h_1^2)$$

and

$$M_p = Z\sigma_{yp} = \tfrac{1}{4}(bh^2 - b_1 h_1^2)\sigma_{yp} \tag{12.12}$$

Upon substitution of Eqs. (12.11) and (12.12) into Eq. (12.10), it is found that

$$f = \frac{M_p}{M_{yp}} = \frac{3}{2}\frac{1 - (b_1 h_1^2/bh^2)}{1 - (b_1 h_1^3/bh^3)} \tag{12.13}$$

The foregoing shows that $f \le \tfrac{3}{2}$. For a beam of rectangular section ($h_1 = 0$), $f = \tfrac{3}{2}$. Thus, if a rectangular beam and an I beam are designed inelastically, the former will be more resistant to complete plastic failure.

*12.7 MOMENT-CURVATURE RELATIONSHIP

As discussed in Sec. 7.8, the procedure for determining the curvature κ is based only upon geometric considerations and applies for a beam of *any* material. Thus Eq. (7.8), $\kappa = d^2v/dx^2$, may still be used in plastic analysis. For a beam of a linearly elastic material, from Eq. (7.11) we have $\kappa = M/EI$.

In the case of an inelastic beam, an expression for the curvature in terms of the bending moment [such as we shall find in Eq. (12.16)] must be derived, as $\kappa \ne M/EI$. Then, upon integrating Eq. (7.8) twice, with adjustments made for the boundary conditions, we can determine the inelastic deflections. A generalized version of the moment-area method (Sec. 10.2), called the *curvature-area method*, can be used more effectively for this purpose. However, computations are usually complicated because of the nonlinear relationships between the displacements and the applied loads. Thus the method of superposition cannot be used in determining the inelastic deflections.

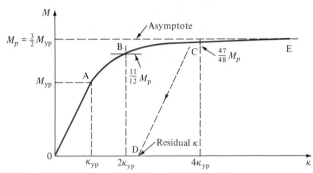

Figure 12.14 Moment-curvature relationship for a rectangular beam.

We now turn to consideration of the moment-curvature relation of an elastoplastic rectangular beam in pure bending (Fig. 12.8). As described in Sec. 7.6, the applied bending moment M produces a curvature which is the reciprocal of the radius of curvature ρ. At the elastic-plastic boundary of the beam (Fig. 12.9b), the yield strain is $\varepsilon_x = \varepsilon_{yp}$. Equation (7.7b) therefore gives

$$\kappa = \frac{1}{\rho} = -\frac{\varepsilon_{yp}}{y_0} \quad \text{and} \quad \kappa_{yp} = \frac{1}{\rho_{yp}} = -\frac{\varepsilon_{yp}}{h/2} \qquad (12.14)$$

Here κ and κ_{yp} correspond to the curvatures of the beam *after* and *at the onset* of yielding, respectively. From Eqs. (12.14), we have

$$\frac{\kappa_{yp}}{\kappa} = \frac{\rho}{\rho_{yp}} = \frac{y_0}{h/2} \qquad (12.15)$$

Substitution of Eqs. (12.15) and (12.8) into Eq. (12.7) results in the required moment-curvature relationship:

$$M = \frac{3}{2} M_{yp} \left[1 - \frac{1}{3} \left(\frac{\kappa_{yp}}{\kappa} \right)^2 \right] = M_p \left[1 - \frac{1}{3} \left(\frac{\kappa_{yp}}{\kappa} \right)^2 \right] \qquad (12.16)$$

Note that Eq. (12.16) is valid only when the bending moment becomes greater than M_{yp}. For $M < M_{yp}$, Eq. (7.11) applies. The variation of M with curvature is shown in Fig. 12.14. We observe that M rapidly approaches the asymptotic value $\frac{3}{2} M_{yp}$, which is the plastic moment M_p. When $\kappa = 2\kappa_{yp}$, or $y_0 = h/4$, eleven-twelfths of M_p has already been reached. Clearly, positions A, B, and E of the curve correspond to the stress distributions depicted in Fig. 12.9. Upon removal of the load at C, for instance, elastic rebound occurs along the line CD, and point D is the residual curvature.

To illustrate the three stages of loading, dimensionless moment is plotted against dimensionless curvature in Fig. 12.15 for a circle, a rectangle, and a typical wide-flange beam. In each case there is an initial range of linear elastic response. This is followed by a curved line representing the region in

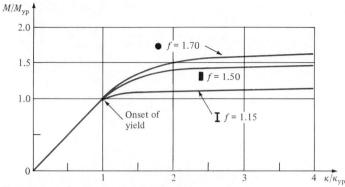

Figure 12.15 Moment-curvature diagrams for different cross-sectional shapes, with shape factor $f = M_p/M_{yp}$.

which the member is partially plastic and partially elastic. This is the region of contained plastic flow. Finally, the member continues to yield with no increase of applied bending moment. Observe the rapid ascent of each curve toward its asymptote as the section approaches the fully plastic condition. At this stage, unrestricted plastic flow occurs and the corresponding moment is the plastic moment M_p. Thus the cross section will abruptly continue to rotate; the beam is said to have developed a *plastic hinge*.

The plastic bending described so far disregards the effects of shear forces. However, these effects are negligible in beams of ordinary proportions (review Sec. 7.7). Hence, in the technical theory of plastic bending, it is assumed that the bending-stress distributions of Fig. 12.9 and the moment-curvature relation, Eq. (12.16), remain valid when shear is present—that is, when the moment varies along the beam axis. The concept of a plastic hinge will be considered further in the next section.

*12.8 LIMIT ANALYSIS OF BEAMS

Plastic, or limit, analysis refers to the procedures for determining limit, or ultimate, loads and for locating the plastic hinges of elastoplastic beams. Generally, limit analysis is simpler than elastic analysis, as it involves consideration of only static equilibrium and a geometric study of a beam (Sec. 1.3). On this basis, safe dimensions can be calculated in what is termed *limit design*. It is clear that such design requires higher than usual factors of safety. Note that beam deflections at the ultimate load can be estimated by using the ultimate load in elastic analysis and multiplying the results by the shape factor f.

The concept of plastic hinges provides a useful limit-analysis method. The rationale for the term *hinge* becomes apparent when describing, for example, the behavior of a simple beam of ductile material subjected to a concentrated load P at midspan (Fig. 12.16a). The moment diagram of the beam is shown in Fig. 12.16b. If $M_{yp} < M_{max} < M_p$, a region of plastic deformation occurs, as shown by the shaded area of Fig. 12.16a. The bending moment at the edge of the plastic region is $M_{yp} = (P/2)(L - L_0)/2$, from which

$$L_0 = L - \frac{4M_{yp}}{P} \qquad (a)$$

As the load P is increased, $M_{max} \rightarrow M_p$, and the plastic region extends inward. When M_{max} becomes equal to M_p, the cross section at midspan becomes fully plastic (Fig. 12.16c). The curvature at the center of the beam becomes unbounded and the beam fails (that is, collapse occurs). The beam halves on either side of the midspan experience rotation in the manner of a rigid bar about a *plastic hinge* under the effect of constant plastic moment M_p. At this instant, $M_{max} = M_p$, so the beam load equals

$$P = \frac{4M_p}{L} \qquad (b)$$

Substitution of Eq. (b) into Eq. (a) provides an expression for the length of the plastic region:

$$L_p = L\left(1 - \frac{M_{yp}}{M_p}\right) = L\left(1 - \frac{1}{f}\right) \qquad (12.17)$$

Interestingly, referring to Fig. 12.15, we find that the foregoing equation leads to $L_p = L/3$ for a rectangular beam and $L_p = 0.13L$ for a wide-flange beam. Thus the hinge actually extends over a finite length of the beam. However, for most purposes, the plastic hinge is assumed to form *at a discrete point* in the beam.

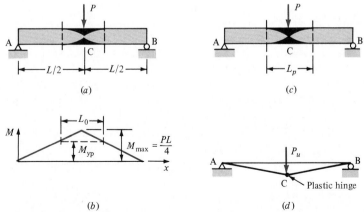

Figure 12.16   (a) Plastic region; (b) moment diagram; (c) plastic hinge; and (d) collapse mechanism.

The configuration indicating the location of the plastic hinge (Fig. 12.16d) or hinges is called a *collapse mechanism*. For a statically determinate beam, the value of the load required to develop a plastic hinge is the *limit*, or *ultimate, load* P_u. Thus by setting $P = P_u$ in Eq. (b), we obtain

$$P_u = \frac{4M_p}{L}$$

Observe that the ratio of the above load to the yield load, $P_{yp} = 4M_{yp}/L$, is equal to the shape factor f. This result is valid *only* for statically *determinate* beams (see Example 12.7 below).

A simple procedure for analyzing statically *indeterminate* beams includes the arbitrary assignment of the value of the plastic moment to the redundant moments. This makes the beam determinate. Then, if we introduce additional plastic hinges at points of high moment, a kinematically admissible collapse mechanism is created. It should be noted that for statically indeterminate beams the ratio of the ultimate load to the yield load varies with the type of beam support and loading (see Example 12.8 and Prob. 12.51). Sometimes more than one collapse mechanism is possible. In such a case, the ultimate load for each possibility must be computed. The mechanism associated with the smallest limit load is the actual collapse mechanism.

EXAMPLE 12.7

A simple beam of elastoplastic material carries a partial uniform load of intensity w_0 (Fig. 12.17a). Determine the ultimate load w_u.

Solution The bending-moment diagram as determined from static equilibrium is shown in Fig. 12.17b. By setting M_{max} equal to M_{yp}, with $w = w_{yp}$, the load at the onset of yielding is

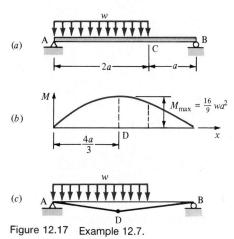

Figure 12.17   Example 12.7.

$$w_{yp} = \frac{9M_{yp}}{16a^2}$$

If we increase the load further, a plastic hinge develops at the section of maximum moment, forming the collapse mechanism shown in Fig. 12.17c. The corresponding ultimate load equals

$$w_u = \frac{9M_p}{16a^2}$$

where M_p is the plastic moment in the beam.

   Note the simplicity of calculating the limit load, which, however, provides no information with respect to deflection.

EXAMPLE 12.8

A beam of ductile material is supported and loaded as shown in Fig. 12.18a. Determine the ultimate load P_u.

Solution   An elastic analysis leads to the moment diagram of Fig. 12.18b. The maximum bending moment occurring at the fixed end is set equal to M_{yp} to obtain the yield load

$$P_{yp} = \frac{16M_{yp}}{3L} \qquad (c)$$

As the load is increased above P_{yp}, a plastic hinge will be developed at A where the moment can attain the value M_p. This, however, does not lead to beam collapse. With the further increase in load, a second plastic hinge will form at C, allowing the beam to undergo large deflection. Thus the member has formed a geometrically admissible mechanism (Fig. 12.18c).

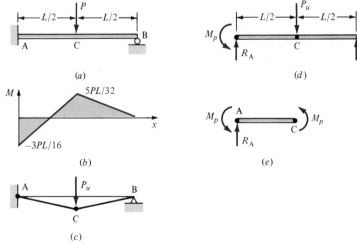

Figure 12.18   Example 12.8.

The limit load can be determined from the conditions of equilibrium. Usually the sense of the plastic moments, determined by inspection, is indicated in the free-body diagrams. Taking moments about B (Fig. 12.18d), we obtain

$$P_u \left(\frac{L}{2} \right) - R_A L + M_p = 0$$

Similarly, from $\Sigma M_C = 0$ (Fig. 12.18e), we have

$$2M_p - R_A \left(\frac{L}{2} \right) = 0$$

Solving these equations, we determine the limit load:

$$P_u = \frac{6M_p}{L} \tag{d}$$

Comparing the results of Eqs. (c) and (d), we find that the ratio of the ultimate load to the yield load is

$$\frac{P_u}{P_{yp}} = \frac{9M_p}{8M_{yp}}$$

Note particularly that there are two reasons P_u exceeds P_{yp}: $M_p > M_{yp}$ and the factor ($\frac{9}{8}$ in this case) owing to the *redistribution of moments* along the length of the statically indeterminate beam.

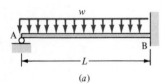

(a)

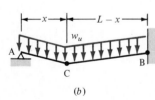

(b)

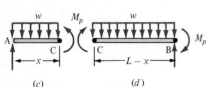

(c)   (d)

Figure 12.19   Example 12.9.

EXAMPLE 12.9

Determine the ultimate load w_u for the uniformly loaded propped cantilever beam of ductile material (Fig. 12.19a).

Solution   The collapse mechanism is shown in Fig. 12.19b. The location of the plastic hinge at C is found from the condition that *no shear* is possible at the point of maximum moment on a continuous curve. Applying the requirement that $\Sigma M_A = 0$ to the free-body diagram of Fig. 12.19c, we have

$$M_p - \tfrac{1}{2} w_u x^2 = 0$$

In like manner, $\Sigma M_B = 0$ (Fig. 12.19d) gives

$$2M_p - \tfrac{1}{2} w_u (L - x)^2 = 0$$

Solving the foregoing equations simultaneously, we obtain

$$w_u = \frac{2M_p}{[(\sqrt{2} - 1)L]^2} \qquad x = (\sqrt{2} - 1)L \tag{e}$$

as the values for the limit load and the location of the plastic hinge C, respectively.

PROBLEMS

Secs. 12.1 to 12.4

12.1 A stepped steel bar ABC is axially loaded until it elongates 15 mm and is then unloaded (Fig. P12.1). For $d_1 = 60$ mm, $d_2 = 50$ mm, $E = 200$ GPa, and $\sigma_{yp} = 250$ MPa, calculate (*a*) the largest value of P and (*b*) the plastic axial deformation of segments AB and BC.

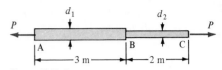

Figure P12.1

12.2 Redo Prob. 12.1 for the case in which $d_1 = 30$ mm, $d_2 = 20$ mm, and $\sigma_{yp} = 300$ MPa.

12.3 Members AB and AC of the truss shown in Fig. P12.3 are fabricated of an elastoplastic material with $\sigma_{yp} = 40$ ksi. If $P_u = 50$ kips and $\alpha = 60°$, determine the minimum cross-sectional areas A_{AB} and A_{BC}.

12.4 Solve Prob. 12.3 for $P_u = 20$ kips and $\alpha = 30°$.

12.5 Determine the ultimate load P_u that can be carried by truss ABC (Fig. P12.3) if each member is constructed of an elastoplastic material with $\sigma_{yp} = 38$ ksi, $A_{AB} = 2A_{BC} = 1.8$ in.², and $\alpha = 40°$.

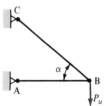

Figure P12.3

12.6 Determine the ultimate load W for the structure shown in Fig. P5.47. Assume that each member is made of mild steel, with $\sigma_{yp} = 250$ MPa, and has the same cross-sectional area, $A = 400$ mm². Use Eqs. (P5.49) and $\alpha = 30°$.

12.7 A rigid bar AB is supported and loaded as shown in Fig. P12.7. The steel wires AC and EF are identical ($E = 30 \times 10^6$ psi and $\sigma_{yp} = 36$ ksi); each has a cross-sectional area of 0.9 in.² and a length of 3 ft. (*a*) Determine the yield load P_{yp} and the corresponding deflection δ_{yp} at B; (*b*) determine the ultimate load and the corresponding deflection δ_u at B; and (*c*) draw the load (P)–deflection (δ_B) diagram.

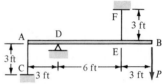

Figure P12.7

12.8 Rework Example 12.2a for the case in which $A_a = 600$ mm², $A_b = 900$ mm², and $a = b = 1$ m.

12.9 Redo Prob. 12.7 for the structure shown in Fig. P12.9.

12.10 As illustrated in Fig. 12.3a, an elastoplastic ($E = 210$ GPa, $\sigma_{yp} = 240$ MPa, and $\alpha = 11.6 \times 10^{-6}/°C$) stepped rod ABC is fixed at both ends and $P = 0$. At a temperature of 25°C the rod is free of stress. Let $A_a = 500$ mm², $A_b = 400$ mm², and $a = 2b = 0.8$ m. If the temperature is raised to 115°C, calculate (*a*) the stress in segment CB and (*b*) the displacement at step C.

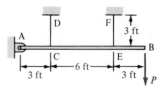

Figure P12.9

12.11 If a load $P = 100$ kips is applied and then removed, determine the residual stress in each bar of the assembly described in Example 12.3.

Sec. 12.5

12.12 A 0.5-m-long circular elastoplastic ($G = 70$ GPa and $\tau_{yp} = 180$ MPa) shaft of diameter 60 mm is twisted until the maximum shearing strain is 9000 μ. Calculate the magnitude of (*a*) the corresponding angle of twist ϕ and (*b*) the applied torque T.

12.13 A 3.6-ft-long circular shaft of 2 in. in diameter is subjected to a torque of 40 kip·in. If the shaft is made of 6061-T6 aluminum (see Table B.1), which is assumed to be elastoplastic, determine (*a*) the radius of the elastic core and (*b*) the angle of twist.

12.14 A 2-m-long circular shaft of 40-mm diameter is twisted such that only a 10-mm-diameter elastic core remains. Assume that the shaft is made of an elastoplastic material for which $\tau_{yp} = 140$ MPa and $G = 80$ GPa. Upon removal of the load, calculate (*a*) the residual stresses and (*b*) the residual angle of twist.

12.15 A hollow shaft has an outside radius c and an inner radius b and is made of an elastoplastic material. (*a*) Show that the *yield torque* and *ultimate torque* of the shaft are, respectively,

$$T_{yp} = \frac{\tau_{yp} J}{c} = \frac{\tau_{yp}}{c}(J_o - J_i) \tag{P12.15a}$$

and

$$T_u = \frac{2}{3}\pi\tau_{yp}(c^3 - b^3) = \frac{4}{3}\tau_{yp}\left(\frac{J_o}{c} - \frac{J_i}{b}\right) \tag{P12.15b}$$

The polar moment of inertia of the shaft is $J = \pi(c^4 - b^4)/2 = J_o - J_i$. (*b*) What is the ratio of T_u to T_{yp} for the case in which $c = 2b$?

12.16 A hollow circular steel tube ($G = 80$ GPa and $\tau_{yp} = 160$ MPa) is subjected to a slowly increasing torque T at its ends. The tube, of length $L = 2$ m, has an inside diameter of 60 mm and an outside diameter of 80 mm. Calculate (*a*) the yield torque T_{yp} and the angle of twist ϕ_{yp}; (*b*) the ultimate torque T_u and the ultimate angle of twist ϕ_u; and (*c*) the residual stresses τ_{res} and the residual rotation ϕ_{res} when the shaft is unloaded. Plot a graph of T versus ϕ.

12.17 The fixed-ended shaft shown in Fig. P12.17 is made of an elastoplastic material for which $G = 80$ GPa and $\tau_{yp} = 240$ MPa. Use $a = 2$ m, $b = 1.5$ m, $d_1 = 100$ mm, and $d_2 = 60$ mm. Assuming that the angle of twist at step C is 0.2 rad, calculate the magnitude of the applied torque T.

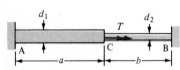

Figure P12.17

12.18 Redo Prob. 12.17 for $d_1 = d_2 = 80$ mm and $a = 2b = 2.2$ m.

12.19 Determine the ultimate load (T_u) for the shaft shown in Fig. P12.17 for $\tau_{yp} = 22$ ksi and $d_2 = 2d_1 = 2$ in., $a = 3$ ft, and $b = 3$ ft.

Secs. 12.6 and 12.7

12.20 A rectangular beam of cross section shown in Fig. P12.20 is made of mild steel with $\sigma_{yp} = 250$ MPa. Use $b = 60$ mm and $h = 80$ mm. For bending about the z axis, calculate (*a*) the yield moment and (*b*) the moment causing a 15-mm-thick plastic zone at the top and bottom of the beam.

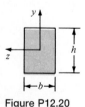

Figure P12.20

12.21   A steel rectangular beam (Fig. P12.20) is subjected to a moment 1.2 times greater than M_{yp}. Determine (a) the distance from the neutral axis to the point at which elastic core ends, y_0 and (b) the residual stress pattern following release of loading.

12.22   Compute the shape factor for a W 10 × 33 beam (see Table B.2).

12.23   Repeat Prob. 12.22 for a W 150 × 30 beam (see Table B.2).

12.24   Redo Prob. 12.22 for an S 10 × 35 beam (see Table B.3).

12.25 through 12.30   Determine the shape factor f for an elastoplastic beam of the cross sections shown in Figs. P12.25 through P12.30.

Figure P12.25

Figure P12.26

Figure P12.27

Figure P12.28

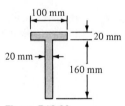

Figure P12.29

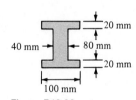

Figure P12.30

12.31   Redo Prob. 12.20 for the beam of the cross section shown in Fig. P12.27. Use $b = 100$ mm and $h = 120$ mm.

12.32   Solve Prob. 12.20 for the beam of the cross section shown in Fig. P12.30.

12.33   A rectangular beam with $b = 3$ in. and $h = 5$ in. (Fig. P12.20) is subjected to a plastic moment M_p. If $\sigma_{yp} = 36$ ksi for this beam of ductile material, calculate, for the case in which the loading has been removed, the residual stresses at the upper and lower faces.

12.34   Rework Prob. 12.33 for the beam of the cross section shown in Fig. P12.27. Use $b = 100$ mm and $h = 120$ mm.

12.35   A beam of the cross section shown in Fig. P12.29 is made of an elastoplastic material for which $E = 200$ GPa and $\sigma_{yp} = 250$ MPa. Calculate the bending moment that produces a bending strain of $-1250\ \mu$ on the lower face of the flange.

Sec. 12.8

12.36 Using Eq. (12.17), determine the value of L_p when a plastic hinge is formed in the beam of Fig. 12.16a. The beam has a hollow circular cross section of $c = 2b$ and is made of an elastoplastic material (Fig. P12.26).

12.37 A beam of elastoplastic material is supported and loaded as shown in Fig. P12.37. Determine the length L_p of the plastic region when the plastic hinge has formed.

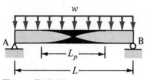

Figure P12.37

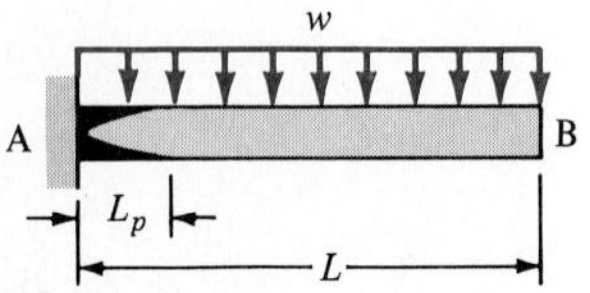

12.38 Rework Prob. 12.37 for the beam shown in Fig. P12.38.

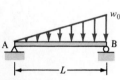

Figure P12.38

12.39 Redo Prob. 12.38 for the case in which an additional vertical load $P = 2wL$ acts downward at end B.

12.40 through 12.48 A beam of ductile material is supported and loaded as shown in Figs. P12.40 through P12.48. Determine the ultimate load in terms of L and M_p.

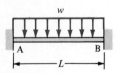

Figure P12.40

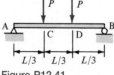

Figure P12.41

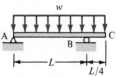

Figure P12.42

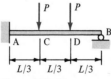

Figure P12.43

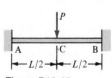

Figure P12.44

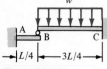

Figure P12.45

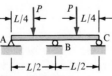

Figure P12.46

Figure P12.47

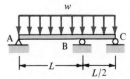

Figure P12.48

12.49   What is the ratio w_u/w_{yp} of the ultimate load to the yield load for the beam shown in Fig. P12.43?

12.50 and 12.51   For the beams of ductile material supported and loaded as shown in Figs. P12.44 and P12.45, determine the ratio P_u/P_{yp} of the ultimate load to the yield load.

CHAPTER
13

BUCKLING OF COLUMNS

13.1 INTRODUCTION

Recall from Sec. 1.2 that the usual objective of mechanics of materials is to investigate the strength, stiffness, and stability of a load-carrying member. We have up to now dealt primarily with the first two qualities. In the same vein, in discussing failure criteria, we have considered only those criteria based upon the attainment of a particular stress, deformation, or energy level within the member. In this chapter we shall consider failure due to elastic instability of a structure. *Stability* is the ability of a structure to support a given load without experiencing a sudden change in configuration. The principal difference between the theories of linear elasticity and linear stability is that, in the former, equilibrium is based upon the undeformed geometry, whereas in the latter, the *deformed* geometry must be considered.

We shall here be concerned only with columns—straight, slender members loaded axially in compression. Such bars are commonly used in trusses and in the framework of buildings. They are also encountered in machine linkages, signposts, supports for highway overpasses, and a wide variety of other structural and machine elements.

Buckling can be defined as the sudden, large, lateral deflection of a column owing to a small increase in an existing compressive load. This response leads to instability and collapse of the member. A wooden household yardstick with a compressive load applied at its ends illustrates the buckling phenomenon. The consideration of material strength (stress level) alone is not sufficient to predict the behavior of such a member.

In Secs. 13.2 through 13.4 we shall describe the critical, or buckling, load, the compressive load that causes the instability, and the effective lengths of columns with various common restraints. The stresses associated with both elastic and inelastic buckling of columns under centric load are considered in Sec. 13.5. Following this, we treat eccentrically loaded columns and then end the chapter with a discussion of column design.

13.2 STABILITY OF STRUCTURES: CRITICAL LOAD

In this section we shall employ equilibrium and energy approaches to examine the concepts of stability and critical load. In this regard, consider the idealized structure shown in Fig. 13.1*a*. Member AB is a rigid bar of length *L,* pinned at B and held in a vertical position by a spring at A of stiffness *k*. The bar supports an axial centric load *P*.

Equilibrium Method. Any momentary lateral force (causing a small displacement δ) will result in rotation of the bar, as shown in Fig. 13.1*b*. If force *P* is small, the system is *stable* and will return to the original position because the *disturbing* moment *P*δ is smaller than the *restoring* moment *k*δ*L,* where both moments are about pin B. But if force *P* is very large and the disturbing moment therefore exceeds the restoring moment, the bar will continue to rotate until the system collapses. Therefore, for a large force, the system is *unstable*.

The borderline between stability and instability occurs when *P*δ = *kL*δ for *any* small displacement δ. This condition is referred to as *neutral equilibrium*. From the foregoing expression, we can define the *critical load* as

$$P_{cr} = kL \qquad (a)$$

Clearly, the system is stable for $P < P_{cr}$ and unstable for $P > P_{cr}$. The bar is in equilibrium only when δ = 0 (Fig. 13.1*c*). Note that A, called the *bifurcation point,* marks the branches of the equilibrium diagram, indicated by two heavy lines in the figure. The horizontal line for neutral equilibrium extends a short distance to the left and to the right of A.

Energy Method. Referring again to Fig. 13.1*b*, we can approximate the work done by *P* as it acts through a distance $L(1 - \cos \theta)$ by the expression

$$\Delta W = PL(1 - \cos \theta) = PL \left(1 - 1 + \frac{\theta^2}{2} + \cdots \right) = \frac{1}{2} PL\theta^2$$

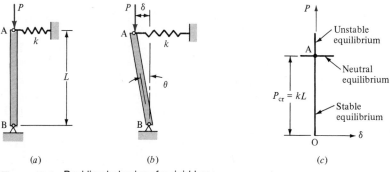

| (a) | (b) | (c) |

Figure 13.1 Buckling behavior of a rigid bar.

The strain energy of the spring as a result of its elongation is

$$\Delta U = \tfrac{1}{2}k(L\theta)^2$$

If $\Delta U > \Delta W$, the system is *stable*. That is, the work done is insufficient to produce a displacement that grows following a momentary force. But if $\Delta U < \Delta W$, the bar is *unstable* because now the work is large enough to result in a displacement that grows subsequent to a lateral disturbance. The transition from a stable to an unstable configuration occurs when $\Delta U = \Delta W$ or $P_{cr} = kL$, as before.

The buckling analysis of columns is usually based upon either the equilibrium or the energy method. The choice of approach depends upon the particulars of the situation under consideration, but elastic stability problems can be treated in a very general way using the energy method.

13.3 BUCKLING OF PIN-ENDED COLUMNS

The behavior of columns is similar to that of the system shown in Fig. 13.1a. In the case of a column, the restoring moment is provided by internal stress resultants rather than an external spring. We now consider a *slender pin-ended column* centrically loaded by compressive forces P at each end (Fig. 13.2a). The member is assumed to be perfectly straight and to be constructed of a linearly elastic material—that is, we have an ideal column.

Refer to Fig. 13.2b in which load P has been increased sufficiently to cause a small lateral deflection. This is a condition of *neutral equilibrium*. The corresponding value of the load is the critical load P_{cr}. Bending moment M at an arbitrary position x equals $-Pv$ (Fig. 13.2c). Employing the relationship $EIv'' = M$, we have

$$EI\,\frac{d^2v}{dx^2} + Pv = 0 \qquad\qquad (13.1)$$

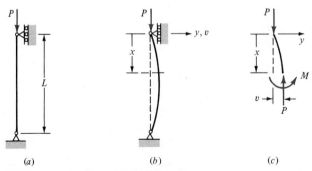

(a) (b) (c)

Figure 13.2 Column with pinned ends.

a linear differential equation with constant coefficients. Setting

$$p^2 = \frac{P}{EI} \tag{13.2}$$

one has

$$\frac{d^2v}{dx^2} + p^2v = 0 \tag{13.3}$$

The general solution of this governing equation is

$$v = A \sin px + B \cos px \tag{13.4}$$

which is a form of the familiar equation of simple harmonic motion. The constants A and B are evaluated from the end conditions:

$$v(0) = 0 \quad \text{and} \quad v(L) = 0$$

The first requirement yields $B = 0$ and the second leads to

$$A \sin pL = 0 \tag{a}$$

The foregoing is satisfied if either $A = 0$ or $\sin pL = 0$. The first of these corresponds to a condition of no buckling and yields a trivial solution. The second case is the acceptable alternative because it is consistent with column deflection. It is fulfilled if

$$pL = \sqrt{\frac{P}{EI}} L = n\pi \quad (n = 1, 2, \ldots) \tag{b}$$

from which

$$P = \frac{n^2\pi^2 EI}{L^2} \tag{c}$$

Only the value for $n = 1$ has physical significance, as it determines the smallest value of P for which a buckling shape can occur under static loading. Thus the *critical load* for a column with a pinned end is

$$P_{\text{cr}} = \frac{\pi^2 EI}{L^2} \tag{13.5}$$

where L represents the original length of the column. The preceding result is attributable to L. Euler (1707–1783), and Eq. (13.5) is thus known as *Euler's formula*; the corresponding load is called the *Euler buckling load*. In Eq. (13.5), EI is the flexural rigidity for bending in the xy plane, taken to be the plane of buckling owing to the restraints imposed by the end connections (Fig. 13.2a). However, if the column is free to deflect in any direction (as in the case of ball-and-socket supports), it will tend to bend about the axis having the *smallest* principal centroidal moment of inertia I.

As defined in App. A, $I = Ar^2$, where A is the cross-sectional area and

r represents the radius of gyration about the axis of bending. Insertion of this relationship into Eq. (13.5) gives

$$P_{cr} = \frac{\pi^2 EA}{(L/r)^2} \tag{13.6}$$

We seek the minimum value of P_{cr}, and consequently the *smallest* radius of gyration should be used in Eq. (13.6). The quotient L/r, called the *slenderness ratio,* is an important parameter in the classification of columns, as we shall see in Sec. 13.5.

Introducing Eq. (*b*) into Eq. (13.4) and recalling that $B = 0$ and $n = 1$, we note that the buckled *shape* or *mode* of the column is

$$v = A \sin \frac{\pi x}{L} \tag{13.7}$$

The value of the maximum deflection, $v_{max} = A$, is undefined. Thus the critical load will sustain *any small* lateral deflection, in agreement with the basic assumptions of beam theory (Sec. 7.7). Because of practical considerations, large deflections are generally not permitted to occur in structures.

EXAMPLE 13.1

A 5-ft-long Douglas fir bar of 2-in. by 4-in. rectangular cross section, pivoted at both ends, is subjected to an axial compressive load. For $E = 1.8 \times 10^6$ psi, calculate (*a*) the slenderness ratio and (*b*) the allowable load, using a factor of safety of $f_s = 1.5$.

Solution   The radius of gyration r of the cross section, from Example A.3, is

$$r = \sqrt{\frac{I}{A}} = \frac{b}{2\sqrt{3}} \tag{d}$$

(*a*) The lowest value of r is obtained when the centroidal axis is parallel to the longer side of the rectangle. We thus have $h = 4$ in. and $b = 2$ in. so that $r = 2/2\sqrt{3} = 0.577$ in. Then $L/r = 60/0.577 = 104$.

(*b*) From Eq. (13.6), the Euler buckling load is

$$P_{cr} = \frac{\pi^2 EA}{(L/r)^2} = \frac{\pi^2(1.8 \times 10^6)(2 \times 4)}{(104)^2} = 13.14 \text{ kips}$$

The largest load the column can support is therefore $P_{all} = 13.14/1.5 = 8.76$ kips.

13.4   COLUMNS WITH OTHER END CONDITIONS

It is evident from the derivation in Sec. 13.3 that the critical load is dependent upon the end restraints. For columns with various combinations of fixed, free, and pinned supports, one need only introduce the appropriate conditions into Eq. (13.4) and proceed as before.

It can be shown that, in each case, the critical load of a uniform column can be expressed as

$$P_{cr} = C\,\frac{\pi^2 EI}{L^2} \qquad (a)$$

where C is a constant dependent upon the end conditions. The values of C for the common cases shown in Figs. 13.3a, b, c, and d are 0.25, 1, 2.04, and 4, respectively. Note that the critical load increases as the degree of constraint increases.

Equation (a) can be made to resemble the fundamental case if L^2/C is replaced by L_e^2, in which L_e is called the *effective length*. Thus the Euler formula can be written in the form

$$P_{cr} = \frac{\pi^2 EI}{L_e^2} \qquad (13.5')$$

As shown in Fig. 13.3, it develops that the effective length is the *distance between the inflection points* on the elastic curves, or *points of zero moment*. This length is often expressed in terms of an effective length factor K: $L_e = KL$ (Fig. 13.3). If we set $I = Ar^2$, the foregoing equation becomes

$$P_{cr} = \frac{\pi^2 EA}{(L_e/r)^2} \qquad (13.6')$$

The quantity L_e/r is referred to as the *effective slenderness ratio* of the column.

The idealized supports shown in Fig. 13.3 seldom occur. Because of uncertainty relative to the fixity of the joints, columns are sometimes taken to be pin-ended. Omitting the case of Fig. 13.3a, where it cannot be applied, this procedure is conservative. Observe that the effective length (and the

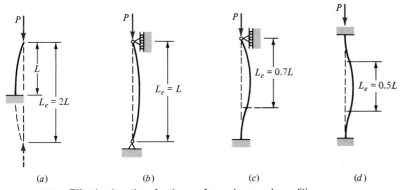

Figure 13.3 Effective lengths of columns for various end conditions.

critical load) for a column fixed at one end and pinned at the other *cannot* be determined by *inspection*, as can be done in the other cases shown in Fig. 13.3. Hence it is necessary to solve the differential equation to obtain P_{cr}, as illustrated in the example that follows.

EXAMPLE 13.2

Determine the critical load for a uniform column fixed at the base, pinned at the top, and subjected to a vertical axial load P (Fig. 13.4a).

Solution   The free-body diagram of the entire column in a slightly deflected configuration is shown in Fig. 13.4b. We note that a statically indeterminate reactive force R is developed at end A in addition to an axial load P. The bending moment at any section is $M = -Pv - Rx$ (Fig. 13.4c). By proceeding as in Sec. 13.2, we obtain

$$\frac{d^2v}{dx^2} + p^2v = -\frac{R}{EI}x \qquad (b)$$

where $p^2 = P/EI$. The general solution of the above differential equation is

$$v = A \sin px + B \cos px - \frac{R}{P}x \qquad (c)$$

in which A, B, and R are constants. The boundary conditions

$$v(0) = 0 \qquad v(L) = 0 \qquad v'(L) = 0$$

are applied to Eq. (c) to give $B = 0$ and

$$A \sin pL = \frac{R}{P}L \qquad Ap \cos pL = \frac{R}{P} \qquad (d)$$

Division of the first expression by the second results in the *buckling equation*:

$$\tan pL = pL \qquad (e)$$

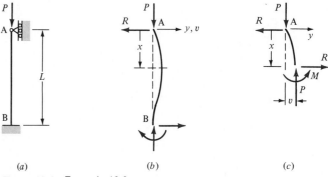

(a)   (b)   (c)

Figure 13.4   Example 13.2.

The smallest nonzero value of pL satisfying Eq. (e), from trial and error, is $pL = 4.4934$. This corresponds to a critical load of

$$P_{cr} = \frac{20.19EI}{L^2} = \frac{2.046\pi^2 EI}{L^2} \tag{13.8}$$

The effective length of the column is determined upon comparison of Eqs. (13.8) and (13.6):

$$L_e = 0.699L \approx 0.7L \tag{f}$$

The foregoing is the distance between the pinned end of the column and the point of inflection in the buckled shape (Fig. 13.3c).

EXAMPLE 13.3

In the structure shown in Fig. 13.5, a rectangular steel bar AB is constructed of a 2-in. by 3-in. section for which $E = 30 \times 10^6$ psi and $\sigma_{yp} = 36$ ksi. What is the buckling load of the column?

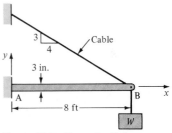

Figure 13.5 Example 13.3.

Solution From statics, the axial force expressed in terms of W is found to be $P = 4W/3$. The properties of the cross section are

$$I_y = \tfrac{1}{12}(3)(2)^3 = 2 \text{ in.}^4$$

$$I_z = \tfrac{1}{12}(2)(3)^3 = 4.5 \text{ in.}^4$$

$$A = (2)(3) = 6 \text{ in.}^2$$

and

$$r_y = \sqrt{\frac{I_y}{A}} = 0.577 \text{ in.} \qquad r_z = \sqrt{\frac{I_z}{A}} = 0.866 \text{ in.}$$

Buckling in xy Plane. The effective length of the column with respect to buckling in the xy plane is $L_e = 0.7L$; therefore

$$\frac{L_e}{r_z} = \frac{0.7(96)}{0.866} = 77.6$$

The Euler formula is applicable in this range. Hence

$$P_{cr} = \frac{\pi^2 EA}{(L_e/r_z)^2} = \frac{\pi^2(30 \times 10^6)(6)}{(77.6)^2} = \frac{4}{3}W$$

from which

$$W = 221.26 \text{ kips}$$

Buckling in xz Plane. The line of action of the compressive force passes through end A, thus causing no moment about the y axis at the fixed support. Therefore $L_e = L$. Then

$$\frac{L}{r_y} = \frac{96}{0.577} = 166.4$$

The buckling load is

$$P_{cr} = \frac{\pi^2 EA}{(L/r_y)^2} = \frac{\pi^2(30 \times 10^6)(6)}{(166.4)^2} = \frac{4}{3} W$$

or

$$W = 48 \text{ kips}$$

The column will obviously fail by lateral buckling when load W exceeds 48 kips. Observe that the critical stress is equal to $P_{cr}/A = (\frac{4}{3})(48 \times 10^3)/(6) = 10.67$ ksi. This compared with the yield strength 36 ksi demonstrates the significance of buckling analysis in predicting the safe working load.

13.5   CRITICAL STRESS: CLASSIFICATION OF COLUMNS

The behavior of an ideal column is often represented on a plot of average compressive stress P/A versus slenderness ratio L_e/r (Fig. 13.6). Such a representation offers a clear rationale for the classification of compression bars. Tests of columns verify each portion of the curve with reasonable accuracy. The range of L_e/r is a function of the material under consideration.

Long Columns.   For a long column, that is, a member of sufficiently large slenderness ratio, buckling occurs elastically at a stress that does *not* exceed the proportional limit of the material. Hence the Euler's load of Eq. (13.6) is appropriate to this case, and the *critical stress* is

$$\sigma_{cr} = \frac{P_{cr}}{A} = \frac{\pi^2 E}{(L_e/r)^2} \tag{13.9}$$

The corresponding portion CD of the curve (Fig. 13.6) is labeled as *Euler's curve*. The *critical* value of slenderness ratio that fixes the lower limit of this curve is found by equating σ_{cr} to the proportional limit (σ_{pl}) of the specified material:

$$\left(\frac{L_e}{r}\right)_c = \sqrt{\frac{\pi^2 E}{\sigma_{pl}}} \tag{13.10}$$

For example, in the case of a *structural steel* with $E = 210$ GPa and $\sigma_{pl} = 250$ MPa, the foregoing results in $(L_e/r)_c = 91$.

From Fig. 13.6 we observe that very slender columns buckle at low levels of stress; they are much less stable than short columns. Use of a higher-strength material does not improve this situation. Clearly, Eq. (13.9) indicates that the critical stress is increased by using a material of higher modulus of elasticity E or by increasing the radius of gyration r. A tubular column, for

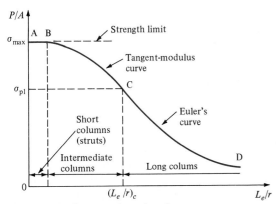

Figure 13.6 Average stress in columns versus slenderness ratio.

instance, will have a much larger value of r than the solid column of the same cross-sectional area. However, there will be a limit beyond which the buckling stress cannot be increased. The wall thickness will eventually become so thin as to cause the member to crumble. This behavior, called *local instability,* is often characterized by a change in the shape of a cross section.

Short Columns. Compression members having low slenderness ratios (for example, steel rods with $L/r < 30$) exhibit essentially no instability and are referred to as *short columns* or struts. For these bars, failure occurs by yielding or by crushing, without buckling, at stresses exceeding the proportional limit of the material. Thus the maximum stress

$$\sigma_{max} = \frac{P}{A} \tag{13.11}$$

is the *strength limit* of such a column, represented by horizontal line AB in Fig. 13.6. It is equal to the yield stress or ultimate stress in compression.

Intermediate Columns. Most structural columns lie in a region between short and long classifications. Such *intermediate-length columns* do not fail by direct compression or by elastic instability. Consequently, Eqs. (13.9) and (13.11) do not apply, and a separate analysis is required. The failure of an intermediate column occurs by *inelastic buckling* at stress levels exceeding the proportional limit. Presented below is one practical approach to determination of inelastic buckling; this approach is known as the *tangent-modulus theory*.

Consider the axial compression of an intermediate-length column, and suppose the loading to occur in small increments until such time as the buckling load P_t is achieved. As increments of load are imposed, the column displays a slight curvature. It is assumed that accompanying the increasing loads and curvature is a continuous increase, or *no decrease,* in the longi-

Figure 13.7 Stress distribution in an intermediate column.

tudinal stress and strain in *every* fiber of the column. Thus the stress distribution is as depicted in Fig. 13.7. The increment of stress $\Delta\sigma$ is due to bending; σ_{cr} represents the value of stress associated with the attainment of critical load P_t. The distribution of strain will display a pattern identical with that shown in the figure.

For small deflections Δv, the stress and strain increments will likewise be small, and as in the case of elastic bending, sections initially plane are assumed to remain plane after bending. The relationship between $\Delta\sigma$ and the change in strain $\Delta\varepsilon$ is therefore $\Delta\sigma = E_t\,\Delta\varepsilon$. Here E_t is the tangent modulus of the material (see Sec. 3.7). Note that within the linearly elastic range $E_t = E$ (Fig. 3.10).

On the basis of the foregoing rationale, all the results for a linearly elastic beam still apply if E is replaced by E_t. Hence the critical stress may be expressed by the generalized Euler buckling formula, or the *tangent modulus formula*:

$$\sigma_{\text{cr}} = \frac{P_t}{A} = \frac{\pi^2 E_t}{(L_e/r)^2} \qquad (13.12)$$

When the tangent moduli corresponding to the given stresses can be read from the compression stress-strain diagrams, the value of L_e/r at which a column will buckle can readily be calculated using Eq. (13.12). If, however, L_e/r is known and σ_{cr} is to be determined, a trial-and-error approach is necessary (see Example 13.5). The behavior of a column for intermediate ratios of L_e/r is represented by the curve BC in Fig. 13.6.

It should be noted that Eqs. (13.9), (13.11), and (13.12) determine the *ultimate stresses* not the working stresses. It is thus necessary to divide the right side of each formula by an appropriate factor of safety, often 2 to 3, depending upon the material, in order to obtain the allowable values. Some typical relationships for *allowable stress* will be introduced in Sec. 13.7.

EXAMPLE 13.4

A 64 by 51 by 6.4-mm steel angle (Fig. 13.8) is to serve as a pin-ended column to support a 40-kN load with a safety factor $f_s = 2$. Assuming that the proportional limit strength $\sigma_{\text{pl}} = 240$ MPa and $E = 210$ GPa, determine the maximum length of the member.

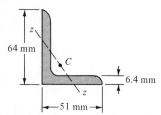

Figure 13.8 Example 13.4.

Solution From Table B.6, we find that the smallest radius of gyration for the principal axis (z) of the angle cross section is $r = 10.8$ mm; the area is 680 mm^2. In this case, the critical load becomes $P_{cr} = 2(40) = 80$ kN and $L_e = L$. Thus $\sigma_{cr} = 80 \times 10^3/(680 \times 10^{-6}) = 117.6$ MPa and we have $\sigma_{cr} < \sigma_{pl}$.

Based upon the assumption that Euler's formula is applicable, we have

$$\sigma_{cr} = \frac{\pi^2 E}{(L/r)^2} \qquad 117.6 \times 10^6 = \frac{\pi^2(210 \times 10^9)}{(L/0.0108)^2}$$

from which $L = 1.434$ m. As the slenderness ratio, $L/r = 1.434/0.0108 = 132.8$, is well into the Euler range, the assumption is valid.

EXAMPLE 13.5

A pipe of 76-mm outside diameter and 3-mm thickness is used as a column of 2-m effective length. Determine the axial buckling load for a material with a stress-strain curve approximated as in Fig. 13.9.

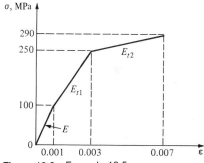
Figure 13.9 Example 13.5.

Solution For a tube cross section (Example A.4), we have

$$A = \frac{\pi}{4}(76^2 - 70^2) = 688 \text{ mm}^2$$

$$r = \tfrac{1}{4}\sqrt{76^2 + 70^2} = 25.8 \text{ mm}$$

Thus $L/r = 77.5$ and the tangent modulus formula is

$$\sigma_{cr} = \frac{\pi^2 E_t}{(L_e/r)^2} = \frac{\pi^2 E_t}{(77.5)^2} = 0.001643E_t \qquad (a)$$

Inasmuch as the value of E_t is different in every range, we must employ a trial-and-error procedure.

For the initial (elastic) range, the slope is $E_t = E = 100$ GPa and Eq. (a) yields

$$\sigma_{cr} = 0.001643(100 \times 10^9) = 164.3 \text{ MPa}$$

As observed from Fig. 13.9, E is not valid above 100 MPa. For the second portion of the stress-strain curve, $E_{t1} = 75$ GPa and therefore

$$\sigma_{cr} = 0.001643(75 \times 10^9) = 123.2 \text{ MPa}$$

This is the critical stress at buckling because the value of E_{t1} lies between 100 and 250 MPa. The solution is therefore

$$P_{cr} = \sigma_{cr}A = (123.2 \times 10^6)(688 \times 10^{-6}) = 84.8 \text{ kN}$$

*13.6 ECCENTRICALLY LOADED COLUMNS

So far we have dealt with the buckling of columns for which the load acts at the centroid of a cross section. We now analyze columns under an *eccentric* load. This case is obviously of great practical significance because of imperfections in load alignment in real situations.

Consider a pinned-end column subjected to compressive forces applied with a small *eccentricity e* relative to the column axis (Fig. 13.10*a*). It is assumed that the member is initially straight and that the material is linearly elastic. With an increase in load, the column deflects, as indicated by the dashed lines in the figure. The bending moment at distance x from the mid-span is $M = -P(v + e)$. For small deflections v, the differential equation for the elastic curve becomes

$$EI \frac{d^2v}{dx^2} + P(v + e) = 0$$

or

$$\frac{d^2v}{dx^2} + p^2v = -p^2e \qquad (a)$$

where $p^2 = P/EI$, as before. The general solution of the foregoing is

$$v = A \sin px + B \cos px - e \qquad (b)$$

Here the first two terms represent the homogeneous solution and $-e$ the particular solution.

To obtain constants A and B, the boundary conditions $v(L/2) = v(-L/2) = 0$ are applied, with the following results:

$$A = 0 \qquad B = \frac{e}{\cos[\sqrt{P/EI}\,(L/2)]}$$

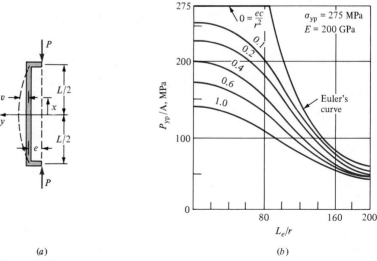

(a)   (b)

Figure 13.10 Graph of the secant formula.

The equation of the deflection curve is therefore

$$v = e \left[\frac{\cos \sqrt{P/(EI)}\, x}{\cos \sqrt{PL^2/(4EI)}} - 1 \right] \qquad (13.13)$$

Expressed in terms of the *critical load* $P_{cr} = \pi^2 EI/L^2$, the midspan ($x = 0$) deflection is

$$v_{max} = e \left[\sec \left(\frac{\pi}{2} \sqrt{\frac{P}{P_{cr}}} \right) - 1 \right] \qquad (13.14)$$

Notice from this expression that, as P approaches P_{cr}, the maximum deflection increases without bound. Thus P should *not be allowed* to reach the critical value obtained in Sec. 13.3 for a column under a centric load.

The maximum compressive stress σ_{max} occurs at $x = 0$ on the concave side of the column. Hence

$$\sigma_{max} = \frac{P}{A} + \frac{Mc}{I} = \frac{P}{A} + \frac{M_{max}c}{Ar^2} \qquad (c)$$

where r represents the radius of gyration and c the distance from the centroid of the cross section to the outer fibers, both in the direction of eccentricity. Substituting $M_{max} = -P(v_{max} + e)$, together with Eq. (13.14), into Eq. (c), we obtain

$$\sigma_{max} = \frac{P}{A} \left[1 + \frac{ec}{r^2} \sec \left(\frac{\pi}{2} \sqrt{\frac{P}{P_{cr}}} \right) \right] \qquad (13.15a)$$

Alternatively,

$$\sigma_{max} = \frac{P}{A} \left[1 + \frac{ec}{r^2} \sec \left(\frac{L}{2r} \sqrt{\frac{P}{AE}} \right) \right] \qquad (13.15b)$$

Equations (13.15), known as the *secant formula*, applies only when the maximum stress remains within the elastic limit of the material. It provides σ_{max} in the column as a function of the average stress (P/A), the *eccentricity ratio* (ec/r^2), and the slenderness ratio (L/r). As Eq. (13.15) is a nonlinear relationship between σ_{max} and P, the maximum stress produced by different axial forces *cannot* be superimposed.

The buckling load specified in Eq. (13.15) is converted to a *limit* or *yield load* P_{yp} by setting $P = P_{yp}$ and $\sigma_{max} = \sigma_{yp}$. By so doing, the secant formula becomes

$$\sigma_{yp} = \frac{P_{yp}}{A} \left[1 + \frac{ec}{r^2} \sec \left(\frac{L_e}{2r} \sqrt{\frac{P_{yp}}{AE}} \right) \right] \qquad (13.16)$$

in which the effective length is used to make the formula applicable to columns with different end conditions. It is necessary to solve this relationship by trial and error to calculate the value of P_{yp}/A for a given column. To facilitate the use of Eq. (13.16), curves such as the one shown in Fig. 13.10*b* can be used to good advantage. This graph is plotted for a steel column of $E = 200$ GPa and $\sigma_{yp} = 275$ MPa (Prob. 13.34). Interestingly, the horizontal line and Euler's curve represent the limit of the secant formula as e approaches zero. The graph shows that the load-carrying capacity decreases with increasing eccentricity.

After finding the limiting load, we can determine the *allowable* axial load P_{all} from

$$P_{all} = \frac{P_{yp}}{f_s} \qquad (d)$$

where f_s is the factor of safety. Equation (13.15) offers an excellent description of column behavior. However, difficulties arise in its effective application because the eccentricity e of the load is seldom known with any degree of accuracy. The development of the secant formula is based upon the assumption that buckling occurs in the xy plane. It is also necessary to analyze buckling in the xz plane, for which Eq. (13.15) does not apply. Note that this possibility relates especially to narrow columns.

EXAMPLE 13.6

A 6-m-long pin-ended steel column of S 200 × 34 section (Fig. 13.11*a*) carries a centric load $P_1 = 400$ kN and an eccentrically applied load $P_2 = 80$ kN (Fig. 13.11*b*). Compute the maximum deflection and stress in the column. Let $E = 200$ GPa.

Solution Loads P_1 and P_2 are statically equivalent to a load $P = 480$ kN acting with an eccentricity $e = 20$ mm (Fig. 13.11*c*). Using the properties of an

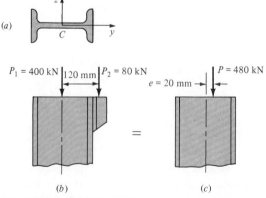

Figure 13.11   Example 13.6.

S 200 × 34 section (see Table B.3), we obtain

$$\frac{P}{A} = \frac{480 \times 10^3}{4370 \times 10^{-6}} = 109.8 \text{ MPa}$$

$$\frac{ec}{r^2} = \frac{eA}{S} = \frac{(0.02)(4370 \times 10^{-6})}{266 \times 10^{-6}} = 0.3286$$

and

$$P_{cr} = \frac{\pi^2 EI}{L^2} = \frac{\pi^2 (200 \times 10^9)(27 \times 10^{-6})}{(6)^2} = 1480.45 \text{ kN}$$

Substitution of $P/P_{cr} = 1/3.08$ and $e = 20$ mm into Eq. (13.14) results in the horizontal deflection at the midpoint:

$$v_{max} = e \left[\sec \left(\frac{\pi}{2} \sqrt{\frac{P}{P_{cr}}} \right) - 1 \right] = 20(1.599 - 1) = 12 \text{ mm}$$

Similarly, Eq. (13.15a) leads to

$$\sigma_{max} = \frac{P}{A} \left[1 + \frac{ec}{r^2} \sec \left(\frac{\pi}{2} \sqrt{\frac{P}{P_{cr}}} \right) \right] = 167.5 \text{ MPa}$$

which is the largest normal stress in the column.

*13.7 COLUMN DESIGN FORMULAS FOR CENTRIC LOADING

In the preceding sections, we assumed the columns to be initially straight, homogeneous prisms. These idealizations are significant in understanding column behavior. However, the design of actual columns must be based upon empirical formulas which reflect the results of numerous laboratory tests. Care must be exercised in using such "special-purpose" formulas. The designer should consider carefully:

1. The specific material to which the formula applies.
2. Whether the formula provides the *allowable* stress or the *ultimate* stress, in which case a factor of safety must be introduced
3. The specific range of slenderness ratios to which the formula applies.

We note that many factors besides those discussed here enter into the design of columns. Thus more specialized compilations should be consulted prior to design of a column for a particular application.

Representative design formulas for *centrically loaded columns* constructed of three different materials are listed in Table 13.1 on page 378. These are specifications recommended by the American Institute of Steel Construction (AISC), the Aluminum Association, and the American Institute of Timber Construction.

Table 13.1   Design Formulas for Centrically Loaded Columns

Structural Steel Columns*

$$\sigma_{\text{all}} = \frac{\sigma_{\text{yp}}}{f_s}\left[1 - \frac{1}{2}\left(\frac{L_e/r}{C_c}\right)^2\right] \qquad \left(\frac{L_e}{r} < C_c\right) \tag{13.17}$$

$$\sigma_{\text{all}} = \frac{\pi^2 E}{1.92(L_e/r)^2} \qquad \left(C_c \leq \frac{L_e}{r} \leq 200\right) \tag{13.18}$$

where

$$C_c = \sqrt{2\pi^2 E/\sigma_{\text{yp}}} \tag{13.19a}$$

$$f_s = \frac{5}{3} + \frac{3}{8}\left(\frac{L_e/r}{C_c}\right) - \frac{1}{8}\left(\frac{L_e/r}{C_c}\right)^3 \tag{13.19b}$$

Aluminum Alloy 6061-T6 Columns†

$$\sigma_{\text{all}} = 19 \text{ ksi} = 130 \text{ MPa} \qquad \left(\frac{L_e}{r} \leq 9.5\right) \tag{13.20}$$

$$\sigma_{\text{all}} = \left[20.2 - 0.126\left(\frac{L_e}{r}\right)\right] \text{ ksi}$$

$$= \left[140 - 0.87\left(\frac{L_e}{r}\right)\right] \text{ MPa} \qquad \left(9.5 < \frac{L_e}{r} < 66\right) \tag{13.21}$$

$$\sigma_{\text{all}} = \frac{51{,}000 \text{ ksi}}{(L_e/r)^2}$$

$$= \frac{350 \times 10^3}{(L_e/r)^2} \text{ MPa} \qquad \left(\frac{L_e}{r} \geq 66\right) \tag{13.22}$$

Timber Columns of Rectangular Cross Section‡

$$\sigma_{\text{all}} = \frac{0.3E}{(L_e/d)^2} \qquad \left(\frac{L_e}{d} \leq 50\right) \tag{13.23}$$

where d is the smallest side dimension of the member. The allowable stress is not to exceed the value of stress for compression parallel to grain of the timber used.

*Source: Manual of Steel Construction, 8th ed., American Institute of Steel Construction, Chicago, 1980.

†Source: Specifications for Aluminum Structures, Aluminum Association, Washington, D.C., 1976.

‡Source: Timber Construction Manual, American Institute of Timber Construction, Wiley, New York, 1974.

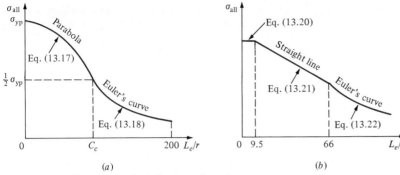

Figure 13.12 Graphs of selected column formulas:
(a) structural steel and (b) aluminum alloy.

Interestingly, the structural steel column formulas given in Table 13.1 are patterned on the schemes illustrated in Fig. 13.12a, in which C_c defines the limiting value of the slenderness ratio between intermediate and long bars. This is assumed to correspond to one-half of the yield stress σ_{yp} of the steel. From Eq. (13.10) we thus have $C_c = L_e/r = (2\pi^2 E/\sigma_{yp})^{1/2}$. It is clear that the use of a variable factor of safety, Eq. (13.19b), provides a consistent buckling formula for intermediate and short columns.

Observe from Table 13.1 that for short and intermediate aluminum columns, σ_{all} is constant and linearly related to L_e/r (Fig. 13.12b). For long columns, a Euler-type formula is used in both cases, as indicated in the graphs. The relationship given for timber columns in Table 13.1 is also a Euler formula, adjusted by a suitable factor of safety.

EXAMPLE 13.7

Determine the diameter d necessary for an aluminum alloy 6061-T6 column to carry an axial load of $P = 100$ kN if the effective length L_e is (a) 1.5 m and (b) 0.4 m.

Solution For the cross section of a solid circular bar, the radius of gyration is $r = d/4$ (see Example A.4). Design of the column requires trial-and-error calculations as follows.

(a) Inasmuch as d is unknown, we assume that $L_e/r \geq 66$ and apply Eq. (13.22) from Table 13.1:

$$\sigma_{all} = \frac{P}{A} = \frac{350 \times 10^9}{(L_e/r)^2} \qquad \frac{100 \times 10^3}{\pi d^2/4} = \frac{350 \times 10^9}{(1.5 \times 4/d)^2}$$

The foregoing yields $d = 60$ mm. The value of L_e/r is 100 for this diameter, thus confirming the use of Eq. (13.22).

(b) Let us again use $L_e/r \geq 66$. Application of Eq. (13.22), following a procedure similar to that used in part a, leads to $d = 31$ mm and hence $L_e/r = 51.6$. This is less than 66 and indicates that Eq. (13.22) is not valid. We now use $9.5 < L_e/r < 66$ and Eq. (13.21) to obtain

$$\frac{P}{A} = \sigma_{\text{all}} = \left[140 - 0.87 \left(\frac{L_e}{r} \right) \right] 10^6$$

or

$$\frac{100 \times 10^3}{\pi d^2/4} = \left[140 - 0.87 \left(\frac{0.4}{d/4} \right) \right] 10^6$$

from which $d = 35.5$ mm. This leads to

$$\frac{L_e}{r} = \frac{(400)4}{35.5} = 45.1$$

Our second assumption, $9.5 < L_e/r < 66$, is correct and $d = 35.5$ mm meets the requirement for this case.

EXAMPLE 13.8

Select the lightest wide-flange steel shape to carry an axial load of 322 kN on an effective length of 4 m. Use $\sigma_{\text{yp}} = 250$ MPa and $E = 200$ GPa.

Solution   Determination of a suitable size for the given shape is facilitated greatly by using tables in the AISC manual. However, we shall employ a trial-and-error procedure here. For the given material properties, Eq. (13.19a) results in

$$C_c = \sqrt{\frac{2\pi^2 E}{\sigma_{\text{yp}}}} = \sqrt{\frac{2\pi^2 (200 \times 10^3)}{250}} = 126$$

as the limiting slenderness ratio.

First Try.   Assume $L_e/r = 0$. Equation (13.19b) then reduces to $f_s = \frac{5}{3}$, and Eq. (13.17) results in $\sigma_{\text{all}} = 250/f_s = 150$ MPa. The required area is

$$A = \frac{P}{\sigma_{\text{all}}} = \frac{322 \times 10^3}{150 \times 10^6} = 2147 \text{ mm}^2$$

From Table B.2, we select a W 150 × 18 section of area $A = 2290$ mm^2 (greater than 2147 mm^2) and with a minimum r of 23.3 mm. The value of $L_e/r = 4/0.0233 = 172$ is greater than $C_c = 126$, which was found above. Hence, using Eq. (13.18), we have

$$\sigma_{\text{all}} = \frac{\pi^2 (200 \times 10^9)}{1.92(172)^2} = 34.8 \text{ MPa}$$

and the allowable load equals $34.8 \times 10^6 (2.29 \times 10^{-3}) = 79.7$ kN. This is less than the design load of 322 kN; therefore a column with either a larger A, a larger r, or both must be selected.

Second Try. Use a W 150 $\times$ 30 section (see Table B.2) for which $A = 3790$ mm^2 and minimum $r = 38.1$ mm. Now $L_e/r = 4/0.0381 = 105$ is less than 126, and Eq. (13.17) is valid. Thus

$$f_s = \frac{5}{3} + \frac{3}{8}\left(\frac{105}{126}\right) - \frac{1}{8}\left(\frac{105}{126}\right)^3 = 1.91$$

and

$$\sigma_{\text{all}} = \frac{250}{1.91}\left[1 - \frac{1}{2}\left(\frac{105}{126}\right)^2\right] = 85.4 \text{ MPa}$$

The allowable load for this section is equal to $(85.4 \times 10^6)(3.79 \times 10^{-3}) = 324$ kN. This is slightly larger than the design load. The lightest column is therefore a W 150 $\times$ 30 section.

*13.8   COLUMN DESIGN FORMULAS FOR ECCENTRIC LOADING

We observed in Sec. 13.6 that the secant formula is a rational formula, and it is also valid for all column lengths. It turns out, however, that it is quite difficult to use in design, although the use of computers facilitates the computations. Of the several methods employed in designing eccentrically loaded columns, the one which seems most simple to apply is known as the *interaction method*. This method is based upon the theory given in Sec. 8.7, in which the maximum stress in a member due to an eccentric compressive load is expressed as

$$\sigma_{\text{max}} = \frac{P}{A} + \frac{Mc}{I} \qquad (a)$$

The first term represents the axial stress developed by the centric load, and the second term is the magnitude of the superimposed bending stress. We have $M = Pe$, where e is the distance between the centroidal axis of the column and the axis through which the load is applied, or the eccentricity (Fig. 13.10a).

In the design of *short columns,* the value of σ_{max} in the foregoing equation is not to exceed the allowable compressive stress σ_{all} of the material:

$$\frac{P}{A} + \frac{Mc}{I} \le \sigma_{\text{all}} \qquad (b)$$

When this expression is divided by σ_{all}, it becomes

$$\frac{P/A}{\sigma_{\text{all}}} + \frac{Mc/I}{\sigma_{\text{all}}} \le 1 \qquad (13.24)$$

The *intermediate* and *long columns,* however, are designed on the basis that the allowable stress will, in general, be different for the two terms in Eq. (13.24). Thus, substituting for σ_{all} in the first and second terms the values of the allowable stress which correspond, respectively, to the centric loading and to the pure bending, we write

$$\frac{P/A}{(\sigma_{\text{all}})_c} + \frac{Mc/I}{(\sigma_{\text{all}})_b} \leq 1 \qquad (13.25)$$

This is referred to as the *interaction formula,* various types of which are in use.

The value of $(\sigma_{\text{all}})_c$ is obtained from one of the axially loaded column formulas (Table 13.1). Note that the *greatest* value of the slenderness ratio of the column should be used to calculate $(\sigma_{\text{all}})_c$, regardless of the plane in which bending occurs. Clearly, $(\sigma_{\text{all}})_b$ is the allowable flexural stress of the material. The AISC specifications permit use of Eq. (13.25) for $(P/A)/(\sigma_{\text{all}})_c \leq 0.15$. Otherwise, in computing the moment M in this equation, the lateral deflection of the column must be included in the moment arm—that is, the second term is modified.

Column design using Eq. (13.25) requires a trial-and-error procedure similar to that employed in Examples 13.7 and 13.8. Any section selected that gives the largest sum (less than unity) of the terms on the left side of the formula is the most efficient section.

EXAMPLE 13.9

A W 250 × 80 steel column (see Table B.2) of 5-m effective length, $A = 10.2 \times 10^3$ mm², $r_y = 65$ mm, $r_z = 111$ mm, and $S_z = 985 \times 10^3$ mm³ supports an eccentric load (Fig. 13.13). Let $E = 210$ GPa, $\sigma_{\text{yp}} = 250$ MPa, and $(\sigma_{\text{all}})_b = 160$ MPa. Using the interaction formula, determine the maximum safe load P.

Solution The largest slenderness ratio of the column is

$$\frac{L_e}{r_y} = \frac{5}{0.065} = 76.9$$

Applying Eq. (13.19a), we have

$$C_c = \sqrt{\frac{2\pi^2 E}{\sigma_{\text{yp}}}} = \sqrt{\frac{2\pi^2(210 \times 10^9)}{(250 \times 10^6)}} = 128.8$$

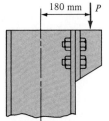
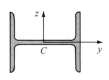

Figure 13.13 Example 13.9.

which is greater than 76.9. Therefore, from Eqs. (13.19b) and (13.17), we have

$$f_s = \frac{5}{3} + \frac{3}{8}\left(\frac{76.9}{128.8}\right) - \frac{1}{8}\left(\frac{76.9}{128.8}\right)^3 = 1.86$$

and

$$(\sigma_{\text{all}})_c = \frac{250}{1.86}\left[1 - \frac{1}{2}\left(\frac{76.9}{128.8}\right)^2\right] = 110.5 \text{ MPa}$$

For the situation described, we have

$$\frac{P}{A} = \frac{P}{10.2 \times 10^{-3}}$$

$$\frac{Mc}{I} = \frac{Pe}{S_z} = \frac{P(0.18)}{0.985 \times 10^{-3}}$$

(c)

Substitution of the given data and Eqs. (c) into Eq. (13.25) yields

$$\frac{P/(10.2 \times 10^{-3})}{110.5 \times 10^6} + \frac{P(0.18)/(0.985 \times 10^{-3})}{160 \times 10^6} \leq 1$$

from which

$$P \leq 493 \text{ kN}$$

The maximum allowable load is thus $P = 493$ kN $\downarrow$.

In conclusion it should be noted that column buckling represents but one case of structural stability.* Other examples include buckling of containers under internal vacuum, external pressure, or axial compression; twist-bend buckling of shafts in torsion; lateral buckling of deep, narrow beams; buckling of circular rings under radial compression; and snap buckling of arches.[†] Treatment of such problems is mathematically complex and beyond the scope of this text.

*See S. H. Crandall, N. C. Dahl, and T. J. Lardner, *Mechanics of Solids,* 2d ed., McGraw-Hill, New York, 1978, chap. 9.

[†]S. P. Timoshenko and J. M. Gere, *Theory of Elastic Stability,* 2d ed., McGraw-Hill, New York, 1961.

PROBLEMS

Secs. 13.1 to 13.5

13.1 Determine the critical load P_{cr} for the system shown in Fig. P13.1. The rigid bar AB is supported by a linear spring of stiffness k at A and a torsional spring of stiffness k_t at B.

13.2 Redo Prob. 13.1 for the case in which the bar is free at B.

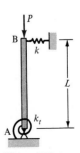

Figure P13.1

13.3 through 13.5 Determine the critical load P_{cr} for the rigid bar-spring system shown in Figs. P13.3 through P13.5, where k and k_t represent the linear and torsional spring constants, respectively.

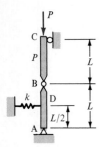

Figure P13.3

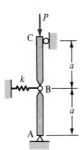

Figure P13.4

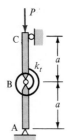

Figure P13.5

13.6 Solve Prob. 13.4, assuming that the two rigid bars AB and BC are replaced by a single rigid bar AC of length $2a$ and that the support at C is removed.

13.7 Calculate the critical load of a wooden yardstick ($E = 1.6 \times 10^6$ psi) of $\frac{1}{8}$ by $1\frac{1}{8}$-in. rectangular cross section.

13.8 The jib crane shown in Fig. P13.8 is of capacity $W = 20$ kN. For $\alpha = 30°$ and a factor of safety of $f_s = 3$, calculate the required minimum cross-sectional area for circular steel bar AB ($E = 200$ GPa).

13.9 Redo Prob. 13.8 for $W = 30$ kN and $\alpha = 40°$.

13.10 Two equal round bars are supported in line with one another and are separated by a small gap δ (Fig. P13.10). Given $\alpha = 12 \times 10^{-6}/°C$, $L = 1$ m, $d = 25$ mm, and $\delta = 1.2$ mm, determine the temperature rise necessary to cause them to (a) just touch and (b) buckle elastically.

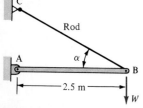

Figure P13.8

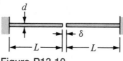

Figure P13.10

13.11 A 60-mm outer-diameter and 50-mm inner-diameter steel pipe ($E = 200$ GPa) 2 m long acts as a spreader member in the assembly of Fig. P13.11. Determine, for a factor of safety $f_s = 2$, the value of Q that will cause buckling of the pipe.

13.12 Rework Example 13.3, given that column AB is supported by a z-directed pin at A.

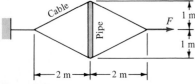

Figure P13.11

13.13   Redo Prob. 13.10 for the case in which $\delta = 0.5$ mm and $L = 0.6$ m.

13.14   Two C 6 $\times$ 10.5 rolled-steel channels (see Table B.4) are bolted back to back, as shown in Fig. P13.14, and used as a pinned-end column 8 ft long. Using $E = 30 \times 10^6$ psi, determine the buckling load for the two members.

Figure P13.14

13.15   Redo Prob. 13.14 for the case in which two L 3 $\times$ 3 $\times$ ¼ steel angles (see Table B.5) are bolted back to back.

13.16   An S 310 $\times$ 74 rolled-steel section (see Table B.3) is used as a column of actual length $L = 4$ m. Given $E = 200$ GPa and a factor of safety $f_s = 2$, determine the critical load P_{cr} for each of the following conditions: (*a*) pinned-pinned; (*b*) fixed-pinned; and (*c*) fixed-free.

13.17   Redo Prob. 13.16 for a W 250 $\times$ 80 rolled-steel column of length $L = 6$ m (see Table B.2).

13.18 through 13.20   Given a factor of safety $f_s = 2$, determine the largest load F that may be applied to the structure shown in Figs. P13.18 through P13.20. Assume that each column is of 50-mm-diameter steel bar ($E = 210$ GPa).

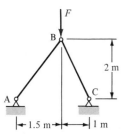

Figure P13.18

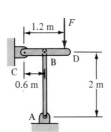

Figure P13.19

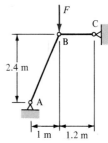

Figure P13.20

13.21   A fixed-ended long column of rectangular cross section 1 by 3 in. is made of aluminum alloy ($\sigma_{pl} = 18$ ksi and $E = 10 \times 10^6$ psi). Determine the minimum length L_c that the column may have and the corresponding critical load P_{cr}.

13.22   A rectangular aluminum ($E = 70$ GPa) tube of uniform thickness $t = 20$ mm (Fig. P13.22) serves as a 6-m-long column fixed at both ends. Calculate the critical stress in the column.

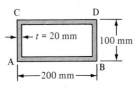

Figure P13.22

13.23   Repeat Prob. 13.22, assuming that the column is pinned at one end and fixed at the other.

13.24   Brace BD of the structure shown in Fig. P2.25 is a steel rod ($E = 200$ GPa and $\sigma_{pl} = 220$ MPa) of square cross section (40 mm on a side). Determine the safety factor against failure by buckling.

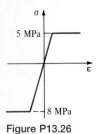

Figure P13.26

13.25 A 6-ft-long column of 4-in. by 2-in. rectangular cross section fits between a rigid ceiling and a rigid floor. Assume the ends to be rounded. For $E = 30 \times 10^6$ psi, $\alpha = 6.7 \times 10^{-6}/°F$, and $\sigma_{pl} = 34$ ksi, calculate the change in temperature that will cause the column to buckle.

13.26 The jib crane of Fig. P13.8 is constructed of a 35-mm-diameter rod BC and a steel tube AB having the following properties: $A = 400$ mm², $I = 8 \times 10^3$ mm⁴, $E = 200$ GPa. The approximate stress-strain curve for both members is as shown in Fig. P13.26. For $\alpha = 45°$, determine the value of W at which failure occurs.

13.27 A 0.5-m-long, 50-mm by 75-mm rectangular column is pinned at both ends. If the stress-strain curve for the material is approximated as shown in Fig. P13.27, determine the buckling load P_{cr}.

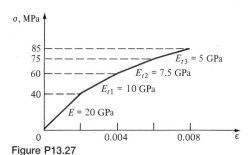

Figure P13.27

13.28 Determine the value of w_0 that may cause the pinned column AB in the structure of Fig. P13.28 to buckle. Assume that the column is constructed of a material for which the stress-strain curve is as shown in Fig. P13.27. Use $L = 0.5$ m.

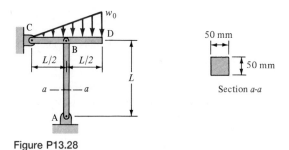

Figure P13.28

13.29 Redo Prob. 13.27, given that one end of the column is fixed and the other is pinned as before.

Sec. 13.6

13.30 A cast-iron pipe of 10-in. outer diameter and $\frac{3}{4}$-in. wall thickness is to serve as a pinned-end column 15 ft long and is to support an eccentric load of 160 kips with a safety factor $f_s = 2$. Using $E = 24 \times 10^6$ psi and $\sigma_{yp} = 33$ ksi, calculate the largest allowable eccentricity e.

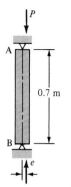

13.31 A pin-ended, 40-mm-diameter rod AB carries an eccentrically applied load $P = 60$ kN, as shown in Fig. P13.31. Given the maximum deflection at the midlength $v_{max} = 0.6$ mm and $E = 200$ GPa, calculate (a) the eccentricity e and (b) the maximum stress in the rod.

13.32 Redo Prob. 13.31, assuming that end B of column AB is fixed.

Figure P13.31

13.33 A 50-mm-diameter steel pinned-end column ($E = 200$ GPa and $\sigma_{yp} = 275$ MPa) is loaded as shown in Fig. P13.31. For a factor of safety $f_s = 3$ with respect to yielding, use Fig. 13.10b to determine allowable load P_{all} for eccentricities e of (a) 5 mm and (b) 3 mm.

13.34 Use Fig. 13.10b to determine the maximum load P_{yp} that can be applied at both ends of an S 310 $\times$ 52 steel link (see Table B.3), given an eccentricity $e = 60$ mm from the center of the section measured along the web. Assume that a stress of 275 MPa is not to be exceeded, $E = 200$ GPa, and $L_e = 5$ m. Check the value found by applying Eq. (13.16).

13.35 Redo Prob. 13.33, this time applying Eq. (13.16).

13.36 A 2-m-long aluminum hollow box column ($E = 70$ GPa) of the cross section shown in Fig. P13.22 is fixed at the base and free at the top. If an eccentric load $P = 150$ kN acts at the middle of side AB (that is, $e = 50$ mm) of the free end, calculate the largest stress σ_{max} in the column.

13.37 Rework Prob. 13.36, assuming that the load acts at the middle of side AC (that is, $e = 100$ mm).

Secs. 13.7 and 13.8

13.38 An S 6 $\times$ 12.5 rolled-steel column (see Table B.3) with pinned ends carries an axial load $P = 20$ kips. What is the longest allowable column length? Use $E = 29 \times 10^6$ psi and $\sigma_{yp} = 36$ ksi.

13.39 Redo Prob. 13.38 for a W 6 $\times$ 20 rolled-steel column (see Table B.2).

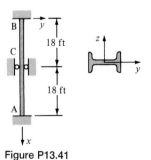

13.40 Determine the allowable axial load P for a W 250 $\times$ 80 rolled-steel column (see Table B.2) for each of three effective lengths: (a) 2 m; (b) 5 m; and (c) 10 m. Use $E = 210$ GPa and $\sigma_{yp} = 250$ MPa.

13.41 Determine the allowable axial load P for a W 14 $\times$ 145 rolled-steel column (see Table B.2) with built-in ends and a length of 36 ft, braced at midpoint C (Fig. P13.41). Use $E = 30 \times 10^6$ psi and $\sigma_{yp} = 40$ ksi.

Figure P13.41

13.42 Redo Prob. 13.41 for an S 20 × 75 rolled-steel column (see Table B.3).

13.43 Rework Prob. 13.40 for a steel pipe column of outer diameter 160 mm and inner diameter 140 mm.

13.44 Determine the smallest allowable diameter d of a steel rod ($E = 29 \times 10^6$ psi and $\sigma_{yp} = 36$ ksi) which may be used to support an axial load $P = 40$ kips for $L_e = 12$ ft.

13.45 Select the lightest wide-flange steel shape (see Table B.2) to support an axial load of 1150 kN on an effective length of 4 m. Use $E = 200$ GPa and $\sigma_{yp} = 250$ MPa.

13.46 A square aluminum alloy 6060-T6 column of length $L = 0.6$ m is pinned at both ends. What is the smallest allowable width a if the member is to support an axial load $P = 250$ kN?

13.47 Redo Prob. 13.46, assuming that the column is pinned at one end and fixed at the other.

13.48 A rectangular aluminum alloy 6061-T6 tube (Fig. P13.22) is used as a column of 5-m effective length. What is the largest permissible axial load P?

13.49 Determine the allowable axial load P for an aluminum alloy 6061-T6 pipe of outer diameter 10 in. and inner diameter $8\frac{1}{2}$ in.; the pipe is used as a column of effective length 12 ft.

13.50 Redo Prob. 13.49, given that the effective length of the column is increased to 20 ft.

13.51 Select the minimum width a of a square timber column ($E = 12$ GPa and $\sigma_{all} = 9$ MPa) of effective length 4 m if it must support an axial load $P = 200$ kN.

13.52 Determine the allowable axial load P for a built-up timber column ($E = 1.6 \times 10^6$ psi and $\sigma_{all} = 2$ ksi) of the cross section shown in Fig. P13.52. The effective length is 15 ft.

13.53 Calculate the allowable axial load P for a 100-mm by 200-mm rectangular timber column ($E = 11$ GPa and $\sigma_{all} = 9$ MPa) if the effective length is (a) 2 m and (b) 4 m.

13.54 A column of 4-m effective length and 100-mm-square cross section is made of timber for which $E = 12$ GPa and $(\sigma_{all})_b = 10$ MPa for compression parallel to the grain. Using the interaction method, calculate the maximum load P that may be safely supported with an eccentricity of $e = 40$ mm.

13.55 Redo Prob. 13.54 for the case in which the column is made of steel. Use $(\sigma_{all})_b = 160$ MPa, $\sigma_{yp} = 250$ MPa, and $E = 200$ GPa.

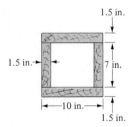

1.5 in.

1.5 in.

7 in.

10 in.

1.5 in.

Figure P13.52

13.56   Solve Prob. 13.54 for the case in which the column is made of aluminum alloy 6061-T6 and the allowable stress in bending is 150 MPa.

13.57   A load of $P = 10$ kips is applied with an eccentricity of $e = 1\frac{1}{2}$ in. to a timber post (Fig. P13.57). Let $E = 1.7 \times 10^6$ psi, $(\sigma_{all})_b = 1.2$ ksi for compression parallel to the grain, and $L = 5$ ft. Employing the interaction method, determine the smallest allowable diameter d of the post.

13.58   An S 4 $\times$ 7.7 steel column (see Table B.3) of 8-ft effective length supports an eccentric load applied $1\frac{3}{4}$ in. from the center of the section measured along the web. Using the interaction formula, determine the maximum safe load P. Let $E = 30 \times 10^6$ psi, $\sigma_{yp} = 36$ ksi, and $(\sigma_{all})_b = 25$ ksi.

13.59   A $2\frac{1}{2}$-in.-diameter aluminum alloy 6061-T6 rod supports a load $P = 8$ kips with an eccentricity $e = 1$ in. (see Fig. P13.57). Employing the interaction method, with $(\sigma_{all})_b = 22$ ksi, determine the longest length L that may be used.

13.60   Redo Prob. 13.59 for the case in which the diameter of the column is decreased to 2 in.

13.61   A W 410 $\times$ 85 steel column of 5-m effective length supports an eccentric load of 400 kN at a distance e from the center of the section measured along the web. Let $(\sigma_{all})_b = 140$ MPa and $E = 210$ GPa. Employ the interaction formula to determine the maximum allowable value of e.

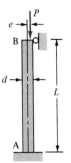

Figure P13.57

APPENDIX A

MOMENTS OF AREAS

A.1 FIRST MOMENTS OF AN AREA: CENTROID

This appendix deals with the geometric properties of cross sections of structural and machine members. Knowledge of the geometric characteristics of a cross section is essential in the study of strength, stiffness, and stability. Properties for most standard shapes encountered in practice are furnished in various handbooks, and the table inside the back cover of this text presents several common cases.

Evaluation of the properties of a plane area begins with the location of the centroid of the area. The *centroid,* or geometric center, represents a point in the plane about which the area is equally distributed. The area A of Fig. A.1 has its *first moments* about the x and y axes, defined, respectively, by

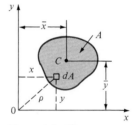

Figure A.1 Plane area.

$$Q_x = \int y \, dA \qquad Q_y = \int x \, dA \qquad (A.1)$$

These quantities are expressed in cubic meters or cubic millimeters in SI units and in cubic feet or cubic inches in U.S. customary units. The centroid of the area A is denoted by C of the coordinates $\bar{x}$ and $\bar{y}$, which are determined by

$$\bar{x} = \frac{Q_y}{A} = \frac{\int x \, dA}{\int dA} \qquad \bar{y} = \frac{Q_x}{A} = \frac{\int y \, dA}{\int dA} \qquad (A.2)$$

If an area possesses an *axis of symmetry,* the centroid must be located on the axis, as the first moment about an axis of symmetry equals zero. To demonstrate this statement, consider the area in Fig. A.2a, which is symmetrical with respect to the y axis. For every element of area dA of abscissa x, there is a corresponding element of area dA' of abscissa $-x$. Therefore, from the first of Eqs. (A.2), we have $\bar{x} = 0$. Similarly, for *two axes of symmetry,* the centroid lies at the intersection of the two axes (Fig. A.2b). Also, from inspection, if an area has no axes of symmetry but does possess a *center of symmetry* O, the centroid coincides with the center of symmetry (Fig. A.2c).

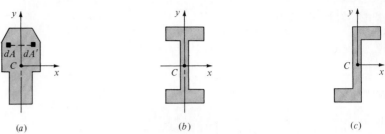

Figure A.2 (a) Area with an axis of symmetry; (b) area with two axes of symmetry; and (c) area with a point of symmetry.

EXAMPLE A.1

Determine ordinate $\bar{y}$ of the centroid of the triangular area shown in Fig. A.3.

Solution A horizontal element of area of length x' and height dy is selected (Fig. A.3). From similar triangles, x' equals $(h - y)b/h$, and

$$dA = \frac{b}{h}(h - y)\,dy$$

Using the first of Eqs. (A.1), we have

$$Q_x = \int y\,dA = \frac{b}{h}\int_0^h (hy - y^2)\,dy = \frac{1}{6}bh^2$$

The second of Eqs. (A.2), with $A = bh/2$, gives

$$\bar{y} = \frac{bh^2/6}{bh/2} = \frac{h}{3} \tag{a}$$

Thus the centroidal horizontal axis of the triangular area is located a distance $\frac{1}{3}$ the altitude from the base of the triangle.

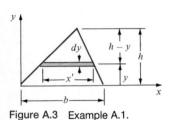

Figure A.3 Example A.1.

In many cases, an area can be divided into simple geometric forms (for example, rectangles, circles, and triangles) whose areas and centroidal coordinates can be readily found. Examples include the cross-sectional shapes depicted in Fig. A.2. When a *composite area* is considered as an assemblage of n elementary areas, the resultant moment about any axis is the algebraic sum of the moments of the component areas. Therefore the centroid of a composite area is located by

$$\bar{x} = \frac{\sum A_i \bar{x}_i}{\sum A_i} \qquad \bar{y} = \frac{\sum A_i \bar{y}_i}{\sum A_i} \tag{A.3}$$

where $\bar{x}_i$ and $\bar{y}_i$ represent the coordinates of the centroids of the component areas A_i ($i = 1, 2, \ldots, n$).

In applying Eqs. (A.3), it is necessary to indicate on a sketch the simple geometric forms into which the composite area is resolved. This is illustrated in the solution of the following numerical problem.

EXAMPLE A.2

Locate the centroid of the unequal-leg section shown in Fig. A.4. The dimensions are given in millimeters.

Solution The area is divided into two rectangles A_1 and A_2 for which the centroids are known. The computation is conveniently carried out in the following tabular form. Note that when an area is decomposed into *only two parts,* the centroid C of the entire area *always lies on the line* connecting the centroids C_1 and C_2 of the components, as indicated in Fig. A.4.

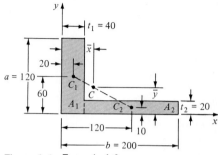

Figure A.4   Example A.2.

No.	A_i, mm²	$\bar{x}_i$, mm	$\bar{y}_i$, mm	$A_i\bar{x}_i$, mm³	$A_i\bar{y}_i$, mm³
1	$40(120) = 4800$	20	60	96×10^3	288×10^3
2	$20(160) = 3200$	120	10	384×10^3	32×10^3
	$\sum A_i = 8000$			$\sum A_i\bar{x}_i = 480(10^3)$	$\sum A_i\bar{y}_i = 320(10^3)$

$$\bar{x} = \frac{\sum A_i\bar{x}_i}{\sum A_i} = \frac{480 \times 10^3}{8000} = 60 \text{ mm}$$

$$\bar{y} = \frac{\sum A_i\bar{y}_i}{\sum A_i} = \frac{320 \times 10^3}{8000} = 40 \text{ mm}$$

The foregoing distances are determined by expressions of the form

$$\bar{x} = \frac{1}{2} \frac{at_1^2 + (b^2 - t_1^2)t_2}{at_1 + (b - t_1)t_2}$$

$$\bar{y} = \frac{1}{2} \frac{a^2t_1 + (b - t_1)t_2^2}{at_1 + (b - t_1)t_2}$$

where a, b, t_1, and t_2 are the dimensions of the area of the section seen in Fig. A.4.

A.2 MOMENTS OF INERTIA

Consider now the *second moment* of an area, or the *moment of inertia,* which is a relative measure of the manner in which the area is distributed about an axis of interest. The moments of inertia of a plane area A about the x and y axes, respectively, are defined by the integrals

$$I_x = \int y^2 \, dA \qquad I_y = \int x^2 \, dA \qquad (A.4)$$

where x and y are the coordinates of the element of area dA (Fig. A.1). Likewise, the *polar moment of inertia* of a plane area A about an axis through O perpendicular to the area is

$$J_O = \int \rho^2 \, dA = I_x + I_y \qquad (A.5)$$

Here ρ is the distance from point O to an element dA, and $\rho^2 = x^2 + y^2$.

The *product of inertia* of a plane area A about the x and y axes is given by

$$I_{xy} = \int xy \, dA \qquad (A.6)$$

We observe from the above that each element of area dA is multiplied by the product of its coordinates (Fig. A.1). The product of inertia of an area about any pair of axes is *zero* when either of the axes is an axis of symmetry. For purposes of illustration, reconsider Fig. A.2a. The products of inertia of element dA and dA' on opposite sides of the axis of symmetry (y) have the same magnitude but opposite sign. Thus $xy \, dA$ cancel one another in the summation of Eq. (A.6).

The foregoing formulations show that I_x, L_y, and J_O are always *positive* quantities, as the coordinates x and y are squared. The moments of inertia have dimensions of length raised to the fourth power, and typical units are meters^4, millimeters^4, and inches^4. The dimensions of the product of inertia are the same as for the moments of inertia. However, the property I_{xy} can be positive, negative, or zero, depending upon the values of the product xy.

The *radius of gyration* is a distance (from a reference axis or a point) at which the entire area of a section may be considered to be distributed for the purpose of calculating the moment of inertia. Thus the radii of gyration of an area about x and y axes, and the origin O (Fig. A.1), are defined as the quantities r_x, r_y, and r_O, respectively:

$$r_x = \sqrt{\frac{I_x}{A}} \qquad r_y = \sqrt{\frac{I_y}{A}} \qquad r_O = \sqrt{\frac{J_O}{A}} \qquad (A.7)$$

Insertion of I_x, I_y, and J_O from Eq. (A.7) into Eq. (A.5) leads to

$$r_O^2 = r_x^2 + r_y^2 \qquad (A.8)$$

The radius of gyration has the units of length.

EXAMPLE A.3

Determine the moments of inertia about horizontal and vertical axes passing through the centroid and the corresponding radii of gyration for the rectangular area of Fig. A.5.

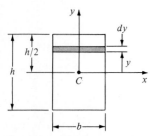

Figure A.5 Example A.3.

Solution   The rectangle possesses two axes of symmetry, and hence the centroid C of the rectangular area coincides with its geometric center (Fig. A.5). We select dA as $b\, dy$. Thus

$$I_x = \int y^2\, dA = \int_{-h/2}^{h/2} y^2 b\, dy = \frac{bh^3}{12} \qquad (a)$$

In a like manner, taking dA as $h\, dx$, $I_y = hb^3/12$. Equations (A.7) then lead to

$$r_x^2 = \frac{bh^3/12}{bh} = \frac{h^2}{12} \qquad r_y^2 = \frac{hb^3/12}{bh} = \frac{b^2}{12}$$

from which

$$r_x = \frac{h}{2\sqrt{3}} \qquad r_y = \frac{b}{2\sqrt{3}} \qquad (b)$$

Observe that the rectangle's product of inertia I_{xy} is zero with respect to centroidal axes x and y because they are the axes of symmetry.

EXAMPLE A.4

Determine the properties about a diameter for the ring-shaped area shown in Fig. A.6.

Solution   Here it is convenient to take dA as $2\pi\rho\, d\rho$. It follows that

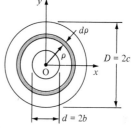

Figure A.6   Example A.4.

$$J_O = \int \rho^2\, dA = \int_b^c 2\pi\rho^3\, d\rho = \frac{\pi}{2}(c^4 - b^4) = \frac{\pi}{32}(D^4 - d^4)$$

Noting the symmetry about both axes, we find that $I_x = I_y$, and Eq. (A.5) becomes $\pi(c^4 - b^4)/2 = 2I_x$. Thus

$$I_x = I_y = \frac{\pi}{4}(c^4 - b^4) = \frac{\pi}{64}(D^4 - d^4)$$

Substituting the foregoing and $A = \pi(c^2 - b^2)$ into Eq. (A.7), we have

$$r = \tfrac{1}{2}\sqrt{c^2 + b^2} = \tfrac{1}{4}\sqrt{D^2 + d^2} \qquad (c)$$

The conditions of symmetry also dictate that $I_{xy} = 0$ about the centroidal axes.

A.3   PARALLEL-AXIS THEOREM

The parallel-axis theorem relates to the moment of inertia of an area with respect to any axis to the moment of inertia around a parallel axis through the centroid. Sometimes called the *transfer formula*, this theorem is useful in determining the moment of inertia of an area composed of several simple shapes.

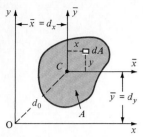

Figure A.7 Parallel axes.

To derive the parallel-axis theorem, consider the area A shown in Fig. A.7. Here $\bar{x}$ and $\bar{y}$ represent the centroidal axes of the area, parallel to the x and y axes, respectively. The distance between the two sets of axes and their origins are d_x, d_y, and d_O. The moment of inertia about the x axis is therefore

$$I_x = \int (y + \bar{y})^2 \, dA = \int y^2 \, dA + 2\bar{y} \int y \, dA + \bar{y}^2 \int dA$$

The first term on the right represents the moment of inertia about the $\bar{x}$ axis, $I_{\bar{x}}$. As y is measured from the centroidal axis $\bar{y}$, $\int y \, dA = 0$. Hence

$$I_x = I_{\bar{x}} + A\bar{y}^2 = I_{\bar{x}} + A\,d_y^2 \tag{A.9a}$$

In a like manner,

$$I_y = I_{\bar{y}} + A\bar{x}^2 = I_{\bar{y}} + A d_x^2 \tag{A.9b}$$

This is the *parallel-axis theorem*: The moment of inertia of an area about any axis is equal to the moment of inertia around a parallel centroidal axis, plus the product of the area and the square of the distance between the two axes.

Similarly, a relationship may be developed connecting the polar moment of inertia J_O of an area about an arbitrary point O and the polar moment of inertia J_c about the centroid of the area (Fig. A.7):

$$J_O = J_c + A d_O^2 \tag{A.10}$$

Finally, it can be shown that the product of inertia of an area I_{xy} about any set of axes is given by

$$I_{xy} = I_{\bar{x}\bar{y}} + A\bar{x}\bar{y} = I_{\bar{x}\bar{y}} + A d_x d_y \tag{A.11}$$

where $I_{\bar{x}\bar{y}}$ designates the product of inertia around the centroidal axes. Note that the parallel-axis theorems, Eqs. (A.9) through (A.11), may be used *only* if one of the two axes involved is a centroidal axes.

For elementary shapes, the integrals appearing in the relationships written so far can usually be evaluated readily, and the geometric properties of the area can be thus determined. (See the table on the inside back cover of this book.) To compute the moments of inertia of a *composite area*, the transfer formula is used rather than integration. The moment of inertia of the original configuration with respect to any axis is the algebraic sum of the moment of inertia of the component areas about the same axis. This approach also applies to the product of inertia.

EXAMPLE A.5

Calculate the moment of inertia of a hollow rectangular area about a horizontal axis through its centroid (Fig. A.8a).

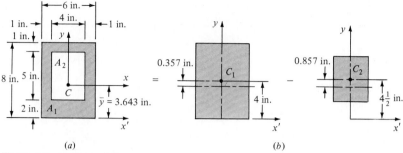

Figure A.8   Example A.5.

Solution

Location of Centroid. The coordinates are selected as shown in Fig. A.8a. The shaded area may be regarded as consisting of the original rectangle A_1 minus square A_2 (Fig. A.8b). The centroid lies on the y axis and its location is given by

$$\bar{y} = \frac{\sum A_i \bar{y}_i}{\sum A_i} = \frac{A_1 \bar{y}_1 - A_2 \bar{y}_2}{A_1 - A_2} = \frac{(6 \times 8)4 - (4 \times 5)(4\frac{1}{2})}{6 \times 8 - 4 \times 5} = 3.643 \text{ in.}$$

Moment of Inertia. Referring to Fig. A.8a, we find that application of the parallel-axis theorem leads to

$$I_x = \sum (I_{\bar{x}} + Ad_y^2) = \sum \left(\frac{1}{12} bh^3 + A\bar{y}^2 \right)$$

$$= \frac{1}{12}(6)(8)^3 + (6 \times 8)(0.357)^2 - \frac{1}{12}(4)(5)^3 - (4 \times 5)(0.857)^2$$

$$= 205.4 \text{ in.}^4$$

This is the centroidal moment of inertia of the entire cross section.

A.4   PRINCIPAL MOMENTS OF INERTIA

It is sometimes necessary to determine the moments and product of inertia $I_{x'}$, $I_{y'}$, and $I_{x'y'}$ about axes x', y' making an angle θ with respect to the original axes when the values for I_x, I_y, and I_{xy} are known (Fig. A.9). For this purpose, the new coordinates of element dA can be expressed by projecting x and y upon the rotated axes:

$$x' = x \cos \theta + y \sin \theta \qquad y' = y \cos \theta - x \sin \theta \qquad (a)$$

Then by definition,

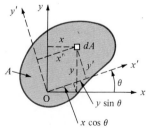

Figure A.9   Rotating axes.

$$I_x = \int y'^2 \, dA = \int (y \cos \theta - x \sin \theta)^2 \, dA$$

$$= \cos^2 \theta \int y^2 \, dA + \sin^2 \theta \int x^2 \, dA - 2 \sin \theta \cos \theta \int xy \, dA$$

Upon introduction of Eqs. (A.4) and (A.6), the above becomes

$$I_{x'} = I_x \cos^2 \theta + I_y \sin^2 \theta - 2I_{xy} \sin \theta \cos \theta \qquad (b)$$

The moment of inertia $I_{y'}$ may be found readily by substituting $\theta + \pi/2$ for θ into the expressions for $I_{x'}$. Similarly, the definition $I_{x'y'} = \int x'y' \, dA$ yields

$$I_{x'y'} = (I_x - I_y) \sin \theta \cos \theta + I_{xy}(\cos^2 \theta - \sin^2 \theta) \qquad (c)$$

These *transformation equations* for the moments and product of inertia may be simplified by introducing the double-angle trigonometric relations:

$$I_{x'} = \frac{I_x + I_y}{2} + \frac{I_x - I_y}{2} \cos 2\theta - I_{xy} \sin 2\theta \qquad (A.12a)$$

$$I_{y'} = \frac{I_x + I_y}{2} - \frac{I_x - I_y}{2} \cos 2\theta + I_{xy} \sin 2\theta \qquad (A.12b)$$

$$I_{x'y'} = \frac{I_x - I_y}{2} \sin 2\theta + I_{xy} \cos 2\theta \qquad (A.12c)$$

In comparing the relationships in this appendix with those in Chap. 4, one observes the following: Moments of inertia $(I_x, I_y, I_{x'}, I_{y'})$ correspond to the normal stresses $(\sigma_x, \sigma_y, \sigma_{x'}, \sigma_{y'})$, the *negative* of the products of inertia $(-I_{xy}, -I_{x'y'})$ correspond to the shear stresses $(\tau_{xy}, \tau_{x'y'})$, and the polar moment of inertia (J_O) corresponds to the sum of the normal stresses $(\sigma_x + \sigma_y)$. Thus the Mohr circle analysis, and all conclusions drawn for stress, apply to these area properties.

By analogy with stress, the *directions* of the principal axes (where $I_{xy} = 0$) are found from Eq. (4.5):

$$\tan 2\theta_p = -\frac{2I_{xy}}{I_x - I_y} \qquad (A.13)$$

This equation has two roots, θ_p' and θ_p''. Similarly,

$$I_{1,2} = \frac{I_x + I_y}{2} \pm \sqrt{\left(\frac{I_x - I_y}{2}\right)^2 + I_{xy}^2} \qquad (A.14)$$

where I_1 and I_2 represent the maximum and minimum *principal moments of inertia,* respectively. Note that, as the product of inertia relative to the axes of symmetry is zero, an axis of symmetry coincides with a centroidal principal axis.

EXAMPLE A.6

Determine the centroidal principal moments of inertia for the Z section shown in Fig. A.10a.

Solution The xy axes are the reference axes through the centroid C (Fig. A.10a). The area is divided into rectangles A_1, A_2, and A_3.

Moments of Inertia. Using the parallel-axis theorem, we have

$$I_x = \sum (I_{\bar{x}} + Ad_y^2) = \sum \left(\frac{1}{12} bh^3 + A\bar{y}^2 \right)$$

$$= 2 \left[\frac{1}{12} (20)(60)^3 + 20(60)(40)^2 \right] + \frac{1}{12} (120)(20)^3$$

$$= 4.64 \times 10^6 \text{ mm}^4$$

$$I_y = \sum (I_{\bar{y}} + Ad_x^2) = \sum \left(\frac{1}{12} hb^3 + A\bar{x}^2 \right)$$

$$= 2 \left[\frac{1}{12} (60)(20)^3 + 60(20)(50)^2 \right] + \frac{1}{12} (20)(120)^3$$

$$= 8.96 \times 10^6 \text{ mm}^4$$

The product of inertia about the xy axes is obtained as described in Sec. A.4:

$$I_{xy} = \sum (I_{\bar{x}\bar{y}} + Ad_x d_y)$$

$$= 0 + (20 \times 60)(40)(-50) + 0 + (20 \times 60)(-40)(50) + 0 + 0$$

$$= -4.8 \times 10^6 \text{ mm}^4$$

Principal Moments of Inertia. Equation (A.13) yields

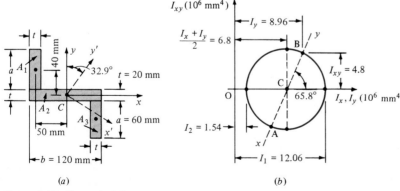

(a)

(b)

Figure A.10 Example A.6.

$$\tan 2\theta_p = \frac{2(4.8)}{4.64 - 8.96} = -2.222$$

$$2\theta_p = -65.8° \quad \text{and} \quad 114.2°$$

Therefore the two values of θ_p are $-32.9°$ and $57.1°$. When we substitute the first of these values, Eq. (A.12a) leads to $I_{x'} = 1.54 \times 10^6 \text{ mm}^4$. The principal moments of inertia are determined from Eq. (A.14):

$$I_{1,2} = \left[\frac{8.96 + 4.64}{2} \pm \sqrt{\left(\frac{8.96 - 4.64}{2}\right)^2 + 4.8^2}\right] 10^6 = (6.8 \pm 5.26)10^6$$

From the foregoing, we find

$$I_1 = I_{y'} = 12.06 \times 10^6 \text{ mm}^4 \qquad I_2 = I_{x'} = 1.54 \times 10^6 \text{ mm}^4$$

The principal axes are denoted by the $x'y'$ axes in Fig. A.10a. It is observed from the figure that I_1 occurs about the axis (y') from which most of the cross-sectional area is *farthest* away.

Mohr's circle provides a convenient, alternative means for transforming the moments of inertia of a plane area. Following a procedure similar to that outlined in Sec. 4.4, we find the principal centroidal moments of inertia to be as shown in Fig. A.10b. As I_{xy} is negative, point A $(4.64, -4.8)$ is located below the horizontal axis and point B $(8.92, 4.8)$ above.

EXAMPLE A.7

Derive expressions for the centroidal moments of inertia and the product of inertia for a Z section (Fig. A.10a). Write the equations for the centroidal principal moments of inertia in a form suitable for computer analysis.

Solution The transfer formula, Eq. (A.9a), leads to

$$I_x = 2\left[\frac{1}{12}ta^3 + at\left(\frac{a + t}{2}\right)^2\right] + \frac{1}{12}bt^3$$
$$= \tfrac{1}{6}ta^3 + \tfrac{1}{2}at(a + t)^2 + \tfrac{1}{12}bt^3 \tag{d}$$

In like manner,

$$I_y = \tfrac{1}{6}at^3 + \tfrac{1}{2}at(b - t)^2 + \tfrac{1}{12}tb^3 \tag{e}$$

and

$$I_{xy} = -\tfrac{1}{2}at(a + t)(b - t) \tag{f}$$

For convenience of programming, we shall let

$$C_1 = \tfrac{1}{6}ta^3 \qquad C_2 = \tfrac{1}{2}at(a + t)^2 \qquad C_3 = \tfrac{1}{12}bt^3$$
$$C_4 = \tfrac{1}{6}at^3 \qquad C_5 = \tfrac{1}{2}at(b - t)^2 \qquad C_6 = \tfrac{1}{12}tb^3$$

and

$$D = -I_{xy} \qquad E = \tfrac{1}{2}(I_x + I_y) \qquad F = \tfrac{1}{2}(I_x - I_y)$$

Equations (d), (e), (A.13), (A.14), and (A.12a) are thus

$$I_x = C_1 + C_2 + C_3 \qquad I_y = C_4 + C_5 + C_6 \qquad \theta_p = \tfrac{1}{2}\arctan\left(\frac{D}{F}\right)$$

$$I_1 = E + (D^2 + F^2)^{1/2} \qquad I_2 = E - (D^2 + F^2)^{1/2}$$

$$I_{x'} = E + F\cos 2\theta - I_{xy}\sin 2\theta$$

We note that if $I_1 = I_{x'}$, then $\theta_{p'} = \theta_p$ and $\theta_{p''} = \theta_p + 90°$, else $\theta_{p'} = \theta_p + 90°$ and $\theta_{p''} = \theta_p$.

PROBLEMS

A.1 through A.6 In terms of dimensions b, h, and t, derive expressions for distances $\bar{x}$ and $\bar{y}$, as needed, to the centroids C of the areas shown in Figs. PA.1 through PA.6. Calculate the values of $\bar{x}$ and $\bar{y}$ for $b = 3$ in., $h = 4$ in., and $t = \tfrac{3}{8}$ in.

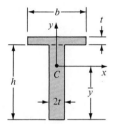

Figure PA.1

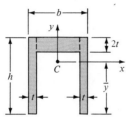

Figure PA.2

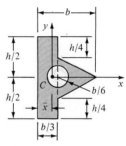

Figure PA.3

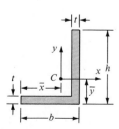

Figure PA.4

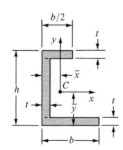

Figure PA.5

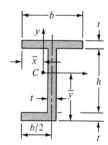

Figure PA.6

A.7 through A.9 In terms of dimensions b, h, and t, derive expressions for the centroidal moments of inertia and the products of inertia of the areas shown in Figs. PA.1 through PA.3. Write the equations for the centroidal principal moments of inertia and their orientations with respect to x in a form suitable for computer analysis; calculate the value of these quantities for $b = 50$ mm, $h = 75$ mm, and $t = 5$ mm.

A.10 through A.12 Redo Probs. A.7 through A.9 for the areas shown in Figs. PA.4 through PA.6.

B

TABLES

Properties of several common engineering materials and selected rolled-steel shapes are presented in the first six tables of this appendix. Table B.7 lists the deflections and slopes of beams with various loadings. The typical values and equations included in these tables are for reference in solving problems in the text.

Data for Tables B.2 through B.6 were compiled from the extensive tables found in the eighth edition of the *Manual of Steel Construction*. The notation used in these tables is as follows:

$$I = \text{moment of inertia}$$
$$S = \text{section modulus}$$
$$r = \sqrt{I/A} = \text{radius of gyration}$$
$$\bar{x} = \text{abscissa of the centroid}$$
$$\bar{y} = \text{ordinate of the centroid}$$
$$\theta_p = \text{angle defining a principal axis}$$

Table B.1 Typical Properties of Selected Engineering Materials* (U.S. Customary Units)

Material	Specific Weight, lb/in.³	Ultimate Strength, ksi			Yield Strength,‡ ksi		Modulus of Elasticity, 10⁶ psi	Modulus of Rigidity, 10⁶ psi	Coef. of Thermal Expans., 10⁻⁶/°F	Ductility, percent Elongation in 2 in.
		Tens.	Comp.†	Shear	Tens.	Shear				
Steel										
Structural, ASTM-A36	0.284	58	· · ·	· · ·	36	21	29	11.5	6.5	30
High strength, ASTM-A242	0.284	70	· · ·	· · ·	50	30	29	11.5	6.5	21
Stainless (302), cold-rolled	0.286	125	· · ·	· · ·	75	· · ·	28	10.6	9.6	12
Cast iron										
Gray, ASTM A-48	0.260	25	95	35	· · ·	· · ·	10	4.1	6.7	0.5
Malleable, ASTM A-47	0.264	50	90	48	33	· · ·	24	9.3	6.7	10
Aluminum										
Alloy 2014-T6	0.101	70	· · ·	42	60	32	10.6	4.1	12.8	13
Alloy 6061-T6	0.098	43	· · ·	27	38	20	10.0	3.8	13.1	17
Brass, yellow										
Cold-rolled	0.306	78	· · ·	43	63	36	15	5.6	11.3	8
Annealed	0.306	48	· · ·	32	15	9	15	5.6	11.3	60
Bronze, cold-rolled (510)	0.320	81	· · ·	· · ·	75	40	16	5.9	9.9	10
Magnesium alloys	0.065	20–49	· · ·	24	11–40	· · ·	6.5	2.4	15	2–20
Concrete										
Medium strength	0.084	· · ·	4	· · ·	· · ·	· · ·	3.5	· · ·	5.5	· · ·
High strength	0.084	· · ·	6	· · ·	· · ·	· · ·	4.3	· · ·	5.5	· · ·
Timber§ (air dry)										
Douglas fir	0.020	· · ·	7.9	1.1	· · ·	· · ·	1.7	· · ·	2.2	· · ·
Southern pine	0.021	· · ·	8.6	1.4	· · ·	· · ·	1.6	· · ·	2.2	· · ·
Glass, 98% silica	0.079	· · ·	7	· · ·	· · ·	· · ·	9.6	4.1	44	· · ·
Nylon, molded	0.040	8	· · ·	· · ·	· · ·	· · ·	0.3	· · ·	45	50
Rubber	0.033	2	· · ·	· · ·	· · ·	· · ·	· · ·	· · ·	90	600

*Properties may vary widely with changes in composition, heat treatment, and method of manufacture.

†For ductile metals the compression strength is assumed to be the same as that in tension.

‡Offset of 0.2 percent.

§Loaded parallel to the grain.

Table B.1   Typical Properties of Selected Engineering Materials*
(SI Units)

Material	Density, Mg/m^3	Ultimate Strength, MPa			Yield Strength,‡ MPa		Modulus of Elasticity, GPa	Modulus of Rigidity, GPa	Coef. of Thermal Expans., 10^{-6}/°C	Ductility, percent Elongation in 50 mm
		Tens.	Comp.†	Shear	Tens.	Shear				
Steel										
Structural, ASTM-A36	7.86	400	· · ·	· · ·	250	145	200	79	11.7	30
High strength, ASTM-A242	7.86	480	· · ·	· · ·	345	210	200	79	11.7	21
Stainless (302), cold-rolled	7.92	860	· · ·	· · ·	520	· · ·	190	73	17.3	12
Cast iron										
Gray, ASTM A-48	7.2	170	650	240	· · ·	· · ·	70	28	12.1	0.5
Malleable, ASTM A-47	7.3	340	620	330	230	· · ·	165	64	12.1	10
Aluminum										
Alloy 2014-T6	2.8	480	· · ·	290	410	220	72	28	23	13
Alloy 6061-T6	2.71	300	· · ·	185	260	140	70	26	23.6	17
Brass, yellow										
Cold-rolled	8.47	540	· · ·	300	435	250	105	39	20	8
Annealed	8.47	330	· · ·	220	105	65	105	39	20	60
Bronze, cold-rolled (510)	8.86	560	· · ·	· · ·	520	275	110	41	17.8	10
Magnesium alloys	1.8	140–340	· · ·	165	80–280	· · ·	45	17	27	2–20
Concrete										
Medium strength	2.32	· · ·	28	· · ·	· · ·	· · ·	24	· · ·	10	· · ·
High strength	2.32	· · ·	40	· · ·	· · ·	· · ·	30	· · ·	10	· · ·
Timber§ (air dry)										
Douglas fir	0.54	· · ·	55	7.6	· · ·	· · ·	12	· · ·	4	· · ·
Southern pine	0.58	· · ·	60	10	· · ·	· · ·	11	· · ·	4	· · ·
Glass, 98% silica	2.19	· · ·	50	· · ·	· · ·	· · ·	65	28	80	· · ·
Nylon, molded	1.1	55	· · ·	· · ·	· · ·	· · ·	2	· · ·	81	50
Rubber	0.91	14	· · ·	· · ·	· · ·	· · ·	· · ·	· · ·	162	600

*Properties may vary widely with changes in composition, heat treatment, and method of manufacture.

†For ductile metals the compression strength is assumed to be the same as that in tension.

‡Offset of 0.2 percent.

§Loaded parallel to the grain.

Table B.2 Properties of Rolled-Steel (W) Shapes, Wide-Flange Sections (U.S. Customary Units)

Designation*	Area, in.^{2}	Depth, in.	Flange Width, in.	Flange Thickness, in.	Web Thickness, in.	Axis x-x I, in.^{4}	Axis x-x r, in.	Axis x-x S, in.^{3}	Axis y-y I, in.^{4}	Axis y-y r, in.
W 24 × 104	30.6	24.06	12.750	0.750	0.500	3100	10.1	258	259	2.91
× 84	24.7	24.10	9.020	0.770	0.470	2370	9.79	196	94.4	1.95
W 18 × 106	31.1	18.73	11.200	0.940	0.590	1910	7.84	204	220	2.66
× 50	14.7	17.99	7.495	0.570	0.355	800	7.38	88.9	40.1	1.65
× 35	10.3	17.70	6.000	0.425	0.300	510	7.04	57.6	15.3	1.22
W 16 × 77	22.6	16.52	10.295	0.760	0.455	1110	7.00	134	138	2.47
× 57	16.8	16.43	7.120	0.715	0.430	758	6.72	92.2	43.1	1.60
× 40	11.8	16.01	6.995	0.505	0.305	518	6.63	64.7	28.9	1.57
W 14 × 145	42.7	14.78	15.500	1.090	0.680	1710	6.33	232	677	3.98
× 82	24.1	14.31	10.130	0.855	0.510	882	6.05	123	148	2.48
× 53	15.6	13.92	8.060	0.660	0.370	541	5.89	77.8	57.7	1.92
W 12 × 72	21.1	12.25	12.040	0.670	0.430	597	5.31	97.4	195	3.04
× 50	14.7	12.19	8.080	0.640	0.370	394	5.18	64.7	56.3	1.96
× 35	10.3	12.50	6.560	0.520	0.300	285	5.25	45.6	24.5	1.54
W 10 × 54	15.8	10.09	10.030	0.615	0.370	303	4.37	60.0	103	2.56
× 45	13.3	10.10	8.020	0.620	0.350	248	4.33	49.1	53.4	2.01
× 33	9.71	9.73	7.960	0.435	0.290	170	4.19	35.0	36.6	1.94
W 8 × 48	14.1	8.50	8.110	0.685	0.400	184	3.61	43.3	60.9	2.08
× 40	11.7	8.25	8.070	0.560	0.360	146	3.53	35.5	49.1	2.04
× 35	10.3	8.12	8.020	0.495	0.310	127	3.51	31.2	42.6	2.03
W 6 × 25	7.34	6.38	6.080	0.455	0.320	53.4	2.70	16.7	17.1	1.52
× 20	5.88	6.20	6.020	0.365	0.260	41.4	2.66	13.4	13.3	1.50
× 16	4.74	6.28	4.030	0.405	0.260	32.1	2.60	10.2	4.43	0.967
× 12	3.55	6.03	4.000	0.280	0.230	22.1	2.49	7.31	2.99	0.918

*A wide-flange shape is designated by the letter W followed by the nominal depth in inches and the weight in pounds per foot.

Source: The American Institute of Steel Construction, Chicago, Ill.

Table B.2 Properties of Rolled-Steel (W) Shapes, Wide-Flange Sections
(SI Units)

Designation*	Area, 10^3 mm²	Depth, mm	Flange Width, mm	Flange Thickness, mm	Web Thickness, mm	Axis x-x I, 10^6 mm⁴	Axis x-x r, mm	Axis x-x S, 10^3 mm³	Axis y-y I, 10^6 mm⁴	Axis y-y r, mm
W 610 × 155	19.7	611	324	19.0	12.7	1290	256	4220	108	73.9
× 125	15.9	612	229	19.6	11.9	985	249	3220	39.3	49.7
W 460 × 158	20.1	476	284	23.9	15.0	795	199	3340	91.6	67.6
× 74	9.48	457	190	14.5	9.0	333	188	1457	16.7	41.9
× 52	6.65	450	152	10.8	7.6	212	179	942	6.4	31.0
W 410 × 114	14.6	420	261	19.3	11.6	462	178	2200	57.4	62.7
× 85	10.8	417	181	18.2	10.9	316	171	1516	17.9	40.6
× 60	7.61	407	178	12.8	7.7	216	168	1061	12	39.9
W 360 × 216	27.5	375	394	27.7	17.3	712	161	3800	282	101.1
× 122	15.5	363	257	21.7	13.0	367	154	2020	61.6	63.0
× 79	10.1	354	205	16.8	9.4	225	150	1271	24.0	48.8
W 310 × 107	13.6	311	306	17.0	10.9	248	135	1595	81.2	77.2
× 74	9.48	310	205	16.3	9.4	164	132	1058	23.4	49.8
× 52	6.65	317	167	13.2	7.6	119	133	748	10.2	39.1
W 250 × 80	10.2	256	255	15.6	9.4	126	111	985	42.8	65
× 67	8.58	257	204	15.7	8.9	103	110	803	22.2	51.1
× 49	6.26	247	202	11.0	7.4	70.8	106	573	15.2	49.3
W 200 × 71	9.11	216	206	17.4	10.2	76.6	91.7	709	25.3	52.8
× 59	7.55	210	205	14.2	9.1	60.8	89.7	579	20.4	51.8
× 52	6.65	206	204	12.6	7.9	52.9	89.2	514	17.7	51.6
W 150 × 37	4.74	162	154	11.6	8.1	22.2	69	274	7.12	38.6
× 30	3.79	157	153	9.3	6.6	17.2	67.6	219	5.54	38.1
× 24	3.06	160	102	10.3	6.6	13.4	66	167	1.84	24.6
× 18	2.29	153	102	7.1	5.8	9.2	63.2	120	1.25	23.3

*A wide-flange shape is designated by the letter W followed by the nominal depth in millimeters and the mass in kilogram per meter.

Table B.3 Properties of Rolled-Steel (S) Shapes, American Standard I Beams
(U.S. Customary Units)

Designation*	Area, in.^{2}	Depth, in.	Flange Width, in.	Flange Thickness, in.	Web Thickness, in.	Axis x-x I, in.^{4}	Axis x-x r, in.	Axis x-x S, in.^{3}	Axis y-y I, in.^{4}	Axis y-y r, in.
S 24 × 100	29.4	24.00	7.247	0.871	0.747	2390	9.01	199	47.8	1.27
× 79.9	23.5	24.00	7.001	0.871	0.501	2110	9.47	175	42.3	1.34
S 20 × 95	27.9	20.00	7.200	0.916	0.800	1610	7.60	161	49.7	1.33
× 75	22.1	20.00	6.391	0.789	0.641	1280	7.60	128	29.6	1.16
S 18 × 70	20.6	18.00	6.251	0.691	0.711	926	6.71	103	24.1	1.08
× 54.7	16.1	18.00	6.001	0.691	0.461	804	7.07	89.4	20.8	1.14
S 15 × 50	14.7	15.00	5.640	0.622	0.550	486	5.75	64.8	15.7	1.03
× 42.9	12.6	15.00	5.501	0.622	0.411	447	5.95	59.6	14.4	1.07
S 12 × 50	14.7	12.00	5.477	0.659	0.687	305	4.55	50.8	15.7	1.03
× 35	10.3	12.00	5.078	0.544	0.428	229	4.72	38.2	9.87	0.980
S 10 × 35	10.3	10.00	4.944	0.491	0.594	147	3.78	29.4	8.36	0.901
× 25.4	7.46	10.00	4.661	0.491	0.311	124	4.07	24.7	6.79	0.954
S 8 × 23	6.77	8.00	4.171	0.425	0.441	64.9	3.10	16.2	4.31	0.798
× 18.4	5.41	8.00	4.001	0.425	0.271	57.6	3.26	14.4	3.73	0.831
S 6 × 17.25	5.07	6.00	3.565	0.359	0.465	26.3	2.28	8.77	2.31	0.675
× 12.5	3.67	6.00	3.332	0.359	0.232	22.1	2.45	7.37	1.82	0.705
S 4 × 9.5	2.79	4.00	2.796	0.293	0.326	6.79	1.56	3.39	0.903	0.569
× 7.7	2.26	4.00	2.663	0.293	0.193	6.08	1.64	3.04	0.764	0.581

*An American Standard Beam is designated by the letter S followed by the nominal depth in inches and the weight in pounds per foot.

Source: The American Institute of Steel Construction, Chicago, Ill.

Table B.3 Properties of Rolled-Steel (S) Shapes, American Standard I Beams (SI Units)

| Designation* | Area, 10^3 mm^2 | Depth, mm | Flange | | Web Thickness, mm | Axis x-x | | | Axis y-y | |
			Width, mm	Thickness, mm		I, 10^6 mm^4	r, mm	S, 10^3 mm^3	I, 10^6 mm^4	r, mm
S 610 × 149	19.0	610	184	22.1	19.0	995	229	3260	19.9	32.3
× 119	15.2	610	178	22.1	12.7	878	241	2880	17.6	34.0
S 510 × 141	18.0	508	183	23.3	20.3	670	193	2640	20.7	33.8
× 112	14.3	508	162	20.1	16.3	533	193	2100	12.3	29.5
S 460 × 104	13.3	457	159	17.6	18.1	385	170	1685	10.0	27.4
× 81	10.4	457	152	17.6	11.7	335	180	1466	8.66	29.0
S 380 × 74	9.5	381	143	15.8	14.0	202	146	1060	6.53	26.2
× 64	8.13	381	140	15.8	10.4	186	151	977	5.99	27.2
S 310 × 74	9.48	305	139	16.8	17.4	127	116	833	6.53	26.2
× 52	6.64	305	129	13.8	10.9	95.3	120	625	4.11	24.9
S 250 × 52	6.64	254	126	12.5	15.1	61.2	96	482	3.48	22.9
× 38	4.81	254	118	12.5	7.9	51.6	103	406	2.83	24.2
S 200 × 34	4.37	203	106	10.8	11.2	27	78.7	266	1.79	20.3
× 27	3.5	203	102	10.8	6.9	24	82.8	236	1.55	21.1
S 150 × 26	3.27	152	90	9.1	11.8	11.0	57.9	144	0.96	17.2
× 19	2.36	152	84	9.1	5.8	9.20	62.2	121	0.76	17.9
S 100 × 14	1.80	102	70	7.4	8.3	2.83	39.6	55.5	0.38	14.5
× 11	1.45	102	67	7.4	4.8	2.53	41.6	49.6	0.32	14.8

*An American Standard Beam is designated by the letter S followed by the nominal depth in millimeters and the mass in kilograms per meter.

Table B.4 Properties of Rolled-Steel (C) Shapes, American Standard Channels
(U.S. Customary Units)

			Flange		Web	Axis x-x			Axis y-y		
Designation*	Area, in.²	Depth, in.	Width, in.	Thickness, in.	Thickness, in.	I, in.⁴	r, in.	S, in.³	I, in.⁴	r, in.	x̄, in.
C 15 × 50	14.7	15.00	3.716	0.650	0.716	404	5.24	53.8	11.0	0.867	0.799
× 40	11.8	15.00	3.520	0.650	0.520	349	5.44	46.5	9.23	0.866	0.778
× 33.9	9.96	15.00	3.400	0.650	0.400	315	5.62	42.0	8.13	0.904	0.787
C 12 × 30	8.82	12.00	3.170	0.501	0.510	162	4.29	27.0	5.14	0.763	0.674
× 25	7.35	12.00	3.047	0.501	0.387	144	4.43	24.1	4.47	0.780	0.674
× 20.7	6.09	12.00	2.942	0.501	0.282	129	4.61	21.5	3.88	0.799	0.698
C 10 × 30	8.82	10.00	3.033	0.436	0.673	103	3.42	20.7	3.94	0.669	0.649
× 25	7.35	10.00	2.886	0.436	0.526	91.2	3.52	18.2	3.36	0.676	0.617
× 20	5.88	10.00	2.739	0.436	0.379	78.9	3.66	15.8	2.81	0.691	0.606
C 8 × 18.75	5.51	8.00	2.527	0.390	0.487	44.0	2.82	11.0	1.98	0.599	0.565
× 13.75	4.04	8.00	2.343	0.390	0.303	36.1	2.99	9.03	1.53	0.615	0.553
× 11.5	3.38	8.00	2.260	0.390	0.220	32.6	3.11	8.14	1.32	0.625	0.571
C 6 × 13	3.83	6.00	2.157	0.343	0.437	17.4	2.13	5.80	1.05	0.525	0.514
× 10.5	3.09	6.00	2.034	0.343	0.314	15.2	2.22	5.06	0.865	0.529	0.500
× 8.2	2.40	6.00	1.920	0.343	0.200	13.1	2.34	4.38	0.692	0.537	0.512
C 4 × 7.25	2.13	4.00	1.721	0.296	0.321	4.59	1.47	2.29	0.432	0.450	0.459
× 5.4	1.59	4.00	1.584	0.296	0.184	3.85	1.56	1.93	0.319	0.449	0.458

*An American Standard Channel is designated by the letter C followed by the nominal depth in inches and the weight in pounds per foot.

Source: The American Institute of Steel Construction, Chicago, Ill.

Table B.4   Properties of Rolled-Steel (C) Shapes, American Standard Channels (SI Units)

Designation*	Area, 10^3 mm²	Depth, mm	Flange		Web Thickness, mm	Axis x-x			Axis y-y		
			Width, mm	Thickness, mm		I, 10^6 mm⁴	r, mm	S, 10^3 mm³	I, 10^6 mm⁴	r, mm	$\bar{x}$, mm
C 380 × 74	9.48	381	94	16.5	18.2	168	133	883	4.58	22.0	20.3
× 60	7.61	381	89	16.5	13.2	145	138	763	3.84	22.5	19.8
× 50	6.43	381	86	16.5	10.2	131	143	688	3.38	23.0	20.0
C 310 × 45	5.69	305	80	12.7	13	67.4	109	442	2.14	19.4	17.1
× 37	4.74	305	77	12.7	9.8	59.9	113	393	1.86	19.8	17.1
× 31	3.93	305	74	12.7	7.2	53.5	117	352	1.62	20.3	17.7
C 250 × 45	5.69	254	76	11.1	17.1	42.9	86.9	338	1.64	17.0	16.5
× 37	4.74	254	73	11.1	13.4	38.0	89.4	299	1.40	17.2	15.7
× 30	3.78	254	69	11.1	9.6	32.8	93	258	1.17	17.6	15.4
C 200 × 28	3.56	203	64	9.9	12.4	18.3	71.6	180	0.82	15.2	14.4
× 21	2.61	203	59	9.9	7.7	15.0	75.9	148	0.64	15.6	14.1
× 17	2.18	203	57	9.9	5.6	13.6	79.0	134	0.55	15.9	14.5
C 150 × 19	2.47	152	54	8.7	11.1	7.24	54.1	95.3	0.44	13.3	13.1
× 16	1.99	152	51	8.7	8.0	6.33	56.4	83.3	0.36	13.4	12.7
× 12	1.55	152	48	8.7	5.1	5.45	59.4	71.7	0.29	13.6	13.0
C 100 × 11	1.37	102	43	7.5	8.2	1.91	37.3	37.5	0.18	11.4	11.7
× 8	1.03	102	40	7.5	4.7	1.60	39.6	31.4	0.13	11.4	11.6

*An American Standard Channel is designated by the letter C followed by the nominal depth in millimeters and the mass in kilograms per meter.

Table B.5 Properties of Rolled-Steel (L) Shapes, Angles with Equal Legs (U.S. Customary Units)

Size and Thickness, in.	Weight, lb/ft	Area, in.^{2}	Axis x-x or y-y				Axis z-z
			I, in.^{4}	r, in.	S, in.^{3}	$\bar{x}$ or $\bar{y}$, in.	r, in.
L 8 × 8 × 1	51.0	15.0	89.0	2.44	15.8	2.37	1.56
× $\frac{3}{4}$	38.9	11.4	69.7	2.47	12.2	2.28	1.58
× $\frac{1}{2}$	26.4	7.75	48.6	2.50	8.36	2.19	1.59
L 6 × 6 × 1	37.4	11.0	35.5	1.80	8.57	1.86	1.17
× $\frac{5}{8}$	24.2	7.11	24.2	1.84	5.66	1.73	1.18
× $\frac{3}{8}$	14.9	4.36	15.4	1.88	3.53	1.64	1.19
L 5 × 5 × $\frac{3}{4}$	23.6	6.94	15.7	1.51	4.53	1.52	0.975
× $\frac{1}{2}$	16.2	4.75	11.3	1.54	3.16	1.43	0.983
× $\frac{3}{8}$	12.3	3.61	8.74	1.56	2.42	1.39	0.990
L 4 × 4 × $\frac{3}{4}$	18.5	5.44	7.67	1.19	2.81	1.27	0.778
× $\frac{1}{2}$	12.8	3.75	5.56	1.22	1.97	1.18	0.782
× $\frac{1}{4}$	6.6	1.94	3.04	1.25	1.05	1.09	0.795
L 3$\frac{1}{2}$ × 3$\frac{1}{2}$ × $\frac{3}{8}$	8.5	2.48	2.87	1.07	1.15	1.01	0.687
× $\frac{1}{4}$	5.8	1.69	2.01	1.09	0.794	0.968	0.694
L 3 × 3 × $\frac{1}{2}$	9.4	2.75	2.22	0.898	1.07	0.932	0.584
× $\frac{1}{4}$	4.9	1.44	1.24	0.930	0.577	0.842	0.592

Source: The American Institute of Steel Construction, Chicago, Ill.

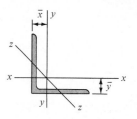

Table B.5 Properties of Rolled-Steel (L) Shapes, Angles with Equal Legs (SI Units)

Size and Thickness, mm	Mass, kg/m	Area, 10^3 mm^2	Axis x-x or y-y				Axis z-z
			I, 10^6 mm^4	r, mm	S, 10^3 mm^3	$\bar{x}$ or $\bar{y}$, mm	r, mm
L 203 × 203 × 25.4	75.9	9.68	37	61.8	259	60.2	39.6
× 19	57.9	7.36	29	62.8	200	57.9	40.1
× 12.7	39.3	5.0	20.2	63.6	137	55.6	40.4
L 152 × 152 × 25.4	55.7	7.1	14.8	45.6	140.4	47.2	29.7
× 15.9	36	4.59	10.1	46.8	92.8	43.9	30.0
× 9.5	22.2	2.8	6.41	47.8	57.8	41.7	30.2
L 127 × 127 × 19	35.1	4.48	6.53	38.2	74.2	38.6	24.8
× 12.7	24.1	3.07	4.70	39.2	51.8	36.3	25.0
× 9.5	18.3	2.33	3.64	39.5	39.7	35.3	25.1
L 102 × 102 × 19	27.5	3.51	3.19	30.1	46.0	32.3	19.8
× 12.7	19	2.42	2.31	30.9	32.3	30.0	19.9
× 6.4	9.8	1.25	1.27	31.8	17.2	27.7	20.2
L 89 × 89 × 9.5	12.6	1.6	1.20	27.3	18.9	25.7	17.5
× 6.4	8.6	1.09	0.84	27.7	13.0	24.6	17.6
L 76 × 76 × 12.7	14	1.77	0.92	22.8	17.5	23.7	14.8
× 6.4	7.3	0.93	0.52	23.6	9.46	21.4	15.0

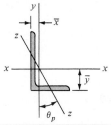

Table B.6 Properties of Rolled-Steel (L) Shapes, Angles with Unequal Legs (U.S. Customary Units)

Size and Thickness, in.	Weight, lb/ft	Area, in.²	Axis x-x				Axis y-y			Axis z-z	
			I, in.⁴	r, in.	S, in.³	$\bar{y}$, in.	I, in.⁴	r, in.	$\bar{x}$, in.	r, in.	tan θ_p
L 8 × 6 × 1	44.2	13.0	80.8	2.49	15.1	2.65	38.8	1.73	1.65	1.28	0.543
× ½	23.0	6.75	44.3	2.56	8.02	2.47	21.7	1.79	1.47	1.30	0.558
L 6 × 4 × ¾	23.6	6.94	24.5	1.88	6.25	2.08	8.68	1.12	1.08	0.860	0.428
× ⅜	12.3	3.61	13.5	1.93	3.32	1.94	4.90	1.17	0.941	0.877	0.446
L 5 × 3 × ½	12.8	3.75	9.45	1.59	2.91	1.75	2.58	0.829	0.750	0.648	0.357
× ¼	6.6	1.94	5.11	1.62	1.53	1.66	1.44	0.861	0.657	0.663	0.371
L 4 × 3 × ½	11.1	3.25	5.05	1.25	1.89	1.33	2.42	0.864	0.827	0.639	0.543
× ¼	5.8	1.69	2.77	1.28	1.00	1.24	1.36	0.896	0.736	0.651	0.558
L 3½ × 2½ × ½	9.4	2.75	3.24	1.09	1.41	1.20	1.36	0.704	0.705	0.534	0.486
× ¼	4.9	1.44	1.80	1.12	0.755	1.11	0.777	0.735	0.614	0.544	0.506
L 3 × 2 × ½	8.5	2.50	2.08	0.913	1.04	1.00	1.30	0.722	0.750	0.520	0.667
× ¼	4.5	1.31	1.17	0.945	0.561	0.911	0.743	0.753	0.661	0.528	0.684
L 2½ × 2 × ⅜	5.3	1.55	0.912	0.768	0.547	0.813	0.514	0.577	0.581	0.420	0.614
× ¼	3.62	1.06	0.654	0.784	0.381	0.787	0.372	0.592	0.537	0.424	0.626

Source: The American Institute of Steel Construction, Chicago, Ill.

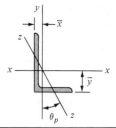

Table B.6   Properties of Rolled-Steel (L) Shapes, Angles with Unequal Legs (SI Units)

Size and Thickness, mm	Mass, kg/m	Area, 10^3 mm²	Axis x-x				Axis y-y			Axis z-z	
			I, 10^6 mm⁴	r, mm	S, 10^3 mm³	$\bar{y}$, mm	I, 10^6 mm⁴	r, mm	$\bar{x}$, mm	r, mm	tan θ_p
L 203 × 152 × 25.4	65.5	8.39	33.6	63.3	247	67.3	16.2	43.9	41.9	32.5	0.543
× 12.7	34.1	4.35	18.4	65.1	131	62.7	9.03	45.6	37.3	33.0	0.558
L 152 × 102 × 19	35.0	4.48	10.2	47.7	102.4	52.8	3.61	28.4	27.4	21.8	0.428
× 9.5	18.2	2.33	5.62	49.1	54.4	49.3	2.04	29.6	23.9	22.3	0.446
L 127 × 76 × 12.7	19.0	2.42	3.93	40.3	47.7	44.5	1.07	21.1	19.1	16.5	0.357
× 6.4	9.8	1.25	2.13	41.2	25.1	42.2	0.60	21.9	16.7	16.8	0.371
L 102 × 76 × 12.7	16.4	2.10	2.10	31.6	31.0	33.8	1.01	21.9	21.0	16.2	0.543
× 6.4	8.6	1.09	1.15	32.5	16.4	31.5	0.57	22.8	18.7	16.5	0.558
L 89 × 64 × 12.7	13.9	1.80	1.35	27.6	23.1	30.5	0.57	17.9	17.9	13.6	0.486
× 6.4	7.3	0.93	0.75	28.4	12.4	28.2	0.32	18.7	15.6	13.8	0.506
L 76 × 64 × 12.7	12.6	1.62	0.87	23.2	17.0	25.4	0.54	18.3	19.1	13.2	0.667
× 6.4	6.70	0.85	0.49	24.0	9.19	23.1	0.31	19.1	16.8	13.4	0.684
L 64 × 51 × 9.5	7.89	1.00	0.38	19.5	8.96	21.1	0.21	14.7	14.8	10.7	0.614
× 6.4	5.39	0.68	0.27	19.9	6.24	20.0	0.16	15.0	13.6	10.8	0.626

Table B.7 Deflections and Slopes of Beams

Load and Support	Maximum Deflection	Slope at End	Equation of Elastic Curve
1.	$-\dfrac{PL^3}{3EI}$	$-\dfrac{PL^2}{2EI}$	$v = \dfrac{Px^2}{6EI}(x - 3L)$
2.	$-\dfrac{ML^2}{2EI}$	$-\dfrac{ML}{EI}$	$v = -\dfrac{Mx^2}{2EI}$
3.	$-\dfrac{wL^4}{8EI}$	$-\dfrac{wL^3}{6EI}$	$v = -\dfrac{wx^2}{24EI}(x^2 - 4Lx + 6L^2)$
4.	$-\dfrac{w_0 L^4}{30EI}$	$-\dfrac{w_0 L^3}{24EI}$	$v = \dfrac{w_0 x^2}{120EIL}(x^3 - 5Lx^2 + 10L^2x - 10L^3)$
5.	$-\dfrac{PL^3}{48EI}$	$\pm\dfrac{PL^2}{16\,EI}$	$v = \dfrac{Px}{48EI}(4x^2 - 3L^2) \qquad (x \le L/2)$

Table B.7 Deflections and Slopes of Beams (*Continued*)

Load and Support	Maximum Deflection	Slope at End	Equation of Elastic Curve
6.	For $a > b$: $$-\frac{Pb(L^2 - b^2)^{3/2}}{9\sqrt{3}EIL}$$ $$x_m = \sqrt{\frac{L^2 - b^2}{3}}$$	$$\theta_A = -\frac{Pb(L^2 - b^2)}{6EIL}$$ $$\theta_B = \frac{Pa(L^2 - a^2)}{6EIL}$$	$$v = \frac{Pbx}{6EIL}(x^2 - L^2 + b^2) \quad (x \le a)$$
7.	$$-\frac{ML^2}{9\sqrt{3}EI}$$	$$\theta_A = -\frac{ML}{6EI}$$ $$\theta_B = \frac{ML}{3EI}$$	$$v = \frac{Mx}{6EIL}(x^2 - L^2)$$
8.	$$-\frac{5wL^4}{384EI}$$	$$\pm\frac{wL^3}{24EI}$$	$$v = -\frac{wx}{24EI}(x^3 - 2Lx^2 + L^3)$$
9.	$$\pm\frac{ML^2}{36\sqrt{12}EI}$$	$$\pm\frac{ML}{24EI}$$	$$v = \frac{Mx}{24EIL}(4x^2 - L^2) \quad (x \le L/2)$$

REFERENCES

American Institute of Timber Construction, *Timber Construction Manual,* Wiley, New York, 1974.

Annual Book of ASTM, American Society for Testing Materials, Philadelphia, Pa.

Baumeister, T., *Mark's Mechanical Engineering Handbook,* 8th ed., McGraw-Hill, New York, 1978.

Crandall, S. H., N. C. Dahl, and T. J. Lardner, *Mechanics of Solids,* 2d ed., McGraw-Hill, New York, 1978.

Dally, J. M., and W. F. Riley, *Experimental Stress Analysis,* McGraw-Hill, New York, 1978.

Faupel, J. H., and F. E. Fisher, *Engineering Design,* 2d ed., Wiley, New York, 1986.

Fleming, J. F., *Structural Engineering Analysis on Personal Computers,* McGraw-Hill, New York, 1986.

Frocht, M. M., "Photoelastic Studies in Stress Concentration," *Mechanical Engineering,* August 1936, pp. 485–489.

Gere, J. M., and S. P. Timoshenko, *Mechanics of Materials,* 3d ed., PWS-Kent, Boston, Mass., 1990.

Hetenyi, M., *Handbook of Experimental Stress Analysis,* Wiley, New York, 1957.

Jacobsen, L. S., "Torsional-Stress Concentrations in Shafts of Circular Cross Section and Variable Diameter," *Trans. ASME,* vol. 47, 1925, pp. 619–638.

Kuhn, P., *Stresses in Aircraft and Shell Structures,* McGraw-Hill, New York, 1956.

McCormac, L. C., *Design of Reinforced Concrete,* Harper & Row, New York, 1978.

Manual of Steel Construction, 8th ed., American Institute of Steel Construction (AISC), Chicago, 1980.

Peterson, R. E., *Stress Concentration Design Factors,* Wiley, New York, 1974.

Pilkey, W. D., "Clebsch's Method for Beam Deflections," *J. Engineering Education,* January 1964, p. 170.

Specifications for Aluminum Structures, Aluminum Association, Washington, D.C., 1976.

Spotts, M. F., *Design of Machine Elements,* 5th ed., Prentice-Hall, Englewood Cliffs, N.J., 1978.

Timoshenko, S. P., and J. M. Gere, *Theory of Elastic Stability,* 2d ed., McGraw-Hill, New York, 1961.

Ugural, A. C., *Stresses in Plates and Shells,* McGraw-Hill, New York, 1981.

Ugural, A. C., and S. K. Fenster, *Advanced Strength and Applied Elasticity,* 2d ed., Elsevier, New York, 1987.

Wahl, A. M., *Mechanical Springs,* McGraw-Hill, New York, 1963.

Young, W. C., *Roark's Formulas for Stress and Strain,* 6th ed., McGraw-Hill, New York, 1989.

Answers to Even-Numbered Problems

CHAPTER 2

2.2 22.5 kN.

2.4 (a) 7.24 mm. (b) 92.8 MPa.
(c) 97 MPa. (d) 116.6 MPa.

2.6 90 kN.

2.8 (a) 10.08 kip·in. (b) 3.133 ksi.

2.10 96 MPa.

2.12 (a) 3.927 kips. (b) 5 ksi.

2.14 $\sigma_{BE} = 25$ MPa; $\sigma_{CE} = 46.7$ MPa.

2.16 $\sigma_{AC} = 0$; $\sigma_{AD} = 6.366$ ksi.

2.18 (a) 22 kN. (b) 6.6 kN·m.

2.20 54.7°.

2.22 (a) $\sigma_{AB} = 6.79$ MPa; $\sigma_{AE} = 6.37$ MPa;
$\sigma_{DE} = -14.9$ MPa. (b) 23.76 MPa.

2.24 (a) 108.9 MPa. (b) $\tau_D = 75.82$ MPa;
$\tau_E = 29.67$ MPa.

2.26 2.79 MPa.

2.28 1.519 ksi.

2.30 29.4 MPa.

2.32 724 mm^2.

2.34 $\sigma_a = 14.7$ MPa; $\sigma_c = 77$ MPa.

2.36 1.2 in.^{2}.

2.38 (a) 3.3 mm. (b) 10.8 mm.
(c) 13.3 mm.

2.40 437.5 mm^2.

2.42 $A_{BC} = 0.808$ in.^{2}. (b) $A_{AB} = 0.4$ in.^{2}.

2.44 12.4 mm.

2.46 $A_{AB} = 104.2$ mm^2; $A_{BC} = 312.5$ mm^2;
$A_{AC} = A_{CD} = 104.2$ mm^2; $A_{BD} =$ any value.

2.48 8.48 mm.

2.50 18.9 mm.

2.52 0.351 in.

2.54 26.2 mm.

CHAPTER 3

3.2 5000 μ

3.4 (a) 2000 μ; (b) 1570 μ.

3.6 $\varepsilon_x = 25$ μ; $\varepsilon_y = 250$ μ; $\gamma_{xy} = -400$ μ.

3.8 5.28×10^{-3} in.

3.10 (a) 0.025y. (b) 2500 μ.
(c) 0.

3.12 1998 μ.

3.14 $\varepsilon_x = \varepsilon_y = -169$ μ; $\gamma_{xy} = -1061$ μ.

3.16 1302 μ.

3.18 (a) $\varepsilon_x = -1500$ μ; $\gamma_{xy} = 0$.
(b) -2700 μ.

3.20 $\varepsilon_{yp} = 1400$ μ; $\varepsilon_f = 0.28$;
$L_f \approx 64$ mm; 28%.

3.22 (a) 0.31. (b) 19.6×10^6 psi.
(c) 7.48×10^6 psi.

3.24 (a) 200 GPa. (b) 255 MPa.
(c) 240 MPa. (d) 450 MPa.

3.26 (a) 69 GPa. (b) 242 MPa.
(c) 275 MPa. (d) 326 MPa.
(e) 20%. (f) 52%
(g) 679 MPa. (h) $E_t = 1.4$ GPa;
$E_s = 10$ GPa.

3.28 474 μ.

3.30 (a) $\Delta L_{AB} = -0.044$ mm; $\Delta L_{BC} = -0.138$ mm; $\Delta L_{CE} = 0.009$ mm.
(b) 105.1 kN.

3.34 0.06°.

3.36 (a) 26.08 ksi. (b) 9.1×10^6 psi.
(c) $\frac{1}{3}$. (d) 3.4×10^6 psi.

3.38 2.00051 in.

3.40 (a) 0.176 mm. (b) -0.014 mm.
(c) -0.0016 mm.

3.42 $\sigma_x = 207.3$ MPa; $\sigma_y = 167.2$ MPa.

3.46 $d_{\min} = 49.978$ mm; $V_{\min} = 2353.629$ mm^3.

3.48 57.6 kips.

3.50 (a) -0.43 mm. (b) -1150 mm^3.

3.52 (a) -0.384×10^{-3} in.
(b) -0.482×10^{-3} in.^{3}.

3.54 (a) -0.013 mm. (b) 0.096 mm.
(c) 0.05 mm. (d) -0.0023 mm.

3.56 (a) 1167 kJ/m^3. (b) 52.5 kJ/m^3.

3.58 (a) 444 kJ/m^3. (b) 44 MJ/m^3.

CHAPTER 4

4.2 $\sigma_{x'} = 84.6$ MPa; $\tau_{x'y'} = 23.1$ MPa.

4.4 $\sigma_{x'} = 10.09$ ksi; $\tau_{x'y'} = -15.83$ ksi.

4.6 $\sigma_{x'} = -11.3$ MPa; $\tau_{x'y'} = 84.5$ MPa.

4.8 $\sigma_{x'} = -16.8$ MPa; $\tau_{x'y'} = 46$ MPa.

4.10 2.89 ksi.

4.12 $\sigma_{x'} = 7.96$ ksi; $\tau_{x'y'} = 593$ psi.

4.14 At $\theta = -30°$: $\sigma_{x'} = 3.5$ ksi;
$\tau_{x'y'} = 4.33$ ksi.
At $\theta = 120°$: $\sigma_{x'} = -1.5$ ksi;
$\tau_{x'y'} = 4.33$ ksi.

4.16 $\sigma_{x'} = -59$ MPa; $\tau_{x'y'} = -17.3$ MPa.

4.18 $\sigma_{x'} = 24.5$ MPa; $\tau_{x'y'} = -6.1$ MPa.

4.20 $\sigma_{x'} = 1.699$ ksi; $\tau_{x'y'} = -1.125$ ksi.

4.22 (a) 400 mm. (b) 120 MPa.

4.24 (a) $\sigma_1 = 6.53$ MPa; $\sigma_2 = -1.53$ MPa;
$\theta'_p = 41.4°\,\circlearrowright$.
(b) $\sigma_1 = 15.11$ MPa; $\sigma_2 = 5.89$ ksi;
$\theta'_p = 6.3°\,\circlearrowright$.
(c) $\sigma_1 = 14.5$ MPa; $\sigma_2 = 5.5$ MPa;
$\theta'_p = 13.3°\,\circlearrowleft$.

4.26 $\sigma_1 = 103.5$ MPa; $\sigma_2 = -73.5$ MPa;
$\theta''_p = 21.4°\,\circlearrowleft$.

4.28 (a) $\sigma_1 = 43.2$ MPa; $\sigma_2 = -83.2$ MPa;
$\theta''_p = 35.8°\,\circlearrowright$.
(b) $\tau_{\max} = 63.2$ MPa; $\sigma = -20$ MPa.

4.30 (a) $\sigma_1 = 19.326$ ksi; $\sigma_2 = 3.674$ ksi;
$\theta'_p = 31.7°\,\circlearrowleft$.
(b) $\tau_{\max} = 7.826$ ksi; $\sigma' = 11.5$ ksi.

4.32 $\sigma_x = 3.62$ ksi; $\sigma_y - 6.62$ ksi; $\theta'_p = 19°\,\circlearrowright$.

4.34 $26.6°\,\circlearrowleft$.

4.36 $\sigma_1 = 17.6$ ksi; $\sigma_2 = 12.4$ ksi; $\theta''_p = 7.5°\,\circlearrowleft$.

4.38 $\theta'_p = \dfrac{\theta}{2} + 45°$; $\sigma_1 = \sigma(1 + \sin\theta)$;
$\sigma_2 = \sigma(1 - \sin\theta)$.

4.40 $(\tau_{\max})_t = 9.7$ MPa; $\sigma' = 9.7$ MPa.

4.42 $\sigma = -12$ MPa; $\theta = 45°\,\circlearrowright$.

4.44 (a) $\sigma_x = 10$ ksi; $\tau_{xy} = -3$ ksi.
(b) $\sigma_1 = 11$ ksi; $\theta'_p = 18.5°\,\circlearrowleft$.

4.64 $\varepsilon_{x'} = 1350\ \mu$; $\varepsilon_{y'} = 350\ \mu$; $\gamma_{x'y'} = -700\ \mu$.

4.66 $\varepsilon_{x'} = -244\ \mu$; $\varepsilon_{y'} = 364\ \mu$; $\gamma_{x'y'} = 434\ \mu$.

4.68 $\varepsilon_{x'} = 550\ \mu$; $\varepsilon_{y'} = -150\ \mu$; $\gamma_{x'y'} = 1789\ \mu$.

4.70 $\varepsilon_{x'} = 336\ \mu$; $\varepsilon_{y'} = 274\ \mu$; $\gamma_{x'y'} = 224\ \mu$.

4.72 $\varepsilon_{x'} = -694\ \mu$; $\varepsilon_{y'} = -226\ \mu$; $\gamma_{x'y'} = -330\ \mu$.

4.74 (a) $\varepsilon_1 = 408\ \mu$; $\varepsilon_3 = -908\ \mu$;
$\gamma_{\max} = 1316\ \mu$.
(b) $(\gamma_{\max})_t = 1316\ \mu$.

4.76 (a) $\varepsilon_1 = 761\ \mu$; $\varepsilon_2 = 39\ \mu$; $\gamma_{\max} = 722\ \mu$.
(b) $(\gamma_{\max})_t = 1104\ \mu$.

4.78 (a) 124.6×10^{-3} mm.
(b) 2.7×10^{-3} mm.

4.80 $\varepsilon_1 = 400\ \mu$; $\varepsilon_2 = -600\ \mu$; $\theta''_p = 26.6°\,\circlearrowright$.

4.84 (a) $\varepsilon_1 = 836\ \mu$; $\varepsilon_2 = -702\ \mu$; $\gamma_{\max} = 1538\ \mu$.
(b) $\sigma_1 = 137.5$ MPa; $\sigma_2 = -99.2$ MPa;
$\tau_{\max} = 118.3$ MPa; $\theta'_p = 32.2°\,\circlearrowright$.

4.86 (a) $\gamma_{\max} = 570\ \mu$; $\theta''_s = 18.9°\,\circlearrowleft$.
(b) $1435\ \mu$.

CHAPTER 5

5.2 (a) 1.13 mm. (b) 3.53 mm.

5.4 28.125 kips.

5.6 0.003 mm.

5.8 $\delta = QL/4AE$.

5.10 (a) 12.12 kN. (b) 0.4 mm.

5.12 0.12 mm $\rightarrow$.

5.14 (a) 1.34 mm $\downarrow$. (b) 0.71 mm $\uparrow$.

5.16 $\delta = \gamma\omega L^3/3gE$.

5.18 1.395 in.

5.20 1.73.

5.22 0.054 in. $\downarrow$

5.24 (a) $R_A = P_B/4$; $R_D = 7P_B/4$.
(b) $0.175P_B/A_1E$.

5.26 $\sigma_s = -13.65$ ksi; $\sigma_c = -1.375$ psi.

5.28 $\sigma_s = -113.3$ MPa; $\sigma_d = -40.8$ MPa.

5.30 $\sigma_b = -79.1$ MPa; $\sigma_s = 142.5$ MPa.

5.32 $A_{AD} = 167$ mm^2; $A_{BE} = 833$ mm^2.

5.34 (a) 0.417 in.
(b) $\sigma_b = 16$ ksi; $\sigma_s = 32$ ksi.

5.36 (a) -215 MPa. (b) 0.15 mm.

5.38 30.5 MPa.

5.40 517 lb.

5.42 0.34 mm $\downarrow$.

5.44 $R_A = \dfrac{\ln\,[2a/(a\,+\,b)]}{\ln\,(a/b)}$; $R_B = P - R_A$.

5.46 $R_A = 79.55$ kN $\rightarrow$; $R_B = 20.45$ kN $\rightarrow$.

5.48 $\sigma_{AC} = -0.506\alpha(\Delta T)E$.

5.50 $\sigma_{max} = 16\alpha E(T_2 - T_1)/3$.

5.52 5.89 kips.

5.54 $\sigma_{x'} = -2.63$ MPa; $\tau_{x'y'} = -3.14$ MPa.

5.56 $\sigma_{x'} = 250$ MPa; $\tau_{x'y'} = -144.3$ MPa.

5.58 $\sigma_{x'} = 294.4$ MPa; $\tau_{x'y'} = -107.1$ MPa.

5.60 19.2 kN.

5.62 (a) 100 MPa. (b) 160 MPa.

5.64 15.56 kips.

CHAPTER 6

6.2 3.046 in.

6.4 0.783.

6.6 $\tau_{AB} = 59.7$ MPa; $\tau_{BC} = 104.5$ MPa;
$\tau_{CD} = 84.9$ MPa.

6.8 $T_C = 21.99$ kip·in.; $T_B = 58.82$ kip·in.

6.10 4.17 kN·m

6.12 10.13 kN·m.

6.14 37.3 GPa.

6.16 $\phi_C = 0.013°$; $\phi_F = 0.067°$.

6.18 (a) 6.52°. (b) 3.67°.

6.20 $T_A = 15.71$ kN·m; $T_B = 22.34$ kN·m;
$T_C = 6.63$ kN·m.

6.22 3.325 in.

6.24 0.086°.

6.26 12.55 kip·in.

6.28 (a) $\phi = 16t_0L^2/\pi d^4G$. (b) 1.33°.

6.30 385.5 N·m.

6.32 184 kip·in.

6.34 $(\tau_{max})_a = 5.21$ ksi; $(\tau_{max})_b = 6.51$ ksi.

6.36 42.27 MPa.

6.38 $T_B = 121$ lb·in.; $T_C = 179$ lb·in.

6.40 (a) $T_A = t_0L/6$; $T_B = t_0L/3$.
(b) 15.01 MPa.

6.42 $\tau_h = 31.03$ MPa; $\tau_s = 27$ MPa.

6.44 $T_C = 1071$ lb·in.; $(\tau_{max})_B = 1939$ psi.

6.46 2 in.

6.48 $d_a = 38$ mm; $d_s = 79$ mm.

6.50 54 mm.

6.52 (a) $d_{AB} = 1.719$ in.; $d_{BC} = 2.139$ in.
(b) 5.27°.

6.54 69 mm.

6.56 4 mm.

6.58 (a) $\tau_{AB} = 40.09$ MPa; $\tau_{AC} = 80.18$ MPa;
$\tau_{BC} = 53.57$ MPa.
(b) 6.84°.

6.60 4.68°.

6.62 (a) $\tau_{AC} = 10.04$ MPa; $\tau_{AB} = 6.69$ MPa.
(b) 0.31°.

6.64 $T = 1.88$ kN·m; $\tau_{max} = 20.51$ MPa.

CHAPTER 7

7.2 $V = \frac{1}{24}w_0L \uparrow$; $M = \frac{1}{48}w_0L^2 \,\backsmile$.

7.4 $V = 14$ kN $\uparrow$; $M = 38$ kN·m$\,\backsmile$.

7.6 $V = 0.2$ kips $\uparrow$; $M = 1.4$ kip·ft$\,\backsmile$.

7.8 $V = 17.5$ kN $\uparrow$; $M = 5$ kN·m$\,\backsmile$.

7.10 $R_A = 1.2$ kips; $M_A = 9$ kips.

7.12 $R_A = 10$ kN; $R_D = 20$ kN.

7.14 $R_A = 5P/2$; $R_D = 7P/12$.

7.16 $R_A = 10.5$ kN; $R_B = 15.75$ kN;
$R_C = 5.75$ kN; $R_D = 11.25$ kN.

7.18 $V_{max} = 4.8$ kN; $M_{max} = 12$ kN·m.

7.20 $R_A = 2P$; $M_A = 5PL/3$.

7.22 $R_A = R_B = w_0L/4$.

7.24 $R_A = R_B = wL/4$.

7.26 $R_A = 4$ kips; $R_B = 4$ kips.

7.28 $R_A = 6.5$ kips; $R_B = 15.5$ kips.

7.30 $R_A = \frac{1}{6}w_0L$; $R_B = \frac{1}{3}w_0L$
$M_{max} = w_0L^2/9\sqrt{3}$.

7.32 $R_A = kL^3/3$; $M_A = kL^4/4$.

7.34 $R_{Ay} = 2$ kN; $R_{Az} = 6.43$ kN;
$R_{By} = 5$ kN; $R_{Bz} = 2.57$ kN.

7.36 $M_{max} = 52.5$ kN·m; $M_B = -10$ kN·m.

7.38 $M_{max} = 10$ kip·ft.

7.40 $M_{max} = -24$ kip·ft; $M_A = 16$ kip·ft.

7.42 $\varepsilon_{max} = 300\ \mu$; $M = 31.5$ N·m.
7.44 1.19.
7.46 7.62 kip·in.
7.48 (a) 1.38 kN/m; (b) 37.4 mm.
(b) 1.74.
7.50 $h/b = \sqrt{2}$.
7.52 −53.9 kN.
7.54 (a) 6255 psi. (b) 9383 psi.
7.56 $\sigma_{max} = 68.9$ MPa, at 2.1 m from A.
7.58 (a) −25.52 MPa. (b) 23.97 MPa.
7.60 238.7 kN·m.
7.62 790 lb/ft.
7.64 (a) 1.77 kips. (b) 15.4 kips.
7.66 (a) 27 MPa. (b) 240.6 MPa.
7.68 16.2 mm.
7.70 279 lb.
7.72 $(\sigma_b)_{max} = 31.9$ MPa; $(\sigma_s)_{max} = 47.1$ MPa.
7.74 $(\sigma_b)_{max} = 37.3$ MPa; $(\sigma_s)_{max} = 18.66$ MPa.
7.76 50.3 N·m.
7.78 $(\sigma_w)_{max} = 1064$ psi; $(\sigma_s)_{max} = 21.8$ ksi.
7.80 29.5 ksi.
7.82 186.6 kN·m.
7.84 593 mm².
7.86 0.57.
7.88 (a) 2.74 MPa. (b) 6.43 MPa.
7.90 (a) $\tau_a = 28.92$ MPa; $\tau_b = 2.64$ MPa; $\tau_c = 14.54$ MPa.
(b) 36.51 MPa.
7.92 2.56 MPa.
7.94 (a) 26.36 kN/m. (b) 996.4 kPa.
7.96 28.4 mm.
7.98 (a) 3.1 MPa. (b) 2.4 MPa.
(c) 1.6 MPa.
7.100 36.5 kN.
7.102 82.34 kN.
7.104 (a) At center. (b) $\frac{4}{3}$.
(c) 84.82 kips.
7.106 16.39 kN.
7.108 1920 lb.
7.110 (a) 33.3 ksi. (b) 92.6 ksi.
7.112 $\tau_{max}/\sigma_{max} = h/3L$.
7.114 (a) 525 mm. (b) 11.43 kN/m.
7.116 230 mm.
7.118 S 200 × 27.

7.120 $b = 156$ mm; $h = 312$ mm.
7.122 $h = 2h_0[(x/L) - (x/L)^2]^{1/2}$.

CHAPTER 8

8.2 $\sigma_1 = 59.73$ MPa; $\sigma_2 = -9.17$ MPa; $\theta_p' = 21.4°\!\circlearrowleft$.
8.4 59.6 kN.
8.6 $\tau_{max} = 88.6$ MPa; $\theta_s'' = 25.8°\!\circlearrowright$.
8.8 $\tau_{max} = 21.34$ MPa; $\theta_s'' = 30.7°\!\circlearrowleft$.
8.10 180 kip·in.
8.12 (a) 1.85 kips. (b) 1.66 kips.
8.14 (a) 192.9 MPa. (b) 141 mm.
(c) 5.7 kN/m.
8.16 (a) 6.08 kN. (b) 5.83.
8.18 4.163 ksi.
8.20 $\sigma_1 = 10.4$ MPa; $\sigma_2 = -19.3$ MPa; $\tau_{max} = 14.9$ MPa.
8.22 $\sigma_1 = 81.5$ MPa; $\sigma_2 = -3.26$ MPa; $\tau_{max} = 42.4$ MPa.
8.24 5.51 ksi.
8.26 4.76 kN.
8.28 $\sigma_1 = 4.08$ ksi; $\sigma_2 = -6.85$ ksi; $\theta_p'' = 37.7°\!\circlearrowleft$.
8.30 15.2 kN.
8.32 (a) 19.5 MPa. (b) 55.6°.
8.34 $\sigma_A = 124.3$ MPa; $\sigma_B = -84.3$ MPa; $\sigma_D = 84.3$ MPa.
8.36 (a) 38.7°. (b) 8.13 ksi.
8.38 (a) $\phi' = 83.2°$. (b) $\sigma_{max} = 20.53$ ksi.
8.40 (a) $\phi' = 80°$. (b) $\sigma_A = -13$ ksi.
8.42 (a) $\phi' = -3.8°$. (b) $\sigma_A = 58.61$ MPa.
8.44 25.4 kN.
8.46 5 kN·m.
8.48 3.34 kN·m.
8.50 $2.42(10^{-3})(PL/t^3)$.
8.52 23.08 kips.
8.54 72.8 mm.
8.56 27 kips.
8.58 (a) $\sigma_B = 46.8$ MPa. (b) $y_n = -4.9$ mm.
8.60 (a) $\sigma_A = 19.78$ MPa. (b) $y_n = -2.13$ mm.
8.62 $\sigma_1 = 42.7$ MPa; $\sigma_2 = -0.54$ MPa; $\tau_{max} = 21.6$ MPa; $\theta_p' = 6.4°\!\circlearrowleft$.

8.64   4.03 kips.

8.66   (a) $\sigma_1 = 6.1$ MPa; $\sigma_2 = -68.8$ MPa;
$\tau_{max} = 37.5$ MPa; $\theta_p' = 16.6°\,\rangle$.
(b) $\sigma_1 = 5.15$ MPa; $\sigma_2 = -88.4$ MPa;
$\tau_{max} = 46.8$ MPa; $\theta_p' = 13.6°\,\rangle$.

8.68   4.5.

8.70   (a) 15.2 ksi. (b) -16 ksi.
(c) 5.47 ksi.

8.72   33.8 mm.

8.76   (a) 43.64 mm. (b) 1.78 MPa.
(c) 2.44 MPa.

8.78 $e = a/2\sqrt{2}$.

8.80 $e = 3b^2h^2t/4I_z$.

8.82   53.3 MPa.

8.84 $e = (h_1^3 t_1/h_1^3 t_1 + h_2^3 t_2)b$.

8.86   (a) 687 N. (b) 669 N.

8.88   1.3.

8.90   6.4.

8.92 $3\frac{3}{8}$ by $6\frac{3}{4}$ in.

8.94   65 mm.

8.96   50 mm.

8.98 $2\frac{1}{4}$ in.

CHAPTER 9

9.2   (a) $v_1 = P(x^3 - 3x^2 L)/12EI$

$$v_2 = \frac{P}{2EI}\left[\frac{1}{3}(x - L)^3 - \frac{5}{8}L^2 x + \frac{1}{4}L^3\right].$$

(b) $5PL^2/16EI\,\nwarrow$.
(c) $3PL^3/16EI\,\downarrow$.

9.4   (a) $v_1 = wax^2(2x - 9a)/12EI$

$$v_2 = \frac{w}{24EI}(-x^4 + 8ax^3 - 24a^2 x^2 + 4a^3 x - a^4).$$

(b) $7wa^3/EI\,\nwarrow$.
(c) $41wa^4/24EI\,\downarrow$.

9.6   (a) $v_1 = \frac{w}{24EI}(-24a^2 x^2 + 8ax^3 - x^4)$

$$v_2 = wa^3(-4x + 2a)/3EI.$$

(b) $4wa^3/3EI\,\nwarrow$.
(c) $10wa^4/3EI\,\downarrow$.

9.14 $\theta_A = 2.82 \times 10^{-3}$ rad $\measuredangle$;
$v_A = 4.24$ mm $\downarrow$.

9.16   (a) $v = \dfrac{w_0}{360EIL}(3x^5 - 15Lx^4 + 20L^2 x^3 - 8L^4 x)$.
(b) $w_0 L^3/45EI\,\nwarrow$.
(c) $5w_0 L^4/768EI\,\downarrow$.

9.18   (a) $v_1 = -\dfrac{M_0 x}{6EIL}(x^2 + 2L^2 - 6aL + 3a^2)$

$$v_2 = -\frac{M_0}{6EIL}(x^3 - 3Lx^2 + 2L^2 x + 3a^2 x - 3La^2).$$

(b) $-\dfrac{M_0}{6EIL}(2L^2 - 6aL + 3a^2)$.

(c) $-\dfrac{M_0}{6EI}\left(\dfrac{9L^2}{4} - 6aL + 3a^2\right)$.

9.20   (a) $v = \dfrac{w_0 L^4}{\pi^4 EI}\sin\dfrac{\pi x}{L}\,\downarrow$.
(b) $w_0 L^3/\pi^3 EI\,\nwarrow$.
(c) $w_0 L^4/\pi^4 EI\,\downarrow$.

9.28   1.4 mm $\downarrow$.

9.30   (a) $v = \dfrac{w_0 x}{360EIL}(-3x^4 + 10L^2 x^2 - 7L^4)$.
(b) $v_C = 5w_0 L^4/768EI\,\downarrow$.
(c) $v_{max} = 0.00652 w_0 L^4/EI\,\downarrow$, at $0.5193L$ from A.
(d) $\theta_B = w_0 L^3/45EI\,\measuredangle$.

9.32   (a) $v_1 = \dfrac{w}{48EI}(3Lx^2 - 8x^3 - L^3)$

$$v_2 = \frac{w}{48EI}[3Lx^3 - 2x^4 - L^3 x + 9L(x - L)^3].$$

(b) $wL^4/192EI\,\downarrow$.
(c) $0.0078wL^4/EI\,\downarrow$, at C.
(d) $wL^3/12EI\,\measuredangle$.

9.34   (a) $v_1 = \dfrac{wx^2}{48EI}(4Lx - 3L^2 - 2x^2)$

$$v_2 = wL^3(-8x + L)/384EI.$$

(b) $3wL^4/384EI\,\downarrow$. (c) $7wL^4/384EI\,\downarrow$.
(d) $wL^3/48EI\,\nwarrow$.

9.36 30.24×10^{-3} rad $\nwarrow$.

9.38 $25wL/64$.

9.40 $R_A = 2.188$ kips; $R_B = 9.816$ kips;
$M_B = 217.39$ kip·in.

9.42 (a) $R_A = 5P/16 \uparrow$; $R_B = 11P/16 \uparrow$;
$M_B = 3PL/16 \circlearrowleft$.
(b) 4.1 mm $\downarrow$.

9.44 (a) $R_A = 3M_0/2L \downarrow$; $M_A = M_0/4 \circlearrowleft$;
$R_B = 3M_0/2L \uparrow$; $M_B = M_0/4 \circlearrowleft$.
(b) 0.

9.46 (a) $R_A = 7w_0L/20 \uparrow$; $M_A = w_0L/20 \circlearrowleft$;
$R_B = 3w_0L/20 \uparrow$; $M_B = w_0L^2/30 \circlearrowleft$.
(b) 2.8 mm $\downarrow$.

9.48 (a) $R_A = R_B = w_0L/4 \uparrow$;
$M_A = 5w_0L^2/96 \circlearrowright$; $M_B = 5w_0L^2/96 \circlearrowleft$.
(b) 3.8 mm $\downarrow$.

9.54 (a) $v = \dfrac{P}{6EI}[x^3 - \langle x - a \rangle^3 - \langle x - L + a \rangle^3 + ax\langle a - L \rangle]$.
(b) 8.4×10^{-3} rad $\angle$.

9.56 (a) $v = \dfrac{w}{384EI}[8Lx^3 - 16\langle x - (L/2) \rangle^4 - 7L^3x]$.
(b) $7wL^3/384EI \searrow$.
(c) $5wL^4/768EI \downarrow$.

9.58 (a) $v = \dfrac{P}{12EI}\left[x^3 - 2\langle x - a\rangle^3 + 3\left(-a^2 + \dfrac{EI}{ak}\right)x - \dfrac{6EI}{k}\right]$.
(b) $\dfrac{P}{4EI}[-a^2 + (EI/ak)]$.
(c) $\dfrac{P}{12EI}[2a^3 + (3EI/k)] \downarrow$.

9.60 (a) $v = \dfrac{M_0}{24EIL}[4x^3 - 12L\langle x - (L/2)\rangle^2 - L^2x]$.
(b) 0.
(c) At $0.7113L$ from A:
$v_{max} = 0.00802M_0L^2/EI \uparrow$.
(d) $M_0L^2/24EI \searrow$.

9.64 (a) $R_A = 57wL/128 \uparrow$; $R_B = 7wL/128 \uparrow$;
$M_A = 9wL^2/128 \circlearrowright$.
(b) 2.3 mm $\downarrow$.

9.66 (a) $R_A = 1.007wa \uparrow$. (b) $2.75wa^4/EI \downarrow$.

9.68 (a) $R_A = 9P/8 \uparrow$; $M_A = 3PL/8 \circlearrowright$.
(b) $3PL^3/128EI \downarrow$.

9.70 (a) $R_A = \dfrac{3M_0}{2L} + \frac{3}{8}wL \uparrow$.
(b) $\dfrac{5L^2}{384EI}(24M_0 + 7wL^2) \downarrow$.

9.72 $19PL^3/48EI \downarrow$.

9.76 8.4×10^{-3} rad $\angle$.

9.78 $v_D = L^2(3PL - 28M_0)/96EI$.

9.80 5.84×10^{-3} rad $\searrow$.

9.84 $R_A = 3M_0/2L \uparrow$; $R_B = 3M_0/2L \downarrow$;
$M_A = M_0/2 \circlearrowright$.

9.86 $R_A = 57wL/128 \uparrow$; $R_B = 7wL/128 \uparrow$;
$M_A = 9wL^2/128 \circlearrowright$.

9.88 $R_A = 49wL/80 \uparrow$; $R_B = 31wL/80 \uparrow$;
$M_A = 9wL^2/80 \circlearrowright$.

9.92 $R_A = 5M_0/4L \uparrow$; $R_B = 3M_0/2L \downarrow$;
$R_C = M_0/4L \uparrow$.

9.94 $R_A = 7wL/16 \uparrow$; $R_B = 5wL/8 \uparrow$;
$R_C = wL/16 \downarrow$.

9.96 $R_A = Pa^3/(L^3 + a^3) \uparrow$;
$R_B = PL^3/(L^3 + a^3) \uparrow$;
$R_H = PL^3/(L^3 + a^3)$;
$M_B = PL^3a/(L^3 + a^3) \circlearrowleft$.

CHAPTER 10

10.2 (a) $R_A = 14.69$ kN; $R_B = 25.31$ kN;
$M_A = 18.75$ kN·m.
(b) $R_A = 27.5$ kN; $R_B = 12.5$ kN;
$M_A = 30$ kN·m.

10.4 (a) $3PL^2/4EI \searrow$. (b) $13PL^3/24EI \downarrow$.

10.6 (a) $3Pa^2/4EI \searrow$. (b) $23Pa^3/12EI \downarrow$.

10.8 (a) $7wa^3/6EI \searrow$. (b) $23wa^4/8EI \downarrow$.

10.10 $\theta_D = 2.36 \times 10^{-3}$ rad $\searrow$; $v_D = 3.5$ mm $\downarrow$.

10.12 (a) 0. (b) $M_0L^2/36EI \downarrow$.

10.14 (a) $7M_0L/48EI \searrow$. (b) $M_0L^2/23EI \uparrow$.

10.16 (a) $5wa^3/96EI \angle$. (b) $5wa^4/32EI \downarrow$.

10.18 6.91×10^{-3} rad $\angle$.

10.20 (a) $4M_0a/3EI \angle$. (b) $7M_0a^2/6EI \uparrow$.

10.22 (a) $7PL^2/48EI \searrow$. (b) $3PL^3/32EI \downarrow$.

10.24 (a) $Pa^2/EI \searrow$. (b) $4Pa^3/3EI \downarrow$.

10.26 2.304×10^6 lb·in.^{2}.

10.28 295.6×10^3 lb·in.^{3}.

10.30 $v_{max} = 0.74wa^4/EI \downarrow$, at $1.472a$ to the left of B.

10.32 $v_{max} = 0.0259PL^3/EI \downarrow$, at $\sqrt{5}\,L/6$ to the right of A.

10.34 $v_{max} = M_0L^2/48\sqrt{2}EI$, at $L/2\sqrt{2}$ to the left of B.

10.36 100×10^3 N·m².

10.38 $0.06815PL^3/EI \downarrow$.

10.40 $PL^3/6Eb_0h^3 \downarrow$.

10.42 $R_A = 3M_0/2L \uparrow$; $M_A = M_0/4 \circlearrowright$; $R_B = -R_A$; $M_B = M_A$.

10.44 $R_A = 3P/4 \downarrow$; $R_B = 7P/4 \uparrow$; $M_A = Pa/2 \circlearrowright$.

10.46 $R_A = w_0L/10 \uparrow$; $R_B = 2w_0L/5 \uparrow$; $M_B = w_0L^2/15 \circlearrowright$.

10.48 $R_A = 3M_0/2L \uparrow$; $R_B = 3M_0/2L \downarrow$; $M_A = M_0/2 \circlearrowright$.

10.50 $R_A = \dfrac{P}{16} + \dfrac{3}{8}\dfrac{\delta_C EI}{a^3}$; $R_B = \dfrac{P}{16} + \dfrac{3}{8}\dfrac{\delta_C EI}{a^3}$; $R_C = \dfrac{P}{8} - \dfrac{3}{4}\dfrac{\delta_C EI}{a^3}$.

10.52 $R_A = 3w_0L/20 \uparrow$; $R_B = 7w_0L/20 \uparrow$; $M_B = w_0L^2/20 \circlearrowright$.

10.54 $0.03402w_0L^4/EI \downarrow$.

10.56   (a) $0.0599wL^3/EI \downarrow$.
(b) $0.041102wL^4/EI \downarrow$.

10.58   (a) $wa^3/64EI \measuredangle$.
(b) $7wa^4/64EI \downarrow$.

10.60 $v_C = 0.00966wL^4/EI \downarrow$.

10.62 $v_C = 0.1016PL^3/EI \downarrow$.

10.64 $v_C = 0.1875wa^4/EI \downarrow$.

10.66 $v_{max} = 0.7656wa^4/EI \downarrow$, at mid-span.

10.68 $v_{max} = 0.13867w_0a^4/EI$.

10.70 $v_{max} = 0.0093M_0L^2/EI \downarrow$, at $L/3$ from the point A.

10.72 $0.025PL^3/EI \downarrow$.

10.74 $\theta_A = 0.07862w_0L^3/EI \measuredangle$.
$v_m = 0.02662w_0L^4/EI \downarrow$.

10.76   (a) $v_{max} = 0.00311w_0L^3/EI \downarrow$, at mid-span;
$\theta_{max} = 0.00923w_0L^3/EI$, at A.
(b) $0.05966w_0L^2 \circlearrowright$.

10.78 $R_A = 0.56P \uparrow$; $R_B = 0.12P \uparrow$; $R_C = 0.44P \uparrow$.

10.80 $R_A = 8P/11 \uparrow$; $R_B = 19P/11 \uparrow$; $M_A = 5Pa/11 \circlearrowright$.

10.82 $R_A = 9P/14 \downarrow$; $R_B = 13P/7 \uparrow$; $R_C = 33P/42 \uparrow$.

CHAPTER 11

11.2 $U_0 = 118 \times 10^{-4}$ in.·lb/in.³;
$U_{0v} = 3.5 \times 10^{-4}$ in.·lb/in.³;
$U_{0d} = 114.5$ in.·lb/in.³.

11.6   (a) 35 ksi. (b) 20.4 in.·lb/in.³.

11.8   71.2 J.

11.10   0.382 in.

11.12   126.4 J.

11.14   5.088 J.

11.16 $w^2L^5/40EI$.

11.18 $w^2L^5/240EI$.

11.20 $M_0^2a/6EI$.

11.22 $w^2L^3/5AG$.

11.24 $w^2L^3/20AG$.

11.26 $PL/2AE$.

11.28 $2Pa^2\left(\dfrac{2a}{3EI} + \dfrac{3}{5AG}\right)$.

11.30 $M_0a/3EI$.

11.32   33 mm $\downarrow$.

11.34 $4PL(2 + \sqrt{2})/AE \downarrow$.

11.36 $10wa^4/EI \downarrow$.

11.38 $8.6wL/AE \downarrow$.

11.40 $wL^3/6EI \measuredangle$.

11.42   (a) $Pa^2(L + a)/3EI$. (b) $Pa(2L + 3a)/6EI$.

11.44   (a) $M_0a(2a + 3L)/6EI \uparrow$.
(b) $M_0(L + 3a)/3EI \measuredangle$.

11.46 $5.828PL/AE \downarrow$.

11.48 $wa^4/48EI \uparrow$.

11.50 $PR^2/EI \measuredangle$.

11.52   (a) $2Pa^3/EI \downarrow$. (b) $3Pa^2/2EI \measuredangle$.

11.54 $3Pa^3/2EI_1 \downarrow$.

11.56   (a) $PL^3/3EI \downarrow$. (b) $PL^2/2EI \measuredangle$.

11.58 $49w_0L^4/3480EI \downarrow$.

11.60 $PR^3/2EI \downarrow$.

11.62 $3M_0a/b(b + 3a)$.

11.64 $R_A = R_B = wL/2 \uparrow$; $M_A = wL^2/12EI \circlearrowright$; $M_B = -M_A$.

11.66 $R_A = 9P/16 \downarrow$; $R_B = P/8 \uparrow$;
$R_C = 7P/16 \uparrow$.

11.68 $R_A = Pb^2(3a + b)/L^3 \uparrow$;
$R_B = Pa^2(a + 3b)/L^3 \uparrow$;
$M_A = Pab^2/L^2 \mathcal{D}$; $M_B = Pa^2b/L^2 \mathcal{D}$.

11.70 $R_A = 5wL/8 \uparrow$; $R_B = 3wL/8 \uparrow$;
$M_A = wL^2/8 \mathcal{D}$.

11.72 $R_A = 5wL/12 \uparrow$; $R_B = 3wL/4 \uparrow$;
$R_C = wL/6 \downarrow$.

11.74 $v_{max} = 3.7$ mm; $\sigma_{max} = 25.7$ MPa.

11.76 $v_{max} = 6.65$ mm; $\sigma_{max} = 41.82$ MPa.

11.78 33.455 in.

11.80 203.6 N.

11.82 (a) 135.5 mm $\downarrow$ (b) 22.77 MPa.

CHAPTER 12

12.2 (a) 94.25 kN. (b) $(\delta_{AB})_P = 0$;
$(\delta_{BC})_P = 10$ mm.

12.4 $A_{AB} = 0.866$ in.^{2}; $A_{BC} = 1.0$ in.^{2}.

12.6 229.9 kN.

12.8 $P_u = 630$ kN; $|\delta_b| = |\delta_a| = 2$ mm.

12.10 (a) -240 MPa. (b) 0.104 mm $\rightarrow$.

12.12 (a) 0.15 rad. (b) 10.114 kN·m.

12.14 (a) 93.5 MPa, at $\rho = 5$ mm; 46.2 MPa,
at $\rho = 20$ mm.
(b) 26.77°.

12.16 (a) $T_{yp} = 11$ kN·m; $\phi_{yp} = 5.73°$.
(b) $T_u = 12.4$ kN·m; $\phi_u = 7.64$.
(c) 24.7 MPa, at $\rho = 30$ mm; 20.4 MPa,
at $\rho = 40$ mm; $\sigma_{res} = 1.18$.

12.18 59.32 kN·m.

12.20 (a) 16 kN·m. (b) 22.875 kN·m.

12.22 1.245.

12.24 1.196.

12.26 $16c(c^3 - b^3)/3\pi(c^4 - b^4)$.

12.28 2.

12.30 1.25.

12.32 (a) 52.875 kN·m. (b) 58.9 kN·m.

12.34 ± 97.8 MPa.

12.36 0.37L.

12.38 $L(1 - \sqrt{1/f})$

12.40 $9\sqrt{3}\ M_p/L^2$.

12.42 $M_p/0.11L^2$.

12.44 $4M_p/L$.

12.46 $128M_p/9L^2$.

12.48 $11.62M_p/L^2$.

12.50 $4M_p/3M_{yp}$.

CHAPTER 13

13.2 k_t/L.

13.4 $ka/2$.

13.6 $ka/2$.

13.8 2033 mm^2.

13.10 (a) 50°C. (b) 58°C.

13.12 108.3 kips.

13.14 105.2 kips.

13.16 101 kN.

13.18 101.7 kN.

13.20 43.41 kN.

13.22 101.8 MPa.

13.24 4.3.

13.26 2.53 kN.

13.28 462 kN/m.

13.30 2.01 in.

13.32 (a) 0.53 mm. (b) 58.6 MPa.

13.34 1029 kN; 1032.5 kN, using Eq. (13.16).

13.36 53.1 MPa.

13.38 116.7 in.

13.40 (a) 1416 kN. (b) 1120 kN.
(c) 465.2 kN.

13.42 201 kips.

13.44 3.26 in.

13.46 49.2 mm.

13.48 192.8 kN.

13.50 207.6 kips.

13.52 75.5 kips.

13.54 14.62 kN.

13.56 141.4 kN.

13.58 9.9 kips.

13.60 74.1 in.

APPENDIX A

A.2 $\bar{x} = 0$; $\bar{y} = 2.585$ in.

A.4 $\bar{x} = 2.218$ in.; $\bar{y} = 1.282$ in.

A.6 $\bar{x} = 1.373$ in.; $\bar{y} = 2.705$ in.

A.8 $\bar{x} = 0$; $\bar{y} = 48.8$ mm;
$I_1 = I_x = 63 \times 10^4$ mm^4; $I_{xy} = 0$;
$I_2 = I_y = 43.5 \times 10^4$ mm^4.

A.10 $\bar{x} = 38.125$ mm; $\bar{y} = 24.375$ mm;
$I_x = 34.9 \times 10^4$ mm^4;
$I_y = 12.7 \times 10^4$ mm^4;
$I_{xy} = 12.3 \times 10^4$ mm^4;
$I_1 = 40.4 \times 10^4$ mm^4;

$I_2 = 7.2 \times 10^4$ mm^4;
$\theta_p' = 24°\,\rotatebox{180}{$\circlearrowright$}$.

A.12 $\bar{x} = 22.97$ mm; $\bar{y} = 48.4$ mm;
$I_x = 77 \times 10^4$ mm^4;
$I_y = 7.6 \times 10^4$ mm^4;
$I_{xy} = 7.1 \times 10^4$ mm^4
$I_1 = 77.7 \times 10^4$ mm^4;
$I_2 = 6.9 \times 10^4$ mm^4;
$\theta_p' = 5.78°\,\rotatebox{180}{$\circlearrowright$}$.

NAME INDEX

SUBJECT INDEX